P9-DFP-070

Coders

Also by Clive Thompson

Smarter Than You Think: How Technology Is Changing Our Minds for the Better

Coders

< The Making of a New Tribe and
the Remaking of the World >

Clive Thompson

PENGUIN PRESS > NEW YORK > 2019

PENGUIN PRESS
An imprint of Penguin Random House LLC
penguinrandomhouse.com

Copyright © 2019 by Clive Thompson
Penguin supports copyright. Copyright fuels creativity, encourages diverse voices, promotes free speech, and creates a vibrant culture. Thank you for buying an authorized edition of this book and for complying with copyright laws by not reproducing, scanning, or distributing any part of it in any form without permission. You are supporting writers and allowing Penguin to continue to publish books for every reader.

Portions of this book first appeared *The New York Times Magazine* and *Wired.*

"In a Station of the Metro," by Ezra Pound, from *Personae*, copyright © 1926 by Ezra Pound. Reprinted by permission of New Directions Publishing Corp.

LIBRARY OF CONGRESS CATALOGING-IN-PUBLICATION DATA
Names: Thompson, Clive, 1968–author.
Title: Coders: the making of a new tribe and the remaking of the world / Clive Thompson.
Description: New York: Penguin Press, [2019] | Includes bibliographical references and index.
Identifiers: LCCN 2018051657 (print) | LCCN 2018056767 (ebook) |
ISBN 9780735220577 (ebook) | ISBN 9780735220560 (hardcover)
Subjects: LCSH: Computer programmers. | Computer programming—Social aspects. | Computer programming—Psychological aspects. | Interpersonal relations. | Information technology—Social aspects.
Classification: LCC QA76.6 (ebook) | LCC QA76.6 .T4496 2019 (print) | DDC 005.1092—dc23
LC record available at https://lccn.loc.gov/2018051657

Printed in the United States of America
10 9 8 7 6 5 4 3 2 1

DESIGNED BY MEIGHAN CAVANAUGH

To Emily, Gabriel, Zev, and my mother

Contents

Coders

< Chapter 1 >

The Software Update That Changed Reality

In the early hours of September 5, 2006, Ruchi Sanghvi rewrote the world with a single software update.

A round-faced, outspoken programmer, Sanghvi was 23 years old when she arrived to work at Facebook. Raised in India, she had long dreamed of growing up to work for her father's company, which lent heavy machinery for the construction of ports, oil refineries, and windmills. But while studying at Carnegie Mellon University, she got intrigued by computer engineering, and then she fell in love with it. It was like constantly solving puzzles: trying to make an algorithm run faster, trying to debug a gnarly piece of code that wasn't working right. The mental chess colonized her mind, and she found herself pondering coding problems all day long. "You're at it for hours, you're not eating, you're not sleeping; it's like you can't stop thinking about it," she tells me.

Sanghvi was, by programming standards, a late bloomer; she was studying alongside kids, nearly all male, who'd been coding since they were nine and playing video games, and they seemed to effortlessly get it. But she kept grinding away, got good grades, then graduated and got hired for her first job in Manhattan, doing math modeling for a derivatives trading desk.

When she arrived in New York, though, she was horrified by the sight of the gray cubicles at the workplace. She wouldn't be having much of an impact on the world here. She didn't want to be a cog in a machine, writing code to support finance work; she hungered to work for a company where the technology itself was the core product, where computer scientists were the main players. She wanted to actually make a product that people used—something tangible, useful. She wanted to do something like Facebook, a site that she'd joined in her last year of college. Now *that* was an addictive bit of software. She'd log in all the time to stay in touch with college friends who'd recently graduated, checking their pages to see if they'd updated anything.

So Sanghvi bailed on Manhattan, quitting the job even before her first day. She fled to San Francisco, where she got a job at Oracle, the database company. And then, one day a college friend invited her to come by the offices of Facebook itself.

It was a tiny firm, serving only college students; everyday folks weren't yet allowed to use Facebook. When she walked up to the office on the second floor above a Chinese-food restaurant, she found a passel of mostly young white men, some who'd recently bailed on Harvard: a 21-year-old Mark Zuckerberg walking around in nearly wrecked sandals, Adam D'Angelo (the guy who'd taught a younger Zuckerberg some coding), and Dustin Moskovitz, Zuckerberg's roommate at Harvard. They worked in a haze of intensity, laptops open on cluttered desks, while playing video games at their nearby dorm-like crash-pad houses, or even while sunning on the roof of the Facebook office. The graffiti artist David Choe was hired around that time to bedeck the walls with murals, one of which depicted "a huge buxom woman with enormous breasts wearing this Mad Max–style costume riding a bull dog" (as early employee Ezra Callahan described it).

They were aggressive about tweaking and changing Facebook, regularly "pushing" new code out to users that would create features like Facebook's famous "Poke," or a "Notes" app that let people write longer posts. They were daredevils; sometimes a new feature would have been written so eagerly and hastily that it produced unexpected side effects, which they wouldn't discover until, whoops, the code was live on the site. So they'd push the code out at midnight and then hold their breath to see whether it crashed Facebook or not. If everything worked, they'd leave; if it caused a catastrophe, they'd frantically try to fix it, often toiling until the early morning, or sometimes just "reverting" back to the old code when they simply couldn't get the new feature working. As Zuckerberg's oft-quoted motto went, "Move fast and break things."

Sanghvi loved it. "It was different, it was vibrant, it was *alive*," she says. "People there were like humming along, everyone was really busy, everyone was really into what they were doing . . . the energy was just so tangible." And as it turns out, Facebook was desperately seeking more coders. It's hard to imagine now, with the company being such a globe-spanning behemoth, but back in 2005 they had trouble attracting anyone to work there. Most experienced software engineers in Silicon Valley thought Facebook was a fad, one of those bits of web ephemera that enjoys a brief and delirious vogue before becoming unspeakably passé. They had no interest in working there. So Sanghvi arrived in a lucky window of opportunity: young enough to have used Facebook and known how addictive it was, but old enough to have actually graduated college and be looking for a coding job. They hired her a week after her visit, as the company's first female software engineer.

Soon, she was given a weighty task. Zuckerberg and the other founders had decided that Facebook was too slow and difficult to use. Back in those early days, the only way to know what your friend was doing was to go look at their Facebook page. It required a lot of active forethought. If someone posted a juicy bit of info—a newly ended relationship, a morsel of gossip, a racy profile photo—you might not see it if you forgot to check their page that day. Facebook was, in effect, like living in an apartment building where you had to keep poking your nose in people's doors to see what was up.

Zuckerberg wanted to streamline things. He'd been carrying around a notebook in which he'd sketched a vision (in his tiny, precise handwriting) for a "News Feed." When you logged in, the feed would be a single page that listed things friends had posted since you last logged in. It'd be like a form of ESP for your social life. As soon as someone posted an update—*Ping!*—it would arrive on the periphery of your vision. The News Feed wouldn't be just a slight cosmetic tweak to Facebook, like a pretty new font or color. It would reconstruct how people paid attention to one another.

And now Sanghvi had to make the News Feed happen. She set to work with a small "pod" of collaborators, including Chris Cox, Matt Cahill, Kang-Xing Jin (known as "KX"), and Andrew "Boz" Bosworth, Zuckerberg's former teaching assistant at Harvard. For nine months they worked intensely, batting ideas around and then clattering away writing code, while Cox blasted James Brown or Johnny Cash from his laptop. Like the other coders, Sanghvi began programming almost around the clock, staying at Facebook until dawn and then staggering home to San Francisco; after nearly crashing her car from lack of sleep, she moved to a house near Facebook's office, from which she'd sometimes wander to work in her pajamas. Nobody minded. All the coders blended socializing and working, playing poker or video games at work; during a video interview in 2005, Zuckerberg chatted while toting a red-cupped beer, and an employee did a keg stand.

It was a boys' club, though for Sanghvi, that wasn't anything new: The world of computer science she'd known had always been a boys' club. There were only a few women in her class of 150 at college. She'd learned to yell back when others started yelling, which, in a roomful of cocky young men, was often. Being loud, and a woman, brought repercussions: "Everyone called me super aggressive," she says. "And that hurt. I don't think of myself as aggressive."

But she kept her head down, grinding on the code, because it was mostly what she cared about—and it was thrillingly fun, weird, and hard. Creating the News Feed required her and the other coders to grapple with philosophically hefty questions about friendship, such as *What type of news do friends* want *to know about each other?* The feed couldn't show everything that every single one of your friends did, all day long. If you had 200 friends posting 10 things each, that was 2,000 items, way more than anyone had time to look at. So Sanghvi and the coders had to craft a set of rules to sift through each person's feed, giving a "weight"—a number that ranked it as more or less important. *How would you weight the relationship between two people?* they'd ask each other, sitting around the Facebook office late at night. *How would you weight the relationship between a person and a photo?*

By mid-2006, they had a prototype working. One night Chris Cox sat at home and watched as the first-ever News Feed message blinked into existence: "Mark has added a photo." ("It was like the Frankenstein moment when the finger moves," he later joked.) By the end of the summer, the News Feed was working smoothly enough that they were ready to unleash it on the public. Sanghvi wrote a public post—entitled "Facebook Gets a Facelift"—to announce the product to the world. "It updates a personalized list of news stories throughout the day, so you'll know when Mark adds Britney Spears to his Favorites or when your crush is single again. Now, whenever you log in, you'll get the latest headlines generated by the activity of your friends and social groups," Sanghvi explained. The changes, she wrote, would be "quite unlike anything you can find on the web."

Not long after midnight, Sanghvi and the other coders pushed the update out to the world. News Feed was live; the team cracked open bottles of champagne and hugged each other. It was this type of moment that got her into computers: writing code that changes people's everyday lives.

There was only one problem: People hated it.

When Sanghvi and the team pushed the code out in those early hours, they clustered around a laptop on her desk to watch the comments from users. She crouched on the ground as Zuckerberg peered down at the screen, clad in a red CBGB's T-shirt, her colleague KX looming in up high behind Zuckerberg. Everyone was vibrating with excitement. "They were thinking," Zuckerberg recalled later, "it was going to be good news."

It was not good news. "This SUCKS" was a typical comment that came scrolling down the screen. Users were in full revolt; many were threatening to leave Facebook or boycott it. Groups had formed with names like "Ruchi Is the Devil." One student, Ben Parr, had created a Facebook group called "Students against Facebook News Feed" that amassed over 250,000 members in barely a day.

What exactly did they loathe so much? "Very few of us want everyone automatically knowing what we update," Parr explained. "News Feed is just too creepy, too stalker-esque." Sure, Facebook had been slow and inefficient before, as Zuckerberg had noted. But Facebook's users, it seems, had grown to rely on that inefficiency. It gave them a small, pleasant measure of secrecy. They could post a new profile photo, decide it was unattractive, and quickly change it back to the old one a few minutes later, knowing that it was unlikely many of their friends saw the change. But now News Feed was like a pushy, nosy robot that was taking your every post and shouting it to the heavens. *Hey, Rita's no longer going out with Jeff! She's single again! Check it out!*

The coders had been right: Their invention really had changed the way people learned about their social circle. But users weren't sure they wanted the machinery of their attention upgraded so quickly, and so dramatically.

The uproar grew all day long, and the next day student protestors were camped out in front of the Facebook building, forcing Sanghvi and the other engineers to sneak in and out through the back door. Online, things were even worse. Fully 1 million Facebook users—10 percent of their entire user base—had joined Facebook groups demanding that the News Feed be turned off.

Staff members started arguing about what to do. Two factions emerged, one in favor of shutting down the News Feed, and the other arguing that it was just an adjustment period. Zuckerberg was part of the second camp. Once the initial shock wore off, he believed, the users would discover they liked it. Sanghvi strongly agreed, though she admits part of her insistence in keeping the News Feed was also fueled by engineering pride. "I'd just spent nine months of my life on this, and there was no damn way I was going to get rid of it," she says.

Zuckerberg's view won the day. But even so, he admitted they'd moved a bit too hastily and needed to meet their irate users halfway. So the Facebook coders hatched a plan to create some extra privacy settings so Facebook users could prevent sensitive updates from appearing on the News Feed. After 48 hours of pell-mell work, they pushed that privacy code out live. Zuckerberg published an apologetic note publicly on Facebook. "We really messed this one up," he admitted.

"[We] did a bad job of explaining what the new features were and an even worse job of giving you control of them." But he was still confident that, in the long run, News Feed would be a hit.

He was right. The feed was unsettling and shocking, but it was also captivating. There was, it turns out, enormous value in seeing a little daily gazette of your friends' doings. As you checked the feed and saw the status updates scroll by, you could begin to build up a nuanced picture of what was going on in your friends' lives. Indeed, the day after News Feed emerged, Sanghvi and the team found that people were spending twice as much time on Facebook than before. They were also forming groups much more quickly. It made sense. If you could see that your friend joined a political cause or fan group for a band, you might think, *Hey, maybe I should do that, too.* Ironically, the whole reason "I Hate News Feed" groups were able to grow so quickly is that they tapped into the power of the feed. (And it wasn't just silly groups that were forming. In the days after News Feed launched, the second-largest group was one focused on calling attention to genocide in Darfur, and the fourth-biggest was to advocate for breast-cancer research.)

Indeed, you could argue that News Feed eventually became one of the most consequential pieces of computer code written in the last twenty years. Its effects can be seen everywhere, fractally, up and down the patterns of our lives. Facebook users learn that their friends have had babies, see snapshots of their cubicles and vacations; they notice stray jokes and click on cat-meme links. Its massive, shared attention pool has made News Feed one of the surest vectors by which a piece of culture goes viral, from a tear-jerky video of a kind act to an outtake by Beyoncé, from the hopeful, pro-democratic beginnings of the Arab Spring to virulent ISIS recruitment videos. News Feed tied people together and propelled a host of acronymized pop-psychology ailments, from TMI to FOMO.

The feed got people to stare at Facebook a lot—on average, 35 minutes a day for each American. It's not hard to see why. The feed's sorting algorithm is designed to give you more of what you like; it pays close attention to everything you do on Facebook—your "likes," your reposts, your comments—the better to find new items to show you, that, the programmers hope, match neatly with your preferences. Giving people mostly what they want to see makes for a terrific business, of course, which is why Facebook made about $40 billion a year in advertising in 2017. But it turned out that Facebook's feed, by concentrating everyone's attention into one funnel, also had some unsettling side effects. It created a central point of failure for civic discourse. If you wanted to seed misinformation, spread rumors, or proselytize hate, the News Feed was a wonderfully efficient tool. By the end of the 2016 US elections and President Trump's first year in power, journalists discovered that all manner of toxic forces—from white supremacists to merchants of political disinfo clickbait—were gleefully gaming the feed, seeding it with stories designed to whip up political hysteria. Worse, it seemed quite likely that the feed was exacerbating America's partisan divide, because it was designed to mostly filter out news that didn't match what you already "liked."

By February 2017, even Zuckerberg appeared to be wondering what sort of creature he'd electrified into existence. He wrote a 5,700-word note that felt like an oblique and defensive apology for Facebook's role in today's political schisms. "Our job at Facebook is to help people make the greatest positive impact while mitigating areas where technology and social media can contribute to divisiveness and isolation," he wrote.

That's a curiously cautious mission statement: *While mitigating areas where technology and social media can contribute to divisiveness.* It's certainly a more measured rallying cry than "Move fast and break things." You could read it, perhaps, as a quiet admission that some things ought to be left unbroken.

... ...

"Software," as the venture capitalist Marc Andreessen has proclaimed, "is eating the world."

It's true. You use software nearly every instant you're awake. There's the obvious stuff, like your phone, your laptop, email and social networking and video games and Netflix, the way you order taxis and food. But there's also less-obvious software lurking all around you. Nearly any paper book or pamphlet you touch was designed using software; code inside your car helps manage the braking system; "machine-learning" algorithms at your bank scrutinize your purchasing activity to help spy the moment when a criminal dupes your card and starts fraudulently buying things using your money.

And this may sound weirdly obvious, but every single one of those pieces of software was written by a programmer—someone precisely like Ruchi Sanghvi or Mark Zuckerberg. Odds are high the person who originally thought of the product was a coder: Programmers spend their days trying to get computers to do new things, so they're often very good at understanding the crazy what-ifs that computers make possible. (What if you had a computer take every word you typed and, quietly and constantly and automatically in the background, checked it against a dictionary of common English words? Hello, *spell-check*!) Sometimes it seems that the software we use just sort of sprang into existence, like grass growing on the lawn. But it didn't. It was created by someone who wrote out—in code—a long, painstaking set of instructions telling the computer precisely what to do, step-by-step, to get a job done. There's a sort of priestly class mystery cultivated around the word *algorithm*, but all they consist of are instructions: Do this, then do this, then do this. News Feed is now an extraordinarily *complicated* algorithm involving some trained machine learning; but it's ultimately still just a list of rules. So the rule makers have power. Indeed, these days, the founders of high-tech companies—the ones who determine what products get created, what problems get solved, and what constitutes a "problem" in the first place—are increasingly technologists, the folks who cut their teeth writing endless lines of code and who cobbled together the prototype for their new firm themselves.

Programmers are thus among the most quietly influential people on the planet. As we live in a world made of software, they're the architects. The decisions they make guide our behavior. When they make something newly easy to do, we do a lot more of it. If they make it hard or impossible to do something, we do less of it. When coders made the first blogging tools in the late '90s and early '00s, it produced an explosion of self-expression; when it's suddenly easy to publish things, millions more people do it. And when programmers invented "file-sharing" tools around the same time, a shudder ran through the entertainment industries, as they watched their lock hold on distribution suddenly evaporate. In fact, they fought back by hiring their own programmers to invent "digital rights management" software, putting it in music and film releases, making those wares trickier for everyday folks to copy and hand out to their friends; they tried to create artificial scarcity. If wealthy interests don't like what some code is doing, they'll pay to create software that fights in the opposite direction. Code giveth, and code can taketh away.

If you look at the history of the world, there are points in time when different professions become suddenly crucial, and their practitioners suddenly powerful. The world abruptly needs and rewards their particular set of skills.

Back during the Revolutionary America of the late eighteenth century, the key profession was law. The American style of government is composed of nothing but laws, of course. So lawyers and legal writers—anyone who could construct legal systems in their head, who could argue persuasively and passionately for a particular framework—were powerful. They had a seat at the table. When you look at the careers of the founding fathers, the majority were trained lawyers (John Adams, Alexander Hamilton, John Jay, Thomas Jefferson), and those who weren't (James Madison) were nonetheless masterful legalists. They were the ones who wrote the rule sets that made America America. They wrote the operating system of its democracy. And the tiniest of their design decisions have had massive, long-standing effects on how the republic evolves. For example, by creating the electoral college the founding fathers inadvertently created a system where—two hundred years later—presidential candidates would discover they really only needed to pay attention to a handful of "swing" states. Once a state votes reliably Democratic or Republican, it effectively vanishes off the political map. No presidential candidate visits it or tries to seduce its voters. A different design decision—say, electing the president by a popular vote across the entire country—would have produced a radically different style of modern election. But the founders created a system that Americans are, absent a rewriting of the Constitution, stuck with.

A hundred years after the American Revolution, a new professional class rose to importance. With the Industrial Revolution roaring onward, the country began urbanizing, and skyscrapers climbed in New York and Boston and Chicago. Suddenly it was crucial to figure out how, precisely, you could get millions of people to live in tight proximity without drowning in their own excrement and while providing them with reasonably clean water and air and some way of moving around. This required a ton of ingenious mechanics, so all of a sudden civil engineers, architects, and city planners were in the driver's seat. Anyone who could work in those fields—subway builders, bridge-builders, park-planners—had an outsize role in determining how the city dwellers of the US would live. And once again, single design decisions would go on to have a huge impact on people's lives. Robert

Moses was a famous city planner in midcentury New York, who, in a flurry of regal activity, built dozens of highways and parks that define how today's Gotham works. Some of his decisions ruined lives. In 1948, he began work on the Cross Bronx Expressway, arguing that it was necessary to help relieve traffic that was going from Long Island to New Jersey. It worked, but at a brutal cost to the mostly black neighborhoods that the expressway sliced through. The expressway quickly attracted so much loud, pollution-belching trucking traffic that it destroyed the property value of any houses nearby, making hundreds of mostly African American families abruptly less wealthy, and living in a much less pleasant neighborhood. (Moses, it seems, also liked to move fast and break things.)

If we want to understand how today's world works, we ought to understand something about coders. Who exactly are the people who are building today's world? What makes them tick? What type of personality is drawn to writing software? How does their work affect us? And perhaps most interestingly, what does it do to *them*?

Nearly every programmer has a similar story about the moment they became mesmerized by coding.

It's often when they write their first, tiny bit of code, telling the computer to say "Hello, World!" Here's what that looks like in Python, a popular language:

```
print ("Hello, World!")
```

Hit Enter to run that code, and the computer prints what you'd expect:

```
Hello, World!
```

Not terribly complicated, is it? Yet the effect on the neophyte programmer is electric and Olympian. "It's this feeling of control," as a coder at Noisebridge, the famous San Francisco hacker space, told me. "I was 13, and I had this machine that came to life and would do whatever I said. And when you're a kid, that feeling is wild. It's like you have a little universe to control, that you create."

The first recorded use of "Hello, World!" was in 1972, when a young computer scientist named Brian Kernighan was writing the manual explaining how to program in the coding language called B. He wanted to show the simplest thing you could get B to do, which was to print a message. As he told me, he'd seen a cartoon of a chick coming out of an egg, saying *Hello, World!*, and liked its funny, quirky ring. So he wrote a simple snippet of B code that displayed that little message. Coders quickly glommed on to Kernighan's witty idea, and ever since then, virtually every guide to a programming language—and there are over 250—begins with that one incantation. "Hello, World!" neatly distilled the existential jolt of coding: the creation of a life-form that lurches into being.

Coding has always had that uncanny hint of thaumaturgy about it. It's a form of engineering, sure. But unlike in every other type of engineering—mechanical, industrial, civil—the machines we make with software are woven from words. Code is speech; speech a human utters to silicon, which makes the machine come to life and do our will. This makes code oddly literary. Indeed, the law reflects this nature of code. While physical machines like car engines or can openers are governed by patent law, software is also governed by copyright, making it a weird sister of the poem or the novel. Yet software is also, obviously, quite *different* from a poem or a novel, because it wreaks such direct physical effects on how we live our lives. (This is part of why some coders think it's been ruinous to regulate code with copyright.) Code straddles worlds, half metal and half idea.

"The programmer, like the poet, works only slightly removed from pure thought-stuff," wrote the software engineer Fred Brooks in 1975. "He builds his castles in the air, from air, creating by exertion of the imagination . . . Yet the program construct, unlike the poet's words, is real in the sense that it moves and works, producing visible outputs separate from the construct itself. It prints results, draws pictures, produces sounds, moves arms. The magic of myth and legend has come true in our time. One types the correct incantation on a keyboard, and a display screen comes to life, showing things that never were nor could be."

That's why the phrase "Hello, World!" is so laden with metaphoric freight. It summons to mind all the religious traditions where a god utters creation into existence: "In the beginning was the Word." (Christian programmers are particular fans of this connection: Robyn Miller, an evangelical who co-programmed the game world *Myst*, would occasionally pause in his coding after crafting something cool and think, "It is good.") But "Hello, World!" also has a creepy side. It reminds you of the unexpected side effects that come from bringing something to life that might escape your control—like Dr. Frankenstein's neglected and vengeful wretch who wound up killing his loved ones, or the uncontrollable, respawning brooms in *The Sorcerer's Apprentice*. And so it is with code. Your News Feed helps friends organize care for a friend sick with cancer; it also helps dank memesters spread utterly bonkers conspiracy theories. This harmonic resonance between magic and code is why teenaged nerd programmers of the early '80s hybridized so fluently with *Dungeons & Dragons* (a game that merged fantasy with probabilistic dice-rolling math), or the spell-uttering wizards of Tolkien's epics. In the '60s, when programmers invented a type of code that would constantly run in the background, they called it a "daemon." When he created the computer language Perl, Larry Wall, the computer scientist, included a function called "Bless." As the coder Danny Hillis once noted, "A few hundred years ago in my native New England, an accurate description of my occupation would have gotten me burned at the stake."

This sense of magical control can be intoxicating and fun; it lends itself frequently to a sort of starry-eyed idealism. And it can also lead, particularly in younger coders—who've yet to be humbled by life and their own screw-ups—to some epic hubris. People who excel at programming, notes the coder and tech-culture critic Maciej Cegłowski, often "become convinced that they have a unique ability to understand any kind of system at all, from first principles, without prior training, thanks to their superior powers of analysis. Success in the artificially constructed world of software design promotes a dangerous confidence." Or as the computer scientist Joseph Weizenbaum noted in 1976, "one would have to be astonished if Lord Acton's observation that power corrupts were not to apply in an environment in which omnipotence is so easily achievable."

Coding isn't easy. It requires sitting alone for hours, trying to mentally inhabit the twisty nuances of a piece of software—how this loop over *here* is triggered by *that* input by the user, unless this *other* subroutine is currently running, in which case *this* function ought to fire up. I've had it variously described to me as "trying to build and memorize the structure of every street in London" and "mentally levitating a massive house of cards into place." This is why it tends to attract introverts and puzzle-solving logical thinkers, the types of folks who are happy sitting at home on Saturday night at 11:00 p.m. creating drivers for decommissioned 1997 webcams just because, well, they found an old 1997 webcam in their drawer and want to get it working, and *damn* it's an interesting problem, and it's certainly more predictable than dealing with people. (This introversion isn't total. In fact, building software these days is increasingly social; it requires teamwork and, frankly, often more time in meetings with colleagues talking about what to do than actually doing it.)

More than introversion or logic, though, coding selects for people who can handle endless frustration. Because while computers may do whatever you tell them, you need to give them inhumanly precise instructions. That "Hello, World!" line of code I showed you earlier? Let's imagine you were typing it in a rush, and typed it this way accidentally . . .

```
print (Hello, World!)
```

. . . so, whoops, you left out the quotation marks. Try to run it, and boom—it crashes. The computer won't run it. And the computer isn't pleasant about it; there's no "I'm really sorry, Clive, something went wrong." There are no niceties. It just spits out an error message like `SyntaxError: invalid syntax`, and it's up to you to figure out what you did wrong. Programming languages are languages, a method of speaking to machines; but to speak to a computer is to speak to the most literal-minded entity on the planet, a ruthlessly prissy grammarian. When we speak to humans, they put a lot of work into helping interpret what we say. Computers don't. They will take every single last one of your smallest errors and grind them in your face, until you fix them. That works its way into your mind and personality, too. When you meet a coder, you're meeting someone whose core daily experience is of unending failure and grinding frustration.

Because code is constantly broken, screwed up, an unholy mess, filled with bugs. Even the stuff you just wrote two minutes ago will probably crash the first time you try to run it. "When you learn to program a computer, you almost never get it right the first time," noted the pioneering computer scientist and educator Seymour Papert, back in 1980. He regarded this experience as the pivot around which all coder psychology turns. You write some code, you try to run it; it fails; so most of your job is figuring out what the hell you just did wrong. Those who can handle that daily vexation thrive. Those who can't, flee. In June 1949, the computer scientist Maurice Wilkes was about to ascend the stairs when he suddenly had the epiphany that "a good part of the remainder of my life was going to be spent in finding errors in my own programs." Seventy years later, all coders live with that moment. Even more fun—and, these days, more common—is the task of detangling errors not in your own work but rather in that of a programmer who was employed by your firm four years ago and who wrote what programmers call "spaghetti code," filled with haphazard formatting, baffling variable names, and a structure as Gordian as that of *Finnegans Wake*. And so you dive in, and slowly, slowly, fix it. Programmers are what I've come to think of as "near Sisypheans." They toil for days in resigned failure, watching the boulder roll back down the hill . . . until one day it abruptly and unexpectedly tips over the crest. And what do they behold on the other side? Another hill.

For decades, TV and film directors always depicted coding as an act of frenetic, mad typing—the software pouring out of the programmer in an ecstatic flow. The truth is far more mundane. If you watch coders at work (and I've watched them for hours), they're most often just sitting there, frowning at the screen, running their hands through their hair in frustration, fidgeting a bit, and then breaking into a tight smile when they make some small, tiny, incremental progress. "Dude, I have no idea how you're going to write this book, because coding is the most boring thing in the world to behold," as one programmer told me. Or as Scott Hanselman, a programmer for Microsoft, told me: "There's no glamour. There's no glamour. You just sit and type."

Interrupt them at your peril. Visit the marketing wing of a high-tech firm, and it's filled with extroverts glad-handing one another and talking nonstop. Saunter over to the coding wing, though, and it is monastically silent, a forest of headphone-clad workers, everyone trying to remain deep in their flow, to hold dozens or scores or thousands of lines of structure in their brain. Once they're in that zone they hate leaving it, because it's so hard to get there in the first place. Clap your hand on their shoulder to ask, *Hey, how are things going?* and you may well provoke a sputtering, simmering rage—because you've broken the magic spell, and it will take them an hour to rebuild that castle in their mind. *Goodbye, world.*

Let's use some code to automate things a bit. What if we wanted our "Hello, World!" program to say hello to a bunch of different people? We could do this:

```
names = ["Cynthia", "Arjun", "Derek",
"Alondra"]
for x in names:
  print ("Hello there, " + x + "!")
```

What we did was to create a "list" of four names, and then we stored that list inside a variable called `names`. Then we wrote a `for` loop, which goes through the list name by name, sticking each one inside the sentence "Hello there, ____!" and printing it. Run that little Python program, and you see . . .

```
Hello there, Cynthia!
Hello there, Arjun!
Hello there, Derek!
Hello there, Alondra!
```

. . . which is more fun. Hey, now the machine is doing some scut work for us. You could give this program a list of 10 names, and that loop will run 10 times; give it 10,000 names, and it'll just as happily print out the greetings 10,000 times. Or 10 million. Or 10 trillion.

Code, in other words, is really good at making things *scale*. Computers may require utterly precise instructions, but if you get the instructions right, the machine will tirelessly do what you command over and over and over again, for users around the world. It is an exhilarating sensation, watching a dutiful robot you've created suddenly being used by millions of people. Many software engineers are thus drawn like moths to the flame of scale. They love to create code that runs not just for themselves, not just for a couple of local friends, but for the whole damn planet, right *now*. Solve a problem once, and you've solved it for everyone.

When Sanghvi and the Facebook engineers were pondering how to make the News Feed, they initially conceptualized it as a sort of personal news service. It would be a modern version of the gazette of eighteenth-century European gentry. An in-the-know servant would deliver a report on the shenanigans of the various society folk you knew. Or, as Sanghvi put it, it would be like a personalized newspaper. "It seems crazy to think that three engineers who had just graduated from college were able to build a personalized newspaper for 10 million users," she says. Such is the deep, vertiginous joy of scale.

It is accompanied by its close cousin, a fierce devotion to efficiency. Since they have, at their beck and call, machines that can repeat instructions with robotic perfection, coders take a dim view of doing things repetitively themselves. They have a dislike of inefficiency that is almost aesthetic—they recoil from it as if from a disgusting smell. Any opportunity they have to automate a process, to do something more efficiently, they will. (Sanghvi even extended this to her nuptials. Her mother wanted to arrange Sanghvi's marriage in India, and Sanghvi agreed, because it struck her as far more efficient than dating—which you could regard, from a purely computer-scientific point of view, as a woefully resource-intensive sorting algorithm. "I was a big fan of arranged marriages," as she said once in a speech. "It appealed to the engineer in me. They're practical, and they had a higher probability of succeeding.")

This instinctive desire to optimize—and scale—is what has led to many collisions between software firms and civic life. Facebook looked at our lives as a problem of inefficient transmission of information. Before Facebook, all day long I was doing (and thinking and reading) things my friends might find intriguing. But I had no way to easily broadcast my life, and they had no way to listen; we had to rely on irregular phone calls, drinks at a bar, conversations on the sidewalk. News Feed was, in essence, a massive optimization of our peripheral vision, on a planetary scale. The same thing goes for Uber, which optimized the experience of hailing cars, or Amazon, which did the same thing for shopping, or the many firms creating just-in-time services with "gig" employees. In each case, though, tech firms that are driven maniacally by a zeal for optimization wreak china-shop havoc with any person or government or community that prizes continuity: drivers and employees who'd rather have reliable jobs than piecemeal gigs, neighbors who lose local stores and jobs when they can't compete with lower-friction online sales. Or a culture that suddenly realizes News Feed, so invaluable in many ways, can also sometimes expose us to each other too much.

Code makes efficiency and scale easy, seductive, almost inevitable. That is also why programmers fit so easily into business building and part of why some slide so frequently into libertarian thinking. Their talents are torqued perfectly for capitalism's central trick, which is, basically, "do something marginally more efficiently than before and then skim off the profit." Mind you, software also, by its very nature, weird-ifies a lot of the assumptions of capitalism. A piece of software is a thing, a machine that does something, so you can own it; but because it's also a form of speech, it can be easily shared. And so it *is* shared, almost incessantly. Coders are remarkably chatty and open about their everyday work, regularly posting their software problems to online forums and spending hours helping other people solve theirs. (A study in the '80s concluded that coders were "less loyal to their employers than to their profession.") Even in the bowels of the most rapaciously capitalistic Silicon Valley firm are coders who spend a lot of time—at their desk, at work—helping solve the bugs of other randos. And there's often a communitarian spirit in the worlds of free and open source software, where coders often contribute work freely, to software that anyone can also use themselves, freely.

The wealthy libertarians of Silicon Valley tend to get a lot of press attention. That's understandable. Because they control a lot of the purse strings and decide what to fund, their obsession with "disruption" dictates a lot of what gets funded in technology. But the politics of rank-and-file coders is rather more diverse than one might imagine: There are arriviste brogrammers who'll admit on the third drink that they voted for Trump while working next to property-is-theft anarchists who live in communal lofts, and traditional California liberal lefties attending JavaScript conferences cheek by jowl with coders who spend their evenings energetically shitposting about feminism on Reddit.

Feminism and diversity are, indeed, sore points in the industry. When it comes to the participation rates for women in the US, software is the rare prestigious, high-income industry that has actually regressed. Women were some of the first-ever coders in the '50s, and they comprised some of the field's first towering figures, such as Grace Hopper, who created the first "compiler," or Adele Goldberg, cocreator of the enormously influential Smalltalk language. In 1983, women were 37.1 percent of computer science majors, but by the 2010s the rate had declined to less than half that, around 17 percent. (On the real-world job market the numbers are the same; in 2015, a tally found that the percentage of women in technical jobs at high-profile places like Google or Microsoft ranges from perhaps the high teens to around twenty.) The racial makeup is not any more diverse. Walk into any start-up, and you'll instantly glimpse what the large-scale data show: Programmers at work are primarily white and Asian men. The percentages of black and Latino coders are in the single-digit range across the country and as low as 1 or 2 percent at top Silicon Valley firms.

This is surpassingly ironic because the software industry has long cherished its self-image as a pure meritocracy. The only thing that matters, in theory, is whether you're good. This reverie can be somewhat understandable, because there's a level at which it is inspired by the hard realities of software. At the level of the machine, code truly does feel meritocratic: Crappily written software crashes, and better-written stuff doesn't. Indeed, that binary clarity is what software engineers love about it; like long-distance running, succeeding at a fiendish coding problem feels like a true measure of yourself. The field of software can also appear more democratically accessible than many other fields, because it's one where self-taught amateurs work alongside people with PhDs. (That certainly isn't true of surgery, law, or aerospace engineering.) And many coders will admiringly tell you of colleagues who seem not just more skilled than themselves, but vertiginously so—elite hackers, Neos in the Matrix who can pour out code and solve puzzles an order of magnitude faster than their mortal peers. So the roots of the mythology go deep. Steeped in a field with these do-or-die specs, it isn't surprising that so many young white men, inexperienced in feeling the daily subtle discrimination faced by underrepresented people of all backgrounds, fall under the sway of the myth; that they believe, genuinely—sometimes with almost a sort of doe-eyed naivete, even when they're being pissy or enraged about it—that the field is a straightforward meritocracy; that if women and minorities aren't thriving, it's because they haven't worked as hard or aren't genetically endowed with the talents.

It's also clear this isn't remotely true. There's plenty of evidence, of course, including studies that use Silicon Valley's much-favored A/B testing. To pick just one revealing experiment, a tech recruiting firm submitted 5,000 résumés to employers, once without names, and once with. When the résumés had no names—and thus employers couldn't know the gender—fully 54 percent of résumés from women were selected for interviews. When the résumés were submitted with names, the number plummeted to 5 percent. And when I interviewed women and minorities at tech firms, many told of every shade and flavor of discrimination, from subtle slurs to outright harassment. The tale of how the software industry narrowed and tightened is a modern cautionary tale about opportunity—and how we might better vouchsafe it.

And the stakes here are surprisingly high. When it comes to product design, culture matters. If a tool is built by a team that's essentially a monoculture, it's going to have serious blind spots, as any first-year MBA student learns. Thus it has been with the code that shapes our lives. Some of the most influential software in recent years has been made by groups of mostly young, mostly white, mostly men who didn't foresee ways their code will affect people who aren't like them. This is, just to pick one example, some part of why Twitter has been the site of so much abuse and harassment. The team of young guys who made it were, demographically, far less likely to have experienced online abuse. They didn't prioritize it early on as an inevitable, looming problem that they would need to address. On the contrary, one staff member dubbed their company "the free-speech wing of the free-speech party." They designed few early safeguards against harassment, and years later, trolls and white supremacists discovered that Twitter was a fabulous way to harass targets.

... ...

Broadening the world of coders, as it turns out, is part of what Ruchi Sanghvi is now trying to do.

When I first met her, she was only 35, but in the hummingbird metabolism of software, that made her an éminence grise. After five years at Facebook, she'd founded her own company, sold it to Dropbox, and worked there as a vice president. After she left Dropbox, she started a venture aimed at inspiring the next generation of coders: South Park Commons. A renovated office building in San Francisco's South of Market neighborhood, it is a hub for young technologists pondering their future.

"It's like a salon," Sanghvi said when I showed up for lunch on a hot summer day. Small clusters of engineers, researchers, and entrepreneurs sat at long tables, plowing through email and reading white papers, beneath a sprawling piece of art on the wall, a series of jagged wooden arches. Sanghvi takes in 30 to 40 members at a time, attracting them via word of mouth, then encourages them to talk, organize lectures, and bat around ideas at the edge of reasonable. It's not really an "accelerator," one of those hothouses where young people kill themselves for three months bootstrapping a firm, in a desperate attempt to attract venture capital. It's more speculative, aimed at convincing members to pick a truly new, weird area to examine. Lately the talk has been heavily about artificial intelligence (AI), and the dark magic of writing algorithms that can learn on their own; at least six of Sanghvi's members have wound up working at Google Brain or the nonprofit OpenAI initiative.

Six start-ups that have come out of the Commons were founded by women. That's an achievement too: Getting more women into critical founder roles means they can deeply influence the trajectory of their firm, and benefit from its success. Sanghvi remembers having to argue over getting a fair share of Facebook's value. When she talked to Mark Zuckerberg, she laid down a demand: "I said, either you bring me up to par on equity or you increase the cash compensation," she tells me. She put it to Zuckerberg in nerd terms: "All I really care about is building stuff," she told him, "and I don't want to have this in the back of my mind while I'm working. I want to be only thinking about building stuff! And *you* don't want me to have that in the back of my mind." (She wasn't complaining specifically about getting less than men, she notes; at the time, she didn't know individually what anyone else was getting.) But it was, in classic coder style, an argument based on efficiency, the mind analogized as a CPU: *You wouldn't want me wasting my scarce brain-processing cycles, would you?* Zuckerberg was convinced. These were terms he could understand.

When I visited South Park Commons, it looked like the next generation of nerds hunting for the same open, greenfield areas that Sanghvi found years ago with Facebook. As she sees it, despite all its systemic problems, the world of software is still (as the author Douglas Rushkoff neatly puts it) a "high leverage point"—a fulcrum where one can shift the world. You just need an opening, a place to get in.

That turns out to be true of software, going all the way back to the beginning of the industry. When you look at the history of code—and who becomes a coder—it's a history of doors that swung open, allowing a new generation to wander in. Each group that showed up changed the fabric of software and left their mark on how we live our lives. It happened in the early '90s and '00s with the web; before that, it happened in the '80s with personal computers; and before then, in the curious, room-sized mystery machines of the '70s and '60s. It's been a story of people who discovered they liked the combination of logic and art that lets you talk to machines.

It's the story of people like Mary Allen Wilkes.

< Chapter 2 >

The Four Waves of Coders

Mary Allen Wilkes did not plan to become a software engineer. Back in the 1950s, she was a Maryland teenager who dreamed of becoming a trial lawyer; she was known for her systematic, probing intelligence. But one day in junior high in 1951, her Geography teacher surprised her with a comment: "Mary Allen, when you grow up you should be a computer programmer!"

Huh? Wilkes had no idea what her teacher was talking about. She had no idea what a programmer was; she wasn't even clear what a computer was. To be fair, relatively few Americans knew what one of these newfangled "digital brains" looked like or how they worked. The first digital computers had been built barely a decade earlier at universities and government labs, and they were a pale shadow of today's computers. They were mostly just high-powered calculators, used by scientists and the military to crack encoded messages from the enemy or to calculate bomb trajectories. Still, Wilkes mentally filed away that comment from her teacher as she went off to Wellesley to earn a BA in philosophy.

As she prepared to graduate four years later, she realized her legal dream would be a hard road. It was 1959, and the field of law was still too sexist to let very many women become trial lawyers. Academic mentors all told her the same thing: Don't even bother applying to law school. "They said, 'Don't do it. You may not get in. Or if you get in, you may not get out. And if you get out, you won't get a job,'" she recalls. Or if she lucked out and got a job, it wouldn't be doing trials: More likely it'd be as a law librarian, a legal secretary, someone processing trust and estates. "And I wanted to be a trial lawyer. But in the 1960s, this was just not going to happen."

Wilkes remembered her junior high school teacher's comment about programming, though. In college she'd heard a bit more about computers and how they were supposedly the machines that held the keys to the future. She knew MIT had some. So on the day of her graduation, she asked her parents to drive her to the MIT campus, where she marched into the employment office. "Do you have any jobs for computer programmers?" she asked.

They did. And as it turns out, they hired her. They were happy to take on an applicant who wandered in with absolutely no experience in computer programming.

That's because in 1959, almost *nobody* had experience in computer programming. The discipline did not yet really exist; there were vanishingly few college courses in it, no majors to take. (Stanford wouldn't create a computer science department until 1965.) Writing software had only recently begun to move toward the style of coding we know today. At its heart, a computer is really just a massive collection of binary switches, each one representing a single bit, a "1" or a "0": *on* or *off.* Bits can do some very fun tricks, though. Putting a bunch of bits in a row lets you represent a number in binary format; the sequence 1101, for example, represents the number 13. You can also build logic statements that help a computer make decisions. For example, there's an

"AND" gate: If switch one *and* switch two are both "on," they activate a third switch. Or there's an "OR" gate: If switch one *or* the other is "on," it triggers a third switch. By connecting a lot of logic gates, you could get a computer to add and subtract numbers very quickly, and perform feats of complicated reasoning upon them. To make it easier to manipulate these binary machines, computer scientists in the late '50s had begun to invent languages like Fortran or COBOL, which let programmers write commands that more slightly resembled English. Wilkes was arriving at the beginnings of this revolution.

As it turns out, Wilkes had an edge: her degree in philosophy. She'd studied symbolic logic, in which you craft arguments and inferences by chaining together AND and OR statements in a similar fashion. It was a style of thinking that dated back to Aristotle, and had risen to prominence in the work of logicians like George Boole and Gottfried Leibniz.

Wilkes quickly became a whiz at writing programs. She worked at first on the IBM 704, one of MIT's massive computers. It required her to write in an "assembly language," which could be pretty abstruse: A typical command would be something like "LXA A,K"—it tells the computer to take the number located in location A of the computer's memory and load it into the "index register" K. Entering the program into the IBM 704 was itself a painstaking affair. There were no keyboards, no screens, as with today's computers. Wilkes wrote the program on paper, handed it to a typist who would punch each command onto punch cards, then an "operator" would feed the stack of cards into a reader. The computer would execute the program and send the results back later as a printed message, typed on a printer.

Worse, there were logjams. There was only one IBM 704 at MIT's Lincoln Lab where she worked, and each computer could run only one program at a time. So programmers had to stand in line, taking turns running their programs or waiting around while others ran theirs.

Programming, Wilkes discovered, required monk-like patience. She'd hand off her code, then cool her heels for hours until the result came back. The massive IBM 704 machine would pulse away, hidden in another room. Programmers were rarely allowed into the inner sanctum to touch the machine itself; the hardware was attended to by a priesthood of technicians who'd scurry around replacing burned-out components. What's more, the computer needed to be sealed in a separate room to mitigate the volcanic heat it produced. The computer rooms were some of the only places in all of MIT that had air-conditioning, without which the computers would quickly melt down.

So Wilkes would wait. Then finally her program would attempt to execute, the printer would clatter to life, and she'd examine the result. Very often her code wouldn't produce the result she wanted. She'd have made a mistake, created a bug that would cause the whole calculation to be thrown amok. Then Wilkes would pore over her lines of code, trying to deduce her mistake, stepping through each line of code in her head and envisioning how the IBM 704 would execute it—turning her mind, as it were, into the computer. Then she'd rewrite the program, have it keyed in on a new batch of punch cards to be fed in again, and then wait, for hours more; then repeat. She learned to be not only precise in her code but parsimonious. Most computers' capacity back then was quite limited; the IBM 704 could handle only about 4,000 "words" of code in its memory. Writing a program was like writing a haiku or a sonnet. A good programmer was concise, elegant, and never wasted a word. They were poets of bits.

"It was like working logic puzzles—big, complicated logic puzzles," Wilkes recalls. She also loved the precision it required; she had a meticulous side that coding satisfied. "I still have a very picky, precise mind, to a fault. I notice pictures that are crooked on the wall . . . I think there's a certain kind of mind that works that way."

Who possessed minds like that? Back in the 1960s, it was frequently women. It's often a surprise to people today, but at MIT's Lincoln Labs in the 1960s, when Wilkes worked there, most of the "career programmers" were female. Indeed, it was often assumed back then that women were naturals at programming. There was also a recent female pedigree; during World War II, some of the first experimental computational machines used for code-breaking at Bletchley Park in the UK were staffed by women, and in the US, the first programmers for the ENIAC computer, calculating ballistics trajectories, were women.

Still, Wilkes noticed gender divides. Part of the reason men weren't full-time programmers is because in the '60s, the sexy, high-glory part of the job was regarded as building the *hardware.* That, engineers felt, was where the devilish challenges lay, as well as the money, with defense contracts pouring in. How do you craft a computer that can read data from its memory faster and faster? How do you fit a computer into a smaller space or get it to use less energy? Solving problems like that got an engineer hired as "research staff" at MIT, with better pay and more vacation time. The actual act of programming the machines—telling the hardware what to do—was, if not exactly an afterthought, seen as a subordinate activity. So at Lincoln Labs, the men gravitated to crafting the circuits for bold new computers. They certainly *could* do programming; they needed to. But they didn't make it their life's goal. Career programmers weren't on the research staff; they served the research staff.

Nonetheless, Wilkes enjoyed the relative comity, the sense of being among intellectual peers, that reigned between the men and women at Lincoln Lab. "We were a bunch of nerds," Wilkes says. "We were a bunch of geeks. We dressed like geeks." (Being the '60s, this still meant Wilkes had to wear heels and a skirt—but she could at least wear a simple blouse; no suit jacket was required.) "I was completely accepted by the men in my group. And it was more interesting than being a secretary."

... ...

In 1961, Lincoln Lab heads assigned her to a prominent new project—helping to design and build the LINC, an audacious bid to develop one of the world's first truly personal computers. It was the brainchild of Wesley Clark, a young computer designer known equally for his vision and his penchant for disobedience; MIT had already fired him twice for insubordination. (He was hired back both times to new positions.) He was intrigued by the emergence of transistors, which could do the same work as vacuum tubes—they could form logic circuits and be the guts of a computer—but they were tiny and low power, so they wouldn't heat up, and they'd be faster to boot. He wanted to make the world's first "personal computer," one that could fit in a single office or laboratory room. No more waiting in line; one scientist would have it all to himself (or, more rarely, herself). Clark wanted specifically to target biologists, since he knew they often needed to crunch data in the middle of an experiment. At that time, if they were using a huge IBM machine, they'd need to stop and wait their turn. If they had a personal computer in their own lab? They could do calculations on the fly, rejiggering their experiment as they went. It would even have its own keyboard and screen, so you could program more quickly: no clumsy punch cards or printouts. It would be a symbiosis of human and machine intelligence. Or, as Wilkes put it, you'd have "conversational access" to the LINC: You type some code, you see the result quickly.

Clark knew he and his team could design the hardware. But he needed Wilkes to help create the computers' operating system that would let the user control the hardware in real time. And it would have to be simple enough that biologists could pick it up with a day or so of training.

Over the next two years, she and a team toiled away, staring at flowcharts, pondering how the circuitry worked, how to let people talk to it. "We worked all these crazy hours, we ate all kinds of terrible food," she recalls. When they had a rough first prototype working, Clark tested it on a real-life problem of biological research. He and his colleague Charles Molnar dragged a LINC out to the lab of neurologist Arnold Starr, who had been trying and failing to record the neuroelectric signals cats produced in their brains when they heard a sound. Starr had put an electrode implant into a cat's cortex, but he couldn't distinguish the precise neuroelectric signal he was looking for. In a few hours, Molnar wrote a program for the LINC that would play a clicking noise out of a speaker, record precisely when the electrode fired, and map on the LINC's screen the average response of the cat to noises. It worked: As the data scrolled across the screen, the scientists "danced a jig right around the equipment."

In early 1964, Wilkes took a break, traveling around Europe for a year. Upon her return, Clark asked her to come back to write the operating system for the LINC, but the lab had moved to St. Louis, and she had no desire to move there. So they agreed to let her work remotely and shipped a LINC across the country to her parents' house in Baltimore, where she was living. They set it up in the living room next to the staircase to the second floor, where it looked like a glimpse of a very weird future: HAL from *2001*, arrived in suburban America. The computer was composed of several big units, including one tall cabinet that stood on a table at the bottom of the stairs, on which magnetic tapes whirred and data glowed on the computer's screen, not much bigger than a piece of white bread; and nearby, a fridge-sized box full of transistorized circuits. Wilkes would sit at a desk wedged between the hardware, writing out the code, working into the wee hours. Like many coders, she was a night owl. Wilkes had become one of the first people on the planet to have a personal computer in her home.

Soon Wilkes had completed the LINC's operating system and written a manual explaining, for complete newcomers, how to program it. (Another pioneering moment: Few people had written a how-to program guide for complete neophytes, because there were few computers aimed *at* neophytes.)

Wilkes was now deeply embedded in the world of coding. She was known as a skilled veteran in this curious new field, and she'd been offered jobs at some of the now-growing number of computer manufacturers around the country.

But Wilkes's original dream of a legal career still gently haunted her. "I also really finally got to the point where I said, 'I don't think I want to do this for the rest of my life,'" she tells me. Computers were intellectually stimulating but socially isolating. "I said, 'I think I need something that's more human interactive. I don't want to spend the rest of my life staring at flowcharts.'"

She applied to Harvard Law School, where her CV stood out as a curiosity: Who was this 30-year-old woman who programmed these odd, newfangled computers? She was accepted, graduated, and went on to spend the next forty years working, as she'd initially craved, in law, including taking cases to trial, teaching at Harvard, and working with the Middlesex DA. She doggedly prepared for each court appearance, walking through possibilities in the lines of questioning the way she'd walked through lines of code. "And I loved it. I absolutely loved it." One of her specialties was, rather appropriately, technology law.

These days, the stereotype of a coder is what you'd see on a show like *Silicon Valley* or *Mr. Robot*: young men, enhoodied. In the US, mostly white, though with some Indian and Asian programmers mixed in; all pretty nerdy. Some of them with vaguely antiestablishment points of view, others out to make a quick million.

The truth is that the question of who becomes a coder has changed over the years. As the industry has developed—as the types of computers that are available have evolved—it has produced several discrete generations of programmers. Mary Allen Wilkes was among the first generation of them, the ones who didn't yet necessarily think of coding as a career and who were part of enormous teams. Computers were the province of institutions, and the ones who were allowed to touch them were institutional.

The next wave of coders, though, were the "hackers" of the '60s and early '70s. They regarded themselves as renegades—the ones who wrested computing away from dour, restrictive institutions.

Much of that culture was born at MIT, when the artificial-intelligence lab acquired some of the early real-time machines of the sort that Wilkes had designed. Equipped with output screens and keyboards, these computers—such as the PDP-1—were often busy during the day with work by grad students in the AI program. But in the evening, they were frequently free, open for interested sorts to find these "conversational" machines.

Quickly, a group of obsessives began to cluster around the lab. One was Bill Gosper, a skinny math prodigy who'd spend hours on the machine creating algorithms to solve problems of math or geometry. (In one of his early achievements, he wrote a routine to solve the "Hi-Q" peg-jumping puzzle game.) Two others were Ricky Greenblatt, an unkempt student who became prolific at pouring out code, and Slug Russell, a boy mesmerized by the possibilities of drawing interactive games on a computer screen. Surrounded by a growing crew of young men—and they were all men—the students would spend all night in the lab, often with the lights turned out, lit by the eerie cathode rays.

They were enthralled by the feeling of being in a direct, intellectual loop with the computer—"the rush of having this live keyboard under you and having this machine respond in milliseconds to what you were doing," as Gosper later told the journalist Steven Levy in Levy's book *Hackers.* They'd have an idea, code it, and instantly see the results; then tweak more and more, watching each idea come alive on-screen. When they started pursuing a new coding challenge, time stood still. "I was really proud of being able to hack around the clock and not really care what phase of the sun or moon it was," Gosper said. Greenblatt would program in 30-hour shifts, eventually so destroying his classroom schedule that he flunked out of MIT. (He took a coding job in a nearby town and kept coming to the AI lab at night.)

But crucially, nobody was telling them what to program or what *not* to program. They were the first group of coders who used the machines to do things that were simply whimsical or creative. They'd get the machines to play music by writing code that turned the vibrating speaker on and off at different frequencies. They'd write programs to calculate chess moves, one of which eventually beat an actual human. Russell wrote *Spacewar!*, one of the world's first graphical computer games, in which two players each piloted a ship around a black hole, trying to shoot the other. Using a $120,000 machine to play a video game would likely have seemed, to computer-makers at the time, a madly frivolous thing to do. But the MIT hackers regarded themselves as liberating programming from its mundane history of mere bean counting and scientific problem-solving. Coding itself, they felt, was a playful, artistic act.

They were establishing, as Levy described it, a "hacker ethic." They believed that there was a hands-on imperative, that everyone in the world ought to be allowed to interact directly with a computer. They also believed in radical openness with code: If you wrote something useful, you should freely share it with others. (This spirit of openness extended to the physical world: When MIT authorities locked cabinets with equipment they needed to fix the computer, they studied lock picking and liberated the equipment.) And they deeply distrusted authority and bureaucratic pecking orders. The white-shirted, buzz-cut IBM types who had kept computers locked behind doors enraged them.

In contrast, they admired great code even if it came from someone with no rank at all: Kids as young as 12 showed up at the AI lab and wound up part of the group. One, David Silver, dropped out of school at age 14 to begin hacking with the group, and he became an expert at programming the AI lab's robots. That wound up annoying the graduate students in AI, who were more philosophically inclined; they thought the most important thing was to have a good theory of how intelligence worked, and they didn't find the coding part as interesting. Silver was precisely the opposite. Like the other hackers, he cared more about actually getting code running—a good "hack," a program that really *did* something, was all that mattered. At one point, he got a robot to push a wallet across the room into a goal. "It sort of drove them crazy," he told Levy. "Because [I] would just sort of screw around for a few weeks and the computer would start doing the thing that they were working on that was really hard. . . . They're theorizing all these things and I'm rolling up my sleeves and doing it. You find a lot of that in hacking in general."

It was often a fiercely anticommercial world. Code was a form of artistic expression, they felt—but it wasn't one they wanted to copyright and make money off of. On the contrary, they believed in freely giving it away and showing it to everyone who wondered, *Hey, how'd you do that?* That's how people were going to learn, right? And that's how these inventions and miracles they were crafting were going to spread to the outside world. This was the ethic that later morphed into "free and open source software"—the act of openly publishing your code and letting anyone repurpose and use it. Famous MIT hackers like Richard Stallman were incensed at corporations that kept their source code secret; he was livid when a group of MIT hackers left to found a firm that produced LISP computers, sold them to MIT, yet wouldn't openly share the code. Stallman responded by launching the free-software movement and beginning work on a full operating system and legal apparatus that enshrined everyone's right to inspect and tinker with the code. The hackers weren't political in the partisan sense of the word; most were only dimly engaged with the raging debates over the Vietnam War. But the emerging politics of software? Now *that* compelled them—at least the ones like Stallman.

But they were also the first generation that began to push women out of the field. Unlike Wilkes's earlier cohort, the core scene of hackers in the MIT lab were exclusively men—often stilted in conversation and living in "bachelor mode," as they put it, with no interest in dealing with anyone except those like themselves. They saw themselves as a priestly class, devoted to their craft above all else: "Hacking," Levy observed, "had replaced sex in their lives." Greenblatt was so famously unshowered and messy that the YMCA kicked him out of its residence. With male hackers sleeping in the lab at night, the environment trended toward that of an all-guy dorm.

The tinkering culture of the hackers, too, could collide with that of MIT's computer scientists, who were trying to use the machines to get important research work done. One of the latter was Margaret Hamilton, a young MIT coder who would later become a famous programmer engineering mission-critical NASA systems, helping to land Apollo missions safely on the moon. Back in those early MIT days, she was trying to run a weather-simulation model, but it kept on crashing. Why? Eventually Hamilton learned it was because the hackers had rejiggered the computer's assembler to suit their desires and hadn't switched it back. They wanted to muck around with pretty cellular automata; she was trying to do weather science. But the hackers simply hadn't appeared to think about the repercussions their tinkering had for other people.

These guys were companionable with each other but mostly uninterested in talking about their own or others' inner lives. "I spent my lifetime walking around talking like a robot, talking to a bunch of other robots," as one of them later said with a sigh.

By the '80s, the nature of computers changed again. The devices were becoming cheaper and cheaper, as a new breed of manufacturer decided it was time to truly bring computers to the masses. Over in Silicon Valley in 1976, Steve Wozniak created the Apple I, one of the first computers that had a radical design element: It could plug into a regular TV. Turn it on, and you could immediately start coding, just like the MIT hackers. Soon plenty of other manufacturers began following Apple's lead, driving the price of computers down to something a middle-class family could afford. In 1981, Commodore released the VIC-20, a plug-and-play machine for $300. The revolution begun by Wilkes had now spread to the wood-paneled basements of America.

Suddenly, teenagers with enough money could essentially *stumble* into the world of programming. That's basically what happened to James Everingham.

Everingham grew up in a Pennsylvania town called Dubois, which was near Punxsutawney, famous for its appearance in the film *Groundhog Day*. In 1981, at age 15, a friend dragged him into a local department store called Montgomery Ward. His friend wanted to see a VIC-20, which had just come out; Everingham couldn't figure out the allure of this nerdy crap. "I don't get these computers," Everingham complained, as his friend cooed over the chunky beige computer. "Why would anybody want those? Like, what do you *do* with it?"

"Watch this," his friend said, and he typed in this program:

```
10 PRINT "JIM"
20 GOTO 10
```

When his friend set the program running, the screen filled up with "JIM" over and over again. Everingham was agog. "I thought it was magic," he tells me. "I just wanted to know *what evil is this.* I was like, 'I must know!' "

It was the language BASIC. I'd posit that BASIC is, historically, the most consequential computer language in history because it dramatically threw open the floodgates to amateurs. Back when Wilkes was hacking away, the Assembly language was pretty cryptic to read and write. It's what's known as a lower-level language, which takes a lot of work to learn to read and master. By the time the MIT hackers arrived, higher-level languages that looked a lot more like Standard English were commonly in use, such as Fortran, aimed at helping everyday scientists and mathematicians use computers to do calculations, or COBOL, designed for businesses. But BASIC was one of the easiest yet. Invented in 1964 at Dartmouth College, it stood for a Beginner's All-purpose Symbolic Instruction Code, and used fairly simple commands that a newbie could readily grasp and wield.

That little program of Everingham's friend shown previously? That was the "Hello, World!" of the VIC-20 generation, the first incantation that most kids tried when they got their mitts on the machine. It's pretty easy to read even if you've never programmed: The first line in the program is numbered "10," and it tells the computer to PRINT the name "JIM." The second line of code, numbered "20," tells the computer to go back to line 10 again. Together, they create a tiny infinite loop that will print and reprint your name until the computer is turned off. This simple little two-line program neatly illustrated the awesome, alien power of computers: Like a robotic genie, it would do what you commanded, precisely and ceaselessly.

There aren't many ways for teenagers to grasp, in such visceral and palpable ways, the fabric of *infinity*. But typing that two-line program gave them a taste of that power. Watching his name scroll by gave Everingham a jolt of the emotion Keats describes so beautifully in his poem "On First Looking into Chapman's Homer": Like stout Cortez, he'd climbed a peak and suddenly beheld an entirely new ocean of possibility, upon which one could sail, seemingly, forever.

And it was an ocean that teenagers found easy to navigate. The computer companies very much encouraged it. Indeed, most of those early '80s computers arrived with a manual explaining, step-by-step, how to write BASIC programs. (I'm Everingham's age, and this is precisely how I first learned to program, too.) The computers also had a decent ability to display graphics and play musical notes, which made them particularly suited for crafting simple computer games—a particularly narcotic allure for 1980s teenagers, who invariably set about trying to remake their own versions of arcade hits like *Space Invaders* or *Pac-Man* or text-adventure games like *Zork*. (It was the fan fiction of the arcade world.) All told, these plug-into-your-TV computers gave birth to a Cambrian explosion of hackers around the world. When you take cheap machines that can do nearly anything you tell them to and hand them over to teenagers with essentially no adult supervision—because their parents had no idea what computers were—you create the infinite-monkeys experiment of software. Soon, teenage coders were cobbling together everything and anything: chatbots that would curse and swear, spellbinding forms of artificial life known as "cellular automata," casino games, little databases and accounting programs, computer music, and endless varieties of games.

Everingham was desperate to join this scene. He was from a lower-middle-class family whose parents couldn't afford to buy him a computer. So he began frantically picking up every spare job he could—mowing lawns, then shoveling snow when winter came—until he'd saved enough, with a contribution from his mother, to get his own VIC-20.

Now he had the machine; how to get games? "I was poor so I really couldn't afford any," he says. He needed free ones. So he began buying computer magazines, like Commodore's own *Run*, that contained the entire code for simple games. As he'd type them out, he began to understand bits of how they'd work. He'd change a few variables and presto: He'd have hundreds of lives in a game of *Space Invaders*! Slowly, through trial and error, he began figuring out how BASIC worked, then writing his own software.

Then, through those computer magazines, Everingham heard of something even cooler: Bulletin Board Systems. On BBSes, you could use a modem to dial into someone else's computer across the country, chat with them, and—best of all—download copies of free software and games. "Free" sounded good to a broke teenager, so again he scrounged and worked odd jobs so he could buy a modem. Everingham began spending hours dialing into BBSes around the country, downloading software and using the downloads to learn more and more BASIC. After a month of this frenetic dialing, though, the telephone bill arrived. His mother came downstairs with it in hand, weeping; he had racked up $500 in charges. Everingham was mortified: "It was more than her *mortgage*," he says.

So now he had a new challenge: how to scam free long-distance phone calls. He started researching, and he discovered how calling-card companies worked. You'd dial into their main 1-800 number and input your account number, which was a six-digit number. So all he needed to do was call the 1-800 number over and over again, iterating through every possible six-digit number, until he stumbled across some that actually worked. That'd be a painful and boring task for a human—but perfect for a computer. (Matthew Broderick's character uses his computer to do this in *WarGames*; so hackers started calling the trick "war dialing.") Everingham quickly wrote a Commodore 64 program that dialed up the 1-800 number for a long-distance firm called "LDX" over and over again, all night long while he slept. When he woke up, it had created a neat list of working long-distance codes he could use. It wasn't legal, and he knew it—but stealing something as intangible as long-distance time didn't feel like a big crime.

Indeed, he even got help from inside the bowels of LDX. One day after talking to a hacker friend in Texas long-distance, he got a call from someone who called himself "Mr. Clean." Mr. Clean worked for LDX and had noticed the illicit activity on the lines, so he traced Everingham's number and rang him. But Mr. Clean, it turned out, wasn't angry. In fact, he wanted to help; he coached Everingham and his friends on how the deep secrets of the LDX phone system worked. Generate the right sound tones, Mr. Clean told them, then play them through a speaker, and you could commandeer any part of the system. Armed with that knowledge, Everingham wrote a Commodore 64 program that could play the right tones. Soon he was not only scamming free long-distance numbers but also writing code that would generate Mastercard and Visa numbers, which he and his friends used once as a prank to call a phone-sex line.

"We were teenagers; you take a stolen credit card, we'd call a porn line, and they had to call you back," he remembers. "So we would conference out to a loop line . . . and there were like eight teenage delinquent kids using a stolen credit card to call back into a loop line. We'd start trying to get 'em to talk dirty and then we'd all laugh and hang up. I mean that's as far as it ever got." He laughs. "Of course this is complete juvenile delinquency." But as he argues: "Juvenile delinquency applied to technology will lead you to success."

Mind you, there still wasn't a clear sense, at this point in coder history, that being a programmer was a particularly lucrative field—or, indeed, even a field at all. Many of these kids tinkering around with BASIC had no clue that software engineering, as a job, existed. As a fortysomething Uber engineer, a contemporary of Everingham's, puts it: "I saw computers as this fascinating thing that I wanted to understand—although I didn't know why. To me, it truly seemed as practical as art history."

This was a big shift in the culture of who became a coder, and why. For the first time, programmers were emerging in living rooms, as teenagers, propelled by the culture of making, acquiring, and sharing software. But given that the video-game scene was primarily one of boys, it began to make coding culture—in this new, more-grassroots phase—ever more male. The same goes for the world of BBSes, where illicit contact with far-flung people on boards was something parents might tolerate or ignore in their sons but more likely forbid in their daughters. If you were one of the kids allowed into that BBS scene, though, then—as Everingham discovered—it was heavily tinged with the antiauthoritarian, open-sharing ethic of the MIT hacker generation. And it was a culture where teenagers learned the value of being connected: Long before the internet, they discovered that powerful knowledge could come from text chats with strangers they'd talk to halfway around the world. Like many boy-coder teens of his cohort, Everingham was spending so much time coding he was flunking his high school classes.

"I was hooked," he says. "I learned that this computer was the ultimate medium of expression and I could create anything that I could think of just out of sheer will. It's a multidimensional paintbrush. I could instruct it to do whatever I wanted."

By age 18, Everingham adroitly realized he'd need to stop his phone- and credit-card-hacking exploits or risk going to adult prison. He doubled down on coding, and began writing and releasing open source software that helped people create interfaces (text boxes, buttons, and the like) for their own programs. Everingham enrolled in Penn State to study computer science but found, as with high school, he'd rather be hacking than sitting in classes. He flunked out. Ironically, though, tech staff at the university discovered Everingham's open source software, and were avid users. When they realized this kid who'd just flunked out was the author, they hired him as a full staff computer scientist. Normally that position required a degree; his superiors had to apply for a waiver. "I was fortunate enough to be there at a time where my skill was rare. It was more out of not brilliance but out of rareness," he notes. They paid him $23,000, and, he said, "I thought I was a king."

By the early '90s, Everingham had built a career in software; he was particularly known for his work on interfaces. Then in 1995, in his early 30s, he was hired to work on the Windows interface for a newfangled product that would transform the world: Netscape, the first popular web browser. The team was led by Marc Andreessen, then a 24-year-old coder; it was a mix of just-out-of-college kids who'd built an experimental browser for their university and more experienced ones like Everingham, who'd been hired to help add structure to the project. The workload was insane. Andreessen knew that several competitors were trying to make a browser and was convinced—correctly—that Netscape needed to be first to market. The team worked nearly around the clock, sleeping on the floor of their offices; they'd shout insults across the room about each other's code. One programmer, vibrating with stress from overwork, once hurled a chair across the room after his computer suddenly rebooted and erased his work.

It was the birth of a new style of software creation. Before, a big software project followed a "waterfall" design: First, you'd figure out what the product was supposed to do, and write a design document laying out each feature in painstaking detail; then coders would spend months, or years, trying to create every last feature. It was a top-down approach: the design spilling from up high down onto the coders. Back then, you had to work extra hard to get rid of any bugs, because once you shipped your software out to customers—on floppy disks—you couldn't easily update or change it. Any bugs in your product were there for a long time, possibly forever. But Netscape was distributed mostly online; many, or most, users got it by downloading it. This changed the entire calculus around how you designed something, and whether bugs were bad. Since it cost almost nothing to distribute the product, the Netscape team could get the product just barely working and then release it—knowing that they could add new features and rerelease it later. Bugs? Sure, your customers would find lots of bugs. But they'd email you about them, transforming your customers, effectively, into free testers; better yet, since you had thousands or millions of customers, odds were high they'd find every significant bug.

This was the beginning of the "Move fast and break things" ethos that Mark Zuckerberg, a decade later, would post on the wall of Facebook's office. Or, as Andreessen's mantra went, "Worse is better." Software that is released quickly—and lets people do something new, however imperfectly and buggily—beats software that takes years to see birth or, worse, never gets released because the engineers spend too long dithering over perfection. "Our code was a mess," Everingham admits. But at Netscape, being messy was a point of pride. If a coder pushed out an update that "broke the build"—ground the browser to a halt—the other Netscape engineers would hang a huge yellow lemon over his chair, as Everingham recalls. It was both a mark of shame (you wrecked the browser because your code was sloppy) and something to which you aspired (you wrecked the browser because you were being daring and trying something new). Everingham says, "If you *never* had the lemon, that was bad. If you had the lemon *all the time*, that was bad. So it became this interesting part of the culture of appreciating the right amount of breakage."

Driven by that hummingbird metabolism, Netscape released four versions of its browser in a single year. The engineers transformed software production into something that was almost like a live performance, a band playing a series of songs and seeing how the audience reacted. The engineers were electrified to watch as everyday people began searching for information online or building their own websites. They'd add a new feature—email in the browser!—and watch it propagate online.

"It was my first time being in a loop with such impact," Everingham says. "I would write a feature, I could see myself in the product, and I could see it changing the world. And that was a drug like no drug I had ever had."

But the crash was to come. After four years, Netscape wound down, hobbled by competition from Microsoft and a corporate acquisition that led to a bungled attempt to rewrite the browser from scratch. Everingham left, convinced he would never experience that sort of success again. For four years, he sank into a depression; Netscape had made him wealthy, but as with young pop stars and lottery winners, sudden riches destroyed his relationships. Coders are no better than the rest of us at coping with life-deforming amounts of money.

"My family acted different, my friends were coming at me different, I had new friends, longtime friends were leaving me. I was like, what the fuck? Everybody wants something," he recalls. Women were coming on to him, which was fun but unsettling; he's a handsome enough guy, but this was new. (With his riches, "suddenly I was Brad Pitt.") Over the next five years he became so jittery about the social problems of his money that he gave a good chunk of it away to family members and charities. But he stayed in software and cofounded a firm that made online–phone call software. It was used heavily by stay-at-home workers in phone-support jobs. One user, though, turned out to be his sister. An inner-ear infection had left her with a vestibular disorder that rendered her homebound. She had spent years with no work but began using the software to become a phone-support worker; eventually she was sufficiently skilled that a firm hired her as a remote manager, Everingham notes, with a five-figure salary.

He began to see a new way to alter lives with code. You could do it flashily, with a browser. Or you could do it quietly, with a product that drew no public plaudits but was equally life-changing. "Changing one person, or changing a small group—and having an impact like that—is just as much of a rush as what I had," he realized.

"I was reprogramming myself."

It would take some years for Everingham to get another job where he'd work on software that was famous, a name on the tip of everyone's tongue. But eventually he'd be hired by the fourth wave of coders—the one that still reigns today. These are the programmers who grew up during the age of the web and mobile phones and used that as their on-ramp to programming.

When the web took off in the mid-'90s, it offered one of the most democratized ways to start coding—because a curious young person could peel back the hood of the web and see how it worked. When you point your browser at a website, the site sends back a long page of code—the HTML, the CSS, the JavaScript—to your browser. The browser runs that code, turning it into the things you see: a list, a picture, a video, a button you click. Back in the '90s, Everingham and the folks at Netscape realized that it would be fun to let people surfing the web see this code, if they wanted to. So they put in a feature that let you view the "source" of a page. If you clicked on it, Netscape would open up a window showing you the raw HTML of the page you were currently browsing.

Pretty soon, people around the world were clicking "view source" and getting a glimpse into how this crazy new world, the web, really worked. It was much like the BASIC revolution on the Commodore 64, except even faster and more widespread. Everingham and his peers in the '80s found it pretty slow going to get their hands on BASIC code to study and learn from; they had to download it from a BBS or buy a tech magazine or book that had printed some programs. There was a long gap between each opportunity to learn something new.

The web collapsed that time frame to zero. Every single web page you visited contained the code showing how it was created. The entire internet became a library of how-to guides on programming. If you wanted, you could cut and paste that code into a new file, change a few elements, and see what happened. If you liked what you'd done, you could put it online using the crude new services for hosting your own websites, like GeoCities. BASIC took programming out from the ivory towers and into teenagers' basements—but the web planted it firmly into the mainstream. Soon, teenagers worldwide were making websites for their favorite bands or video games, bedecked with hallucinogenically weird typography and graphics.

One of those teenagers making sites was Mike Krieger. As a middle schooler living in São Paulo, Brazil, he loved video games and had learned a bit of BASIC. "I'd spend my entire summers planted in front of the computer," he tells me. By age 11, he was so entranced by the web that he and a friend spent all their time tinkering with HTML. At school, they would hand in their book reports as custom websites. "We were, like, of course, super nerdy. It was like, 'Why didn't this guy just write a book report like everyone else?' "

Krieger didn't think of himself as a programmer; he dreamed of being a journalist or documentary filmmaker, roaming around São Paulo and reporting on its corrupt politics. He'd met Kátia Lund, the codirector of *City of God*, and she'd given him career advice. "Don't study journalism," Lund told Krieger, "study a subject you want to make a film about."

But in 2004, the summer before he went to college, Krieger became increasingly fascinated by the world of open source software, where hundreds or even thousands of coders would collaborate on building apps that anyone could use or modify. One popular tool was the email app Thunderbird. One night, while blasting his favorite band, Weezer, and reading the Thunderbird discussion boards, Krieger discovered that an American corporate executive had a complaint. The executive used Thunderbird to read several different email accounts—personal and work—and sometimes he'd get the two mixed up, accidentally sending a work email from his personal account, and vice versa. He wished Thunderbird had a color-coding system to show which emails belonged to which accounts.

Krieger was intrigued. "This wasn't like curing cancer or anything, but this guy has a problem," he told me. Could he figure it out and create some new Thunderbird code—a plug-in that created color coding? Krieger began "spelunking," as he put it, hunting around online for any Thunderbird plug-ins he could find. He figured he could use the same approach he took to learning HTML: See how other people did things, and learn from that. Gradually he began to piece his plug-in together. It took weeks to get one tiny part working, something as small as, say, getting his code to display the number of email accounts a user possessed. But when he did, it was a burst of adrenaline, an air-pumping moment of success so fun it made him willing to tolerate the next 30 hours of head-scratching bafflement. He discovered that coding was like playing a video game, where you'd beat a series of "mini-bosses"—small achievements—along the way to beating the "final boss." He explained, "Like, your goal is to open the Temple—Indiana Jones–style—and you need to get the right four symbols in the right place. And even getting *one* is like, woo, I'm on the right track!"

After three weeks, the whole plug-in was done. He posted it online and emailed the executive who'd originally complained about his problem. It turned out to be Greg Brandeau, an executive vice president at Pixar. Brandeau was so happy with Krieger's plug-in that he invited Krieger to a film premiere, "if you ever come to the States," as Krieger recalls. For a teenage kid in Brazil to be solving corporate problems for a major US executive was intoxicating. (And four years later, he joined Brandeau for the launch of *Wall-E*.)

In 2004, Krieger arrived at Stanford to study a quirky curriculum called Symbolic Systems, a mash-up of computer science, psychology, artificial intelligence, cognitive science, philosophy, and linguistics. (He discovered it via a zagging path: Google ran a social network called Orkut that was a complete failure in the US but wildly popular in Brazil, and for some reason the Symbolic Systems program had an Orkut page, so Krieger saw it. *This sounds incredible!* he thought.) While at Stanford, Krieger took a class with B. J. Fogg, who taught him about "persuasive computing": how the design of software could nudge people into new, hopefully useful behaviors. Krieger loved the concept. He wasn't just interested in programming machines; he liked thinking about people's emotional lives, and programming *that*. With another student, Tristan Harris, he built an app called Send the Sunshine, which encouraged people to try and cheer up friends; if the app detected you were standing in a sunny area, it would suggest you send a sun-soaked photo to a friend who was currently in a rainy, cloudy part of the world. It was clunky, but his aspirations got him noticed. As Fogg later wrote of him: "Many thousands of people can write code. But only a relative few can get the psychology right."

When Krieger graduated, he worked for Meebo, a firm that ran chat apps for hundreds of websites. He watched the seasoned Meebo coders handle "scale," the hair-raising challenges that came when one of your apps suddenly got hot and millions of people began using it, with load that threatened to crash your site. What really intrigued him now, though, was less websites than the newest coding frontier, the iPhone. It was a comparatively new field, ready for someone to figure out the question: What was this gizmo good for? Krieger plunged in and, much as he had with Thunderbird, tinkered with the platform, reading others' code, figuring out what worked, building little experiments. One was an augmented-reality app, where you could look through the camera and it would show you information about the world around you. "It'd show you an overlay of what crimes had happened around you, on your phone. So it was actually pretty terrifying," he says. " 'There was an arson ten feet away!' "

One night at a San Francisco coffee shop he ran into Kevin Systrom, a former fellow student from Stanford. Systrom was building a website designed to let friends share their nightlife experiences, called "Burbn." (Not, as the duo would later admit, the catchiest name.) Systrom had raised $500,000 in investment to develop it, but he needed help. "Mikey," as Systrom called him, got excited at the possibility and quit Meebo; soon, they were ensconced in cafés, sketching on paper how Burbn should work and what it should look like. In nightlong coding jags, they crammed it full of features—users could check in to show where they were, set up plans with friends for future dates, and get gamified Burbn "points" for meeting with friends. It was a lot of features; so many that users were increasingly confused by the overstuffed, tofurky-like nature of Burbn, which did tons of things in a mediocre fashion but nothing remarkably well. "Feature creep," Krieger was discovering, is a particular challenge of the coder personality. It's much more fun to create *new* things, to get an idea at 10:00 p.m. for something to add to your app—*a mapping animation! emailable alerts!*—and to excitedly bang it until 4:00 a.m., pushing new code into the world. It's not so much fun to refine and improve things that exist. Programmers are constantly tempted to turn their tools into Swiss Army knives. Burbn languished, with a small user base.

There was, however, one thing people liked to do: Upload photos. This was all the more remarkable because Burbn made it a painful chore to do so. (You had to email the photo to a Burbn address, and a script would attempt to slurp it out and post it on your feed.) Yet users kept doing it, working hard to find cool-looking snaps of their lives and get them in front of their friends. This, Systrom and Krieger realized, was interesting. Why did people enjoy it so much? Possibly because there weren't many apps that focused on sharing photos. Facebook had photos, but they got drowned in the slurry of the News Feed. Meanwhile, apps like Hipstamatic were hot on the iPhone; these let you "filter" photos to look retro, like old Polaroids, but they didn't have any feed so your friends could see them. That's when they realized their users were telling them something. They decided to scrap nearly all of Burbn and strip it down to nothing but photo sharing, with some clever filters to make pictures pop and enabling comments on the pictures.

After eight weeks of work, they had the new app done—and had settled on a name: Instagram. In a genius bit of retro styling, it was a portmanteau of "telegram" and retro '70s-style "instant" photography. And it did only one, simple thing. "See the world as it happens through another's eyes," as they described it.

… … …

When I first met Krieger and Systrom, they were crouched in a hallway charging their phones at the tech conference South by Southwest. Systrom, a tall, reserved sort, and Krieger, who often sports an easy, goofy grin, were both bleary with exhaustion. "We're kind of a mess," they joked. Launched a year and a half earlier, Instagram had been a sudden, rocketing hit. But as Krieger reminded me, while it looked like an overnight success, it had been years in the making. He'd spent years polishing his iPhone-app-making skills; Systrom, a fan of old-style analog cameras, had been obsessing over photo apps since his days interning for Google (and rebuffing Mark Zuckerberg's attempts to hire him in the early days of Facebook). Part of what made Instagram such a hit was that photos are a universal language: It was the first social network where you could follow—and be followed—by people around the world who didn't speak or write your language. By the time I met Systrom and Krieger, Instagram still had only about a dozen employees, but they'd amassed almost 30 million users and over 60 pictures were being uploaded every second.

Now they had a problem peculiar to coding success: scale. Making a piece of web software work for a handful of people is easy. Making it work for a few thousand gets harder; if a lot of people try to send or receive pictures from your database all at once, signals can get crossed. If you have millions? It was, Krieger discovered, like managing traffic in downtown Manhattan.

"It's terrifying," he told me years later, as he poured a complex, bespoke coffee for me when I visited him at Instagram's offices in Menlo Park, California. "I was, for a while, one of two people who was running the entire city. And if there's a fire over there, you gotta go put it out. And while *that* fire's happening, it causes *this* traffic jam! It's an organism way more than it is a knowable system. And it's totally not deterministic. Because you're not in control of who's using your site." Because the site was international, they'd discover a sudden crunch of network traffic at any hour of the day: In the morning, the East Coast of the US started posting pictures, and at 2:00 in the morning it was Korea and Japan. They'd go to bed at 2:00 a.m., setting their alarms for 8:00 a.m., "but know that we are going to get up at 3:00, because something's going to burn down." After two years, Facebook, realizing this trend of instant photo-sharing wasn't going away, bought Instagram for $1 billion.

While Krieger and I wandered through Instagram's sunny offices, we ran into someone he'd recently hired—James Everingham. He'd been brought in, it turns out, to help prevent chaos. As Instagram's phalanx of coders had grown to over 150, and the teams had begun to fragment into different groups (one doing the iPhone app, one doing the Android app, one managing the website), they had become unruly in size. Krieger was no longer sure he could keep everyone on schedule. He realized he needed an experienced hand, someone who'd been to the crazy scaling-up rodeo several times, who could structure their organization so it didn't spin out of control. "Jim was here for three weeks, just having meetings and listening and learning, then he came to me and said, okay, your structure is a *mess*," Krieger says, and laughs.

"I'm fifty-one, so I'm the oldest guy here," Everingham said. "They have these hashtag Instagram T-shirts; I'm going to get one that says #instagrandpa or something."

In many ways, the world of programming had changed wildly since he first peered at that VIC-20 back in the early '80s. For the first three generations of programmers—the folks like Mary Allen Wilkes, the MIT hackers, and the teenage Everingham—code was something they did mostly because it was interesting. There was a raw intellectual challenge in getting the machine to do, well, *anything*. They found the act of coding thrilling on its own terms. But Krieger's generation was the first to grow up in the shadow of the millionaires created by the likes of Netscape, Yahoo, and Google—and their enormous social impact. Sure, politics, law, and business are powerful, but if you want to *really* remold the contours of society? Write code. It didn't hurt that, as kids in the '90s, they'd all watched the leather-clad adepts of movies like *Hackers* or *The Net* or *The Matrix*. They grew up in a world where hackers weren't just outcast nerds anymore but superheroes who wielded powers the normies couldn't grasp.

This latest cohort is also increasingly being confronted with the unexpected side effects their creations visit upon society. Instagram, for example, is appreciated by many for having inspired new heights of photographic creativity. But many critics have worried about how it creates a culture of relentlessly performing a selfied "perfect" version of oneself, and mental-health experts have fretted that it also wreaks havoc with the self-esteem of young people, particularly young women. Some research has found that women's exposure to glam shots on Instagram creates "increased negative mood and body dissatisfaction." And anorexic women have also taken to Instagram, as they have to nearly every form of social media, to post "thinspiration" photos designed to glamorize their disorder, a behavior that sends shudders through health experts. Starting in 2012, Instagram has tried to combat the problem by banning hashtags like #thinspo or #thigh gap, and diverting anyone who searches for them to a pop-up offering a connection to eating-disorder services. But this merely provoked a classic social-media cat-and-mouse game; as researchers found, anorexic users shift to new coinages like #thygap or #thynspo. (Krieger and Systrom left Instagram in September 2018, citing a desire to explore new ideas; Everingham moved that spring to a division of Facebook that develops blockchain technology.)

Neither Krieger nor Systrom actively *set out* to erode anyone's self-esteem, of course. They loved photography, dug code, and aimed to unlock the latent energy of a world where everyone's already carrying a camera 24/7. But social software has impacts that the inventors, who are usually focused on the short-term goal of simply getting their new prototypes to work (and then scale), often fail to predict.

And frankly, the money was deforming decisions—what code gets written and why. By the time Instagram exploded, the finances of big start-ups were pretty well known. You gave away your tool for free, then tried to get millions of people to check it nonstop, to bring in a geyser of ad money. Apps like Instagram certainly did that—but almost too well, *dangerously* well, in the eyes of some more skeptical programmers.

Tristan Harris, that friend of Krieger's at college who collaborated on the Send the Sunshine app, had also become a successful denizen of Silicon Valley, but one who was much more dismayed by the social effects he'd help wreak. After graduating from college, Harris had sold a company to Google and gone to work for the search giant. But he grew disillusioned by how the designers of social apps were too often optimizing only for incessant "engagement," or compulsive use. "The job of these companies is to hook people, and they do that by hijacking our psychological vulnerabilities," as he told *1843* magazine. Or: "Never before in history have basically fifty mostly men, mostly twenty to thirty-five, mostly white engineer designer types within fifty miles of where we are right now, had control of what a billion people think and do when they wake up in the morning and turn their phone over . . . Who's the Jane Jacobs of this attention city?" No one, he feared. His colleagues were sharp and clever but, he argues, unprepared to think about the society-wide implications of what they do. For his part, Krieger noted that Instagram and Facebook had, by August 2018, released tools so users could track the time they spent on the apps, set an amount of time to spend per day, and set an alert if they exceeded that amount. Managing time spent is, he added, "an issue we care about deeply."

The Niagara of money in Silicon Valley, though, has been bringing in a new breed of coder that cared considerably less. By the late '00s, the example of companies like Instagram—minting so many multi-millionaires by making something that Facebook or Google might buy—caught the attention of a new breed of striver, one that we might call the Ivy League frat boys. In previous generations, they'd head for Wall Street, hungry to make quick millions. After the financial collapse of 2008, they needed a new field and found it in Silicon Valley. These "brogrammers" who began to show up, some straight from the frat house, were even concerned about the impact of their code; they weren't even necessarily interested in the intellectual thrill of the work, frankly. For them, the industry was purely a route to power. It's safe to say that overall this new population has made less of a contribution to coding than to the overall culture of Silicon Valley.

If you really want to understand at a high level the way that those who code think, it's useful to construct, from the ground up, what they actually do all day long. The type of people who get intrigued by the field are the ones who enjoy building, tinkering, playing with logic. But the ones who stay and thrive are the ones who can survive something much more mundane and grinding:

The bugs.

< Chapter 3 >

Constant Frustration and Bursts of Joy

Dave Guarino needed a vacation, so he took off to Fiji. That was precisely when the bug began to wreak havoc.

Guarino is a voluble and witty programmer who works as a director and developer for Code for America. It's a nonprofit that hires public-minded nerds and puts them to work revamping the creaky high-tech systems of governments. In this case, the system was food stamps. In early 2013, Guarino's three-person team at Code for America had been hired by the city of San Francisco to make it easier for low-income residents to apply online for public benefits, including food assistance.

And god knows, the government needed help. Its existing food-stamp website was a sludgy, painful mess. To apply for assistance, you had to fill in information on over 30 separate web pages, none of which were designed for the dimensions of mobile phones. This made it particularly hard for low-income Californians, most of whom used inexpensive phones with skimpy data plans to access the Net. California's clunky system had become a real problem; it was so hard to use that barely 50 percent of the people who were eligible for food stamps in California had applied for them. Guarino and his team were keen to fix the problem.

The problem was, they couldn't actually go in and recode the government website from the ground up. They weren't asked to do that, and in any case it would entail an expensive gut renovation of some vintage HTML and old databases.

So they hit upon an ingenious hack. They'd take the existing gnarly site and add a sleek robotic layer on *top* of it, hiding the old one.

To start, they coded a new, simple website. It was only one crisp page that would load briskly on mobile phones; it asked the new food-stamp applicant to enter his or her basic information. Then, behind the scenes, the coders would do some magic. They'd take the info that applicant had entered and hand it over to an automated script. *That* script would navigate over to the old, unwieldy government site and laboriously enter in the new applicant's info, field by field. In essence, the Code for America team had created an invisible daemon—an algorithmic helper—that would do all the hard work. The applicant would never have to deal with the details. They'd just type their info into the clean mobile-phone site, and presto, they'd be entered into the system.

It wasn't pretty. But it worked! "It was the hackiest shit imaginable," jokes Alan Williams, one of Guarino's teammates. But if you spend enough time around complex old systems, "you realize that the fastest, shortest way to a deep, solid, elegant integration is just by putting another layer of dirt on top." In the world of coding, this sort of fix is what's known as an "encasement strategy"—fixing a hairy, complex system by boxing it in. But sometimes it is the only option you have; and in this case, it was wildly successful. Within weeks, new applicants were streaming in. Within two years, there were so many sign-ups that GetCalFresh (as the app was called) would be heralded as one of the most useful new software projects of the California government. Guarino and his team were justifiably proud of their work. Sure, they could make more money devising apps that help millennials share cat GIFs. But it was far more satisfying to sling code that truly helped the needy.

Indeed, they got another idea for another feature that they thought would improve the system even more.

But this, alas, is where they created a ferocious bug.

Their idea was to make it simpler for food-stamp users to check how much money was left on their card. California's old system required users to call a phone line and wade through a ponderous voice-tree of options. "Screw *that*," Guarino said. Voice trees are an insult to humanity; everyone knows that. So instead he wrote some Ruby back-end code that created a phonebot. If you sent it your card number by texting it in (or typing it into an online form), the phonebot would check your balance and text it back to you. Much more pleasant for users! To promote it, they took out Google ads.

By now, Guarino needed some downtime. He's a twitchy, wired-up guy to begin with, who frequently picks nervously at his beard and talks in a rapid-fire pattern; after months of nonstop work, he needed a vacation. "Okay guys," he told the team. "I'm going to Fiji. I'm not bringing my laptop. No phone. No nothing. I'm just going to be there hanging out. You guys deal with stuff." *Okay, cool*, they said. *Have a good time!*

A few days after Guarino left, Williams and his team came into work to a disaster unfolding. The phone-checking system had run amok.

On a typical day, it would be used only about 100 times. But now it was getting 5,000 texts in barely a few hours. *What the* fuck's *going on?* they wondered. At this rate, the line was going to ring up massive charges, or, worse, get cut off by the mobile-phone provider. Why were people suddenly pinging it so rapidly?

Their first theory: It was being used for fraud. One government rumor had it that the Russian mob had been using food-stamp cards to steal money from the government. "The Russian mob must have found our service and they're using it in bulk to check the balance on all these cards! That must be what's going on!" they presumed. Worst of all, they had no way of contacting Guarino, who was the expert on that code. "So they're freaking out and I'm off the grid," Guarino says with a laugh. "They didn't know how the system works—I was the only back-end developer. I did all the code. So they were like, 'Oh God! I don't know what's going on!' "

Guarino returned from Fiji a few days later, and quickly set about trying to quash the bug; he paged frantically through logs of server activity. And finally, after hours, he spotted what had gone wrong.

It turns out a user had made a mistake. Someone out there had used the service to find their balance, as is normal. But instead of inputting their card number—which is what they were supposed to do—the user had accidentally sent in *the number of the phonebot service itself.* So the software got stuck in a loop. "The service was texting itself back and forth, back and forth, back and forth," Guarino says.

It was, he admits, ultimately his mistake, a flaw in how he'd written the code for the textbot. He could have easily written a rule checking to make sure that someone didn't accidentally text the bot its own phone number. But it never occurred to him that a real live person would ever do that.

"Users," he says ruefully, "will find a way." You might think you've stamped out your bugs, but they find new ones.

What type of personality, what type of psychology, makes someone good at programming? Some traits are the obvious ones. Coders tend to be good at thinking logically, systematically. All day long you're having to think about *if-then* statements or ponder wickedly complex ontologies, groups that are subgroups of subgroups. (Philosophy students, it seems, make excellent coders: I met philosophy majors employed at Kickstarter, start-ups, and oodles of other firms.) Coders are curious, relentlessly so, about how things work. When the pioneering coder Grace Hopper was a child, she destroyed so many clocks trying to open them up that her parents restricted her to just one to dismantle and rebuild.

But if you had to pick the central plank of coder psychology, the one common thread in nearly everyone who gravitates to this weird craft?

It's a boundless, nigh masochistic ability to endure brutal, grinding frustration.

That's because even though they're called "programmers," when they're sitting at the keyboard, they're quite rarely writing new lines of code. What are they actually doing, most of the time? Finding bugs.

What exactly is a bug? A bug is an error in your code, something mistyped or miscreated, that throws a wrench into the flow of a program. They're often incredibly tiny, picky details. One sunny day in Brooklyn, I met in a café with Rob Spectre, a lightly grizzled coder who whipped out his laptop to show me a snippet of code written in the language Python. It contained, he said, a single fatal bug. This was the code:

```
stringo = [rsa,rsa1,lorem,text]
_output_ = "backdoor.py"
_byte_ = (_output_) + "c"
if (sys.platform.startswith("linux"))
  if (commands.getoutput("whoami")) != "root":
```

```
        print("run it as root")
        sys.exit() #exit
```

The bug? It's on the fourth line. The fourth line is an "if" statement—if the program detects that `(sys.platform.startswith ("linux"))` is "true," it'll continue on with executing the commands on line five and onward.

The thing is, in the language Python, any "if" line has to end with a colon. So line 4 should have been written like this . . .

```
    if (sys.platform.startswith("linux")):
```

That one tiny missing colon breaks the program.

"See, this is what I'm talking about," Spectre says, slapping his laptop shut with a grimace. "The distance between looking like a genius and looking like an idiot in programming? It's *one character wide*."

Even the slightest mistake in an instruction can produce catastrophe. One morning in early 2017, there was a massive collapse of Amazon Web Services, a huge cloud-computing system used by thousands of web apps, including some huge ones, like Quora or Trello. For over three hours, many of those big internet services were impaired. When Amazon finally got things up and running again and sent in a team to try and figure out what had gone wrong, it appeared that the catastrophe had emerged from one of their systems engineers making a single mistyped command.

The word *bug* is deceptive. It makes bugs sound like an organic process—something that just sort of *happens* to the machine, as if through an accident of nature. One early use of the term was in 1876, when Thomas Edison complained about malfunctioning telegraph equipment he was developing. ("Awful lot of bugs still," as he wrote in his notebook later while working on glitchy incandescent lights.) The phrase entered the lore of programming after 1947, when Harvard engineers discovered that their Mark II was on the fritz because a moth had flown inside the huge machine, and, perhaps seeking the warmth of its internal components, got pinned inside a relay and prevented it from closing. They taped the bug to the log book, scribbling "first actual case of bug being found" next to it.

In reality, though, bugs are rarely an accident of happenstance. They are the fault of the programmers themselves. The joy and magic of the machine is that it does precisely what you tell it to, but like all magic, it can quickly flip into monkey's-paw horror: When a coder's instructions are in error, the machine will obediently commit the error. And when you're coding, there are a lot of ways to screw up the commands. Perhaps you made a simple typo. Perhaps you didn't think through the instructions in your algorithm very clearly. Maybe you referred to the variable `numberOfCars` as `NumberOfCars`—you screwed up a single capital letter. Or maybe you were writing your code by taking a "library"—a piece of code written by someone else—and incorporating it into your own software, and *that* code contained some hidden flaw. Or maybe your software has a problem of timing, a "race condition": The code needs Thing A to take place before Thing B, but for some reason Thing B goes first, and all hell breaks loose. There are literally uncountable ways for errors to occur, particularly as code grows longer and longer and has chunks written by scores of different people, with remote parts of the software communicating with each other in unpredictable ways. In situations like this, the bugs proliferate until they blanket the earth like Moses's locusts. The truly complex bugs might only emerge after years of coding, when your team suddenly realizes that a mistake they made on the first few weeks of work is now interfering with more recent programming.

As Michael Lopp, the vice president of engineering at Slack, once noted: "You are punished swiftly for obvious errors. You are punished more subtly for the less obvious ones."

Coding is, in a profound way, less about making things than about fixing them. The pioneering computer scientist Seymour Papert had a koan: *No program works right the first time.* Spectre discovered this the hard way in his early jobs. Raised in a working-class family in an 800-person Kansas town, he taught himself programming using library books in the early '90s, on computers he built from parts scavenged during dumpster diving outside his high school. After graduating with a degree in history, English, and philosophy, he went to work building Flash web pages for a car-dealer's website—which eventually led to a gig working for a game company in San Francisco. It felt like the big leagues; he'd told his grandmother as a kid in the '90s how he dreamed of working for a Silicon Valley company.

The reality was much bleaker. The company had created server software to run online games in which, ideally, millions of players would play happily together. But when the company launched its first game, the server tech turned out to be a bug-ridden catastrophe. "They had a pile-of-shit server, just, like, the worst code base you've ever seen," Spectre recalls. "We couldn't get more than a hundred people onto the game at the same time." The original designers had, early on, made several critically bad decisions in the design of the code, as Spectre saw it. They had created 160 different software applications spread across 23 different servers, and each could overwrite the memory of the other ones—which is a bit like trying to get a team of bickering colleagues to coordinate when they all (a) hate each other yet (b) can exert mind control on the contents of each other's brains. "It was," he says, "a nightmare to debug."

Spectre settled in, and his team began to quash as many bugs as they could. They got a lot fixed. But others were so obscure, caused by errors that overlapped on errors, embedded deep inside a structure they couldn't revise—mending it would have been, to use an old expression, like rebuilding a plane in midair as they were flying it.

One bug in particular drove them to despair: the "magic number five bug." It was a game service—a piece of code that would usually appear in the game's combat systems—that would suddenly shut down, causing a domino-toppling set of errors that would crash the entire game. They poked and prodded the code to figure out why it was behaving this way. When they're debugging, coders will sometimes use debugging tools that report on what different parts of the code were doing—what was in their memory—at the moment of failure. Or sometimes they'll code "print" statements in different subroutines, so the software will leave a trail of messages; the coder reads the messages, like Sherlock Holmes examining clues littered around a parlor room.

Finally, they were finally able to see what was causing the bug. It was a tiny memory error: The number *five*, written in hexadecimal notation, was getting written to a piece of memory that was already occupied. But try as they might, they could never figure out what other part of the software was sending in that bad number five. "No one ever did. No one *ever* did," Spectre says.

This explains a lot about the mental style of those who endure in the field. "The default state of everything that you're working on is fucking broken. Right? Everything is broken," he says with a laugh. "The type of people who end up being programmers are self-selected by the people who can endure that agony. That's a special kind of crazy. You've got to be a little nuts to do it."

Nearly every coder who works on big, complex software will tell you a version of this. The dictates of working with ultraprecise machines, so brutally intolerant of error, can wind up rubbing off on the coder.

Jeff Atwood has observed this closely. In 2008, he cofounded Stack

Overflow, a wildly popular board where programmers pose questions when they're stuck and receive help, often within seconds. Stack Overflow is a stew of the best and worst of coder behavior; answers can be remarkably generous, with programmers painstakingly solving each other's errors. But comments can also be snippily dismissive if a question seems naive or, horror of horrors, not precise enough. ("You are confusing me. Elaborate your question," is a typical Spock-like reply to a newbie's question.)

Why can coders be so snippy? Atwood asks, rhetorically. He thinks it's because working with computers all day long is like being forced to share an office with a toxic, abusive colleague. "If you go to work and everyone around you is an asshole, you're going to become like that," Atwood says. "And the computer is the ultimate asshole. It will fail completely and spectacularly if you make the tiniest errors. 'I forgot a semicolon.' Well guess what, your spaceship is a complete ball of fire because you forgot a fucking semicolon." (He's not speaking metaphorically here: In one of the most famous bugs in history, NASA was forced to blow up its Mariner 1 spacecraft only minutes after launch when it became clear that a bug was causing it to veer off course and possibly crash in a populated area. The bug itself was caused by a single incorrect character.)

"The computer is a complete asshole. It doesn't help you out at all," Atwood explains. Sure, your compiler will spit out an error message when things go wrong, but such messages can be Delphic in their inscrutability. When wrestling with a bug, you are brutally on your own; the computer sits there coolly, paring its nails, waiting for you to express what you want with greater clarity. "The reason a programmer is pedantic," Atwood says, "is because they work *with* the ultimate pedant. All this libertarianism, all this 'meritocracy,' it comes from the computer. I don't think it's actually healthy for people to have that mind-set. It's an occupational hazard! It's why you get the stereotype of the computer programmer who's being as pedantic as a computer. Not everyone is like this. But on average it's correct."

Atwood ran Stack Overflow for four years, trying to keep the community civil and productive, and was frequently successful; it's an invaluable resource for developers. But dealing with coder sensibility nonstop wore him down, and in 2012 he left the firm. "I had to step back. I had peered into the abyss. I was super burned out, I was super cranky, I was getting into fights with people," he says. "And one of the main things was I was getting sick of programmers. And I understood it deeply!" He *was* precisely like them, too. He's an introvert; as a child, his parents would be holding a party downstairs while he was holed up in his bedroom with a manual on the C programming language. These days, he's doing much less coding, and far more management, working with people. "I'm kind of glad I don't program anymore," he says, and sighs.

There's a flip side to dealing with the agonizing precision of code and the grind of constant, bug-ridden failure. When a bug is finally quashed, the sense of accomplishment is electric. You are now Sherlock Holmes in his moment of cerebral triumph, patiently tracing back the evidence and uncovering the murderer, illuminating the crime scene using nothing but the arc light of your incandescent mind.

My friend Max Whitney has been a programmer for over two decades, but she still remembers the first time she fixed a truly fiendish bug. She was working as a programmer for New York University, and students were reporting some trouble logging in to the university's main web portal. Specifically, they would sometimes discover that they were logged into someone else's account. Whatever was going on?

At first, they noticed that a large number of complaints came from students who were logging into NYU's portal while using computers at the Kinko's copy center around the corner. Maybe Kinko's was somehow to blame? But then Whitney saw reports of the same login bug from computers located on-campus. It became clear that the culprit was the university's login system itself. Unfortunately, that login code had been written years earlier by a staff programmer who no longer worked for NYU. Since Whitney couldn't ask him to help debug his code, she sat down to scrutinize it line by line with the help of another expert programmer.

Reading someone else's code can be a baffling task. That's because there's rarely a simple, obvious way to write a piece of code. Idiosyncrasies abound. Different coders have very different styles. If you asked four different programmers to write a pretty basic algorithm—say, one that figures out and prints the first 10,0000 prime numbers, for example—you'll likely get four different approaches that were structured and looked a bit different. Even something as simple as picking the names for one's variables can be a source of bitter argument between coders. Some prefer to use extremely short, one-letter variables (`x = "Hello, World!"`), arguing it keeps their code more compact and thus easier to glance at. Others prefer to use more descriptive variable names (`greetingToUser = "Hello, World!"`), pointing out that it'll be easier, a year later when the code is crashing, to look at a variable like `greetingToUser` and know what it means. When code gets particularly lengthy, or if something is particularly dense, coders are usually encouraged to leave little comments in their code that explain exactly what the heck is going on, so that some poor soul years hence will have guidance in sifting through the thicket. But often when coders are working fast, or under pressure, they don't "document" their code very much; and even with comments, frankly, figuring out the flow and logic of a piece of code can still be a brow-furrowing affair. (In well-functioning firms, no code is put into production until it's undergone "code review," with colleagues looking it over—not just to make sure it works, but that it's sufficiently readable by others.) One estimate suggests that coders spend 10 times the amount of time parsing lines of software than they do writing them. This is another reason coders can be so snippish and judgy about the style of their colleague's code. They know they may eventually need to read it.

This is the situation in which Whitney found herself. She and her colleague pored over the login code for hours, slowly figuring out how it worked, like an electrician patiently following the tangled wires that someone else had laid down in an apartment. *Hmmm*, *this* section triggers *that* chunk of code, which would get that *other* function to start up. . . .

Then suddenly, they saw it. When they finally had enough of the code's structure loaded into their minds, they could see the bug.

The problem began the moment someone connected to NYU's network. When students logged in, the system gave them a random temporary ID number for that session. To generate the random ID number, the program would "seed" its random-number generator using the timestamp, the exact instant that the student logged in. But what if two students just coincidentally logged in at precisely the same second? They'd be issued the same quasi-random number; oops! To prevent this, the programmer added another "seed" number, the IP address of the computer that the student was using. NYU had tons of IP addresses, so the programmer figured there was no chance any two students logging in would have precisely the same timestamp and precisely the same IP address. Right?

Nope. Years later, NYU and Kinko's switched over to a new technology that funneled lots of computers through just one or two IP addresses. They did this to handle the explosive growth of internet use on campus, but nobody realized it might interfere with the old login system, written so many years previously. But it did. Suddenly it became possible for two people to log in at the same time *and* have the same IP address. The two users would be assigned the exact same session ID, and presto: One of them would be logged into the other person's account, able to see the other person's email and notes.

In a flurry, Whitney wrote some code to test to see if their diagnosis was correct. It was. They'd figured it out. It'd take more weeks of slogging to actually fix the bug, but at least the mystery was solved.

And she was suffused with a drug-like euphoria, a feeling of mastery and accomplishment that rendered her aglow. "It was wonderful," she recalls. "I walked the halls of Warren Weaver Hall, up and down the little H-shaped hall, just going, *I am a golden god! I am a golden god!*"

She wanted to savor the moment, because she knew it wouldn't last.

"I knew that the moment I sat down again, I was gonna find the next thing that was broken," she says, and sighs. Sure, lots of the code at NYU worked fine. Most of it did, probably some of which she had written herself. But you didn't spend much time pondering the stuff that worked. Indeed, by definition, if it's working, you're usually ignoring it. "The actual thing that a programmer spends their time on is all the shit that's broken. The entire activity of programming is an exercise in continual failure.

"The programmer personality is someone who has the ability to derive a tremendous sense of joy from an incredibly small moment of success."

Part of what's so thrilling about a programming "win" is how abruptly it can emerge. "Code can quickly change states; it goes from not working at all to working, in a flash," as Cal Henderson, the CTO and cofounder of Slack, once told me.

This provides a narcotic jolt of pleasure, one so intense that coders will endure almost any grinding frustration just to taste it again, however briefly. And part of the thrill is, I suspect, the sheer unpredictability of when a piece of code will suddenly cooperate, suddenly begin functioning. When you're chasing a bug, success could be a day from now. Or it could be fifteen seconds from now. Who knows? This sporadic nature of success is eerily similar to the psychology behind casinos. When you're playing a one-armed bandit in Las Vegas, you have no idea when you're going to win. Payouts happen irregularly. But it is this very irregularity that makes slot machines so compulsive, as Natasha Dow Schüll writes in her study of casinos, *Addiction by Design*. Because we know that something really good *might* happen at any instant, we get caught in the loop of chasing it. Casino patrons wind up spending hours in a "state of suspended animation that gamblers call the zone," in which the payouts arrive as bursts of biochemical delight in their brain chemistry. Coders very often told me that they were "addicted" to coding; I suspect this is part of the reason why.

"I think everybody that does this for a career or does this long term has an unnatural, and almost unhealthy, satisfaction with problem solving," Spectre says. "Because if you're getting slammed in the face with failure all the time, that endorphin kick when you get it right has to feel—I mean, it has to feel really good to make up for the pain, right? It has to feel *amazing*," he says. "For me, it always did feel great in a way that is probably unhealthy."

The rapid fluctuation between grinding frustration and abrupt euphoria creates whiplashing self-esteem in programmers. Catch them in the low moments, when nothing is functioning, and you'll find the most despondent, self-flagellating employee on earth. But if you randomly come back an hour later—during which, hello, a three-week-long problem has been figured out—you might find them crowing and preening, abruptly transformed into the most arrogant, grandiose person you've ever met. Two years ago the programmer Jacob Thornton decided to rewrite his podcast-creation app Bumpers using a JavaScript framework called React, and the six-week process was a neck-snapping sine curve of highs and lows.

"Lately my coding style has been some sort of sociopathic oscillation between fits of crippling self doubt and an extreme kanye-like god complex," he wrote in a blog post, "where I'm either marching around my apt alone all day crying aloud or I'm calling my mother to let her know her 30 year old son is 'f***ing the game up (in a good way).' "

Programmers are the ones who create the flaws in their own code, of course. But it's the users who reveal the flaws; the users who, poking around and clicking on things, uncover all the inadequacies of the coders' work. This can often wind up planting a deep, vibrating misanthropy in some programmers' souls. Because, hey, your software would work fine *if all those stupid users didn't keep on doing dumb things that break it, right*?

Dealing with unpredictable behavior of users turns coders into "pessimistic, paranoid lunatics," as Blake Ross, the cocreator of the Firefox browser, once wrote. It happens gradually, because every time they create a piece of software they discover that people do things with it that they, the creators, never predicted. His description is funny enough that it's worth quoting at length:

> It starts when we're 8 and coding our very first program. "What's your favorite color?" it asks, sweetly, twirling a lock of Visual Basic around its finger. You type in your answer, the screen changes color accordingly, and boom—time to show off to family.
>
> Then Aunt Jody calls.
>
> "Honey, it froze on me. 'Color.exe has crashed.' I don't know what that means."
>
> You take a look at her entry. She entered: 2.
>
> "I thought it asked how many favorite colors I had?"

But how could you . . . but what does it even mean to have more than one favori . . . ok, fine. No big deal. You add a sliver of code to stop people from typing numbers into the box.

Next you post your program to the Internet. Thirty seconds later, you receive another crash report. That user entered: fart.

You can patch this, too, but you'd really like to understand it first. Was this just, somehow, another honest mishap? You send the user an email: "Why? Why would you enter 'fart'?"

He writes back: "blue." This is the moment you realize that some people just want to watch the world burn. And nothing is ever the same again. . . .

Mercifully, you graduate high school to bigger and better things: quant trading at Renaissance; digital cochair for Ross Perot '92. The sharper you get, the more important the work. But the more valuable the work, the craftier—and more determined—your adversaries. Every attack is more novel than the last.

By the time you land an engineering gig at Apple, you are a twitchy, tinfoily mess.

Of course, this blame game goes both ways. One can just as easily argue that it is the job of software developers to figure out what their "users" will do. This is precisely correct, and the problem we users face is that there are lots of developers with a stunning lack of imagination, so they don't even bother trying. But even a developer who's trying hard to inhabit a user's mind space can't predict everything. Machines may be linear, but humans are unpredictable.

So mature software firms have evolved an array of strategies to cope. Maybe they'll do tons of user testing, giving a thousand users early access to an "alpha" version of their app, then observing what breaks. Or maybe they'll go the Netscape route and just release the whole damn free program as a bug-ridden mess, letting users report what crashes, fixing as they go. Or maybe, if the developers are being paid millions to write a database for a big firm like a bank, they'll embed some of themselves with bank employees, the better to figure out how these mysterious humans will use their code, and to evolve the software in response to what the customer needs, in what's known as "agile" development. (The very adjective "agile" is telling: It's meant to suggest a team that can nimbly change direction, but it also brings to mind creators acrobatically bending and weaving their bodies to accommodate the unpredictable hazards of their users' needs, like ninjas gymnastically leaping to avoid a mesh of laser trip wires.)

In fact, frequently in much software development today at big firms, the decisions aren't always directed by the coders. The people determining what the software ought to do—what its features should be, what itch it should be scratching—are the project managers, working with designers and user-interface experts. They know that leaving coders to design the actual software could lead to a deep form of nerditis: You'd wind up with something that is usable only if you're a programmer who's completely fluent in typing text commands into computers.

That's because this is how many coders prefer to operate, given their own druthers. Frequently when they're interacting with the computer—or writing little scripts and tools to help accomplish a quick task—they use the "command line," the blinking cursor on the text-only screen, rather than a software-ified dialogue box. They're talking directly to the machine. Some of this is for functional reasons; it's remarkably more efficient and flexible to get the computer to do *precisely* what you want that way. But it can also wind up breeding a sort of disdain for everyday, nontechie users. Once you're used to the clicking precision of the command line, then the whole world of mouse pointers and arrows and icons—of being "user-friendly"—can start to seem slow and childish. Indeed, the phrase "user-friendly" is itself hilariously revealing: It implies that most software is "user-*hostile*," until someone can wrest the design from the coders' hands and get it in the mitts of someone who actually understands how real humans function. So that's what good designers and project managers and user-interface experts will typically do: Figure out how the software ought to work, round up users and interview them to find out how the tool ought to feel—then create a design for what the coders should build. A good project manager can figure out how the software on-screen (and behind the scenes) should flow and behave so that it works with, and not against, the behavior of everyday users.

Left to their own devices, a great many coders will simply be too estranged from everyday life to figure out this sort of "soft stuff." For them, software is often a psychological battleground between those who make it and those who use it.

Indeed, for many programmers, a profound allure of coding is that it's a refuge *from* the unpredictability of humans, from their grayscale emotions and needs.

A while ago, I met Michael, a guy who at the time had recently switched jobs into coding and found it deeply satisfying, in part for this reason. A quiet, precise 32-year-old, he'd initially studied mechanical engineering, intending to work in the nuclear industry. But the prospect of committing to a long study only to enter a field with few available jobs wasn't very appealing, so he got a job at a consultancy that analyzed the performance characteristics of buildings. His job was to write up the findings in reports.

But the work felt meaningless, unattached to reality. The models he used were so complex, and relied on so many assumptions fed into them, that he didn't feel they were comprehensible; he also wasn't convinced his work had any impact. As he produced thick, turgid reports, he started feeling like he was living in a Dilbert cartoon.

"You do get disillusioned, working in the corporate world, because it's all bullshit," Michael told me bluntly, when we met on a cool spring morning, walking along the High Line, an elevated railway park in Manhattan. "I'd work for two weeks on a paper, and it's, like, 'No one's ever going to read this. I *know* no one's going to read this!'"

He'd learned the basics of coding in college, and done a bit of it in his job. But he decided to go in deeper, so in his spare time he started programming more and more. He discovered that he loved the feeling of making software. In 2016, he quit his job to work full time as a programmer, creating a start-up app with a friend. Their product died, and Michael decided, as a sort of dare to himself, to write "an app every day," each one themed off a holiday or a major event of the day. For Valentine's Day he made a bot that would generate a love letter from a picture of your partner. For Christmas he made a bot that you could text with a picture of a loved one; it used AI to analyze the person's surroundings and recommend a gift to buy on Amazon. (When a friend sent the bot a picture of a musician playing guitar, it recommended a retro-style record player.) Michael riffed on political events, too: When the opioid crisis was in the news, he began working on a messaging app that could refer people to the nearest addiction-help services. When he woke up on Martin Luther King Jr. Day he was dismayed to see people posting hostile messages on Twitter. "I thought of making a bot to yell at them," he says, "but then it might just be my Martin Luther King bot yelling at a Russian bot."

"It gets addictive," he said, as we climbed the stairs to his apartment in the West Village, where he tossed his dark coat onto the couch and sat down at his MacBook, beneath two large, lovely, and nerdy pieces of artwork: pictures composed of tiny blue typed-up words from *Moby-Dick* and *War and Peace*. "It turns out I love solving problems for people! Having this impact—you get an idea, and boom, it's there, it's working, it's helping someone out."

Over the next few hours, he sipped a coffee and puzzled away at the code for his app of that day. It would be a Twitterbot that would recognize whenever someone tweeted a photo of sculptures in the High Line park, and then tweet back info about the artist. He trained an AI image-classifier on photos of the sculptures; he plinked out code to check Twitter for #highline tagged pictures. By the time he finally got things working, he'd been hunched over his keyboard for twelve hours, and his back was aching. The hours could be far longer than at his previous job.

But coding had one pleasure his old job didn't: a sense of clarity, of proof that his work actually was valid. His former life as a consultant was all about persuasion, convincing people, massaging PowerPoint presentations, trying to win arguments. It was shades of truth and Dale Carnegie, trying to curry favor with higher-ups so they'd shine favor on you. But with programming—getting the machine to do your bidding—persuasion was unnecessary. "The code works or it doesn't," he says. If your code is buggy, you can't bullshit the computer. It doesn't care whether you *seem convincing* or not. You simply have to hack away until you get the code right. The more software he wrote, the more Michael found it was imbuing himself with an odd new sense of confidence. He didn't worry as much about whether he had valuable skills, whether he was contributing to the world. He could prove it, with running programs that were solving people's problems.

"Learning to code is hard, but you get the self-esteem of 'I built this, and it works,'" he told me. "Nobody can *tell* me it's good or not. It actually works. Either the code compiles and works or it doesn't."

This love of objective, tangible results underpins most forms of engineering. In *Shop Class as Soulcraft*, the author Matthew B. Crawford had an epiphany, and a journey, similar to Michael's. Crawford worked for a conservative political think tank but quit because the work seemed empty. In the world of punditry, your value came from your status, how many people you could convince you were brilliant, the "chattering *interpretations*" you made to vindicate your value. Crawford fled to the world of motorcycle repair, where he discovered that it was enormously more spiritually fulfilling to wrestle with unforgiving physics. When you won—when you fixed the motorcycle—you had, as with coding, straightforward proof of your skills. A successful craftsperson, he writes, does not need to boast about his work, "He can simply point: the building stands, the car now runs, the lights are on."

This is the same sort of objective pride that coders often take when they've got something working. On the flip side, it's also part of what can render them arrogant and dickish. You can convince yourself that what you're doing is *real* work, and everyone else—the folks in marketing or sales or management—are just pencil pushers who don't truly produce anything. (*Ooooh*, the marketing folks say they've increased the firm's "mind share." The coders will sit, unimpressed, arms crossed, asking: Can anyone really *prove* that's true?) From the perspective of a really linear-minded coder, people who are involved in so-called soft-skills work are constantly hiding behind PowerPoint presentations because they can't actually demonstrate that they've accomplished anything of merit. They're just slinging buzzwords. This attitude can be a particular hazard among younger programmers fresh out of school, who haven't yet realized the value of all their nontechnical colleagues—the others who are frantically selling and organizing and keeping the company from falling apart. ("I remember as a young programmer going to staff meetings, my eyes glazing over as some big-shot leader with whom I would never directly work showed chart after chart of numbers that I believed to be completely irrelevant to me," as Chad Fowler, a longtime coder, wrote ruefully in his book *The Passionate Programmer*. "My teammates sat together, looking like a row of squirming children on a long car ride. None of us understood what was being presented, and none of us cared. We blamed what we felt was a complete waste of time on the incompetent managers who had called the meeting. Looking back on it, I realize how foolish we were.")

One front-end coder I know used to do a lot of visual art but found she did it less now that she was programming for a living. Why? She so deeply enjoyed the binary jolt of joy when a program started working. "I think part of the reason I'm not doing just art anymore is it just doesn't satisfy me the same way," she said. "There is no 'aha' moment with a painting," no instant when it suddenly begins "working."

I noticed this effect myself. I'm in my late 40s but haven't done much programming since I was kid. But when I started researching this book, I began teaching myself a bit more and quickly discovered something dangerous: It was much more satisfying to program than to write.

One week, for example, I decided to write a Python program that would help archive links I'd posted on Twitter. I often tweet about scientific or high-tech news and usually regret, months later, not having saved all that stuff in one easy-to-find place. So I decided to create a script that would log into my Twitter account every morning at 8:30 a.m., scrape out all my tweets of the last 24 hours, and analyze them: It would make a list of any tweets that had links in them, and ignore all the others, where I was just @-replying to people. Then it would email me a nicely formatted list of all the links, with the text of my tweet alongside.

Fortunately, some of the hard work was done for me: I could use Tweepy, a piece of open source Python code, to help me sift through my recent tweets and pull out the ones that had links. But even so, Tweepy could be a little baffling to use for a neophyte like me. When

I wrote code to query my tweets, I'd get back a huge "object," which appears as a long list of formatted data that looked like this:


```
_json={u'follow_request_sent': False,u'has_
extended_profile': True, u'profile_use_
background_image': True,
u'default_profile_image': False, u'id': 661403,
u'profile_background_image_url_https':
u'https://pbs.twimg.com/profile_background_
images/3908828/pong.jpg' . . .
```


. . . if you can imagine that going on for three or four solid pages, yikes. Essentially, it was my job to write an algorithm in Python that would step through every item in that huge object, looking to see if it possessed a URL I cared about. I made a dozen stupid mistakes—mostly the naive ones of a neophyte, like getting basic Python syntax wrong.

But by the end of the evening, success loomed. I was scraping the right information out. I hooked the bot up to a Gmail account, and set it up to run on a daily basis. When it finally emailed me the first correct gazette of my links, it looked like this . . .

```
"Isn't Baldwin a well-known pervert?" Inside
J. Edgar Hoover's FBI files on James Baldwin–
http://lithub.com/a-look-inside-james-baldwins-
1884-page-fbi-file/
```

```
Tarot cards were an invention of occult-obsessed
Paris in 1781:
https://aeon.co/essays/tarot-cards-a-tool-of-cold-
tricksters-or-wise-therapists
```

```
Three hours of music played with one single note
on a piano—D, at seven different octaves:
https://www.nytimes.com/2017/06/16/arts/music/
listen-to-three-hours-of-music-from-a-single-note.
html?_r=0
```

. . . and I let out a little whoop of joy.

Over the next few days, I set about writing several other Python Twitter tools for myself. I wrote a daily bot to archive tweet-threads between me and a few friends. I wrote another to download tweet-storms from other folks that I found interesting. And after a couple days I began to realize I hadn't done a lick of work on my book . . . because it was far more fun to write code.

Obviously, I enjoy writing! I've been a journalist for 25 years. Some of the joys of publishing are the same as the joys of coding. They both have the joy of creation ex nihilo; where there was once nothing, now there is something. But with coding, I noticed the same thing Michael found. My achievements at coding—modest as they were—felt more objectively solid, more tangible, than my achievements at writing.

The quality of writing is notoriously subjective. This is something that tortures all scribes. Let's consider the column I published last month in *Wired* magazine. Is it a good column? Well, that depends on who you ask. As the writer, I can tell you how satisfied I was with it; maybe it seemed "serviceably okay" to me, or maybe I felt I knocked it flaming out of the ballpark. Or you could ask readers or poll their reaction. Did it get a lot of shares on Twitter or Facebook? Did it produce a surge of traffic on the *Wired* site? Those are crude measures of popularity and don't necessarily reflect, in some abstract and Platonic sense, the column's *good*-ness. In writing, there is no objective measure as to one's quality or efficacy. A writer's reputation is built off the approval of others, which is what makes it feel like such a dodgy, slithy enterprise. When you're writing you're constantly second-guessing yourself, wondering how other people will receive it.

But the code I wrote? There's no question there: It *works.* Days, months, after I got the Python script running, it was still dutifully emailing me the tweet summary every day, an obedient robot. Is it the cleanest, most elegant code? Eh, hardly. But does it work—does it solve my problem? There's no question: Absolutely! In contrast, I have no such simple affirmations of the efficacy of my journalism. You cannot really ask whether any particular *Wired* column I write is "working"; writing does not possess similarly binary pass-fail parameters.

What's more, most coding is about breaking a big, hard task down into small pieces. You don't write one big honking program at one go; you write little chunks of code, small subroutines—functions, modules—that, when chained together, accomplish the big task. If you were cooking breakfast, you could think of each action you do as a function: Cracking the eggs is function (`crackeggs()`), buttering the toast is another (`buttertoast()`). Chain them all together in a logical flow, and that's your main program.

Generally when you're coding, you're focusing on creating one function at a time, and after you've written it you're supposed to test it to make sure it works. If you spent days writing a long program without testing its various components, it would almost certainly fail when you tried to run it, and it might be very hard to figure out which part (or parts) of the program was the problem. So you go piece by piece, testing as you go. You get a little "win" every time a function passes its test.

The upshot is that this can produce a remarkably soothing sense of progress. Research by the Harvard professor Teresa M. Amabile and researcher Steven J. Kramer has found that employees are happiest at jobs where they experience "the power of small wins"—regular, daily, visible progress. They hate it when they have only a fuzzy sense of accomplishment. But if, every day, they experience some concrete moment of success, they're thrilled. Coding gave me precisely that sense of linear, regular achievement. Writing rarely did. Producing a book always feels more like piloting a boat across a foggy lake. I know I'll eventually arrive at my destination, but the voyage will be filled with nail-biting doubt about my bearings.

My friend Saron Yitbarek, a programmer who founded CodeNewbie (a podcast and online community of new programmers) and runs the Codeland conference, also made a midcareer shift into programming after years of doing everything from marketing to journalism and research science. She, too, craved the concrete feel of making tangible progress. And she, like me, discovered the narcotic pleasures of solving everyday problems, watching a nonfunctioning bit of code suddenly spring to life: "There's something about those many little wins, I think, that make coding extremely satisfying in a way that other things weren't. If I'm solving a simple little problem like finding a glitch, a bug, I'm like—'Oh, that was done! All right, *that's* done, now it's a little bit better.' I can see it taking shape. It's like sculpting."

Faced with the pleasures of coding and the pain of writing, I started opting for the former. Any time my writing got difficult or existentially painful, I abandoned it—and pecked away at some code instead. It didn't matter what type of programming; anything would do, no matter how seemingly useless or pointless or simply exploratory. It was just fun to coax the machine to do something with precision. I'd spend hours at a site called Project Euler, which gives coders little puzzles and challenges you to solve them with algorithms. ("By listing the first six prime numbers: 2, 3, 5, 7, 11, and 13, we can see that the 6th prime is 13. What is the 10,001st prime number?") Or I'd page idly through coding blogs, until I found a fun new "library"—a module of code prewritten, that you can use to help create a new program of your own.

(*Heyyy, a new way of doing dataviz in JavaScript using info hoovered out of Google Sheets! Maybe I should play around with* that *for a few hours.*) I craved those moments of absolute clarity and success, the moment when the program came to life, doing precisely what I asked it to.

Gabriella Coleman is an anthropologist friend of mine who has closely studied hacker culture for years. As a writer herself, she noticed something interesting about the act of coding versus scribbling.

"You don't have 'coder's block' the way you have 'writer's block,'" she told me. "Of course some writers really love writing. But you hear so much about the agony of writing, the procrastination. Whereas I find with programmers, you don't find the same agony. If anything, it's, like, 'Yeah, I want to get back to coding!'"

Of course, there are obvious dangers to this seductive, Manichean world of the machine. Spend enough time in a world where pure logic and rigorous structure reign; where persuasion doesn't work, and can't; and you can wind up developing habits of mind that feel half machine themselves.

< Chapter 4 >

Among the INTJs

That was a bad move," Bram Cohen told me.

It was 2004, and I was sitting in Cohen's house outside of Seattle, hunched over a board game. Cohen, 29 years old at the time, had become famous in the previous year for inventing BitTorrent, a protocol that let people share massive files online with lightning speed. Hollywood and the TV industry were deeply alarmed by Cohen's software; they'd seen the music industry get blindsided only years earlier by the rise of people sharing MP3s online. For a while, they'd been immune to the problem, because digitized TV shows and movies were such huge files that it was too slow for fans to share those. That was true—until BitTorrent came along. Now Cohen was famous for creating a new tool that would "Napsterize" TV, and *Wired* had asked me to write a profile of him.

Cohen was one of the coder-iest coders I'd ever met. He wore his hair in a shoulder-length mop, sported a half shave, and loped about the house in a gray shirt with a dragon design. His work area was a room on the first floor, and behind his desk was an enormous plastic bin filled with dozens of Rubik's Cube–style "twisty puzzles"; he twiddled them with twitchy intensity, solving them and rescrambling as he pondered how to make BitTorrent run incrementally faster. Cohen was, I discovered, obsessive about puzzles and games. He was designing his own twisting 3-D puzzles, one of which was going to be produced soon for sale. (The goal of a good puzzle, he said, is to make it always feel like it's *just* about to be solved, when it isn't.) And he had a huge collection of board games, including one he'd bought just a few days earlier, Amazons. In Amazons, each player takes turns dropping more and more tokens on a grid, trying to box in the other until their opponent cannot move. He trounced me several times in a row. Hunched over the board, he described—his voice precise, methodical—his philosophy of the perfect game.

"The best strategy games," he explains, "are the ones where you put a piece down and it stays there for the whole game. You say, 'Okay, I'm staking out this area.' But you can't always figure out if that's going to work for you or against you. You just have to wait and see. You might be right. You might be wrong." He liked ones that were pure logic—no luck involved at all.

I'd first met Cohen in 2002, when I attended CodeCon, a tiny convention for hackers he'd organized that had one rule: Exhibitors could only present functioning, running code. "I'm sick of hand waving and vaporware, and people telling me what they're *going* to do," he told me; "I want them to just *do* it." We were sitting backstage in a dingy chaise lounge, with him gesticulating energetically in his black leather jacket. Silicon Valley had just gone through the dot-com collapse, which had seen the implosion of hundreds of companies whose mandates he regarded with contempt. Shipping pet food online? Flash-driven designer-clothes shopping e-malls? What sort of technical challenge was *that*? Cohen had been coding since he was 6 years old ("I entered a coding competition when I was 12 and won—or, wait, no, I didn't win, but I did really well.") He vowed he wouldn't work on such dot-com drivel any more, or, frankly, anything that never shipped. He'd worked for several firms that had collapsed before producing a product, or after producing one that no one really used. "I wanted to get a product in people's hands," he said. He wanted to solve a big, enormous, skull-crushing problem.

Cohen is an extraordinarily good programmer. I wound up seeing him on and off for the next fifteen years, and when I was writing this book his coding peers would all nod approval at the mention of his name: He's astonishing, they'd all say. So would Cohen himself, who was matter-of-fact about his abilities and work ethic. Why bother lying about something that's a fact? "I kicked ass," he said, casually, when I asked how he'd done in an interview at Google back when he was younger. They hadn't offered him a job, though he said it didn't matter; he didn't want to work for anyone again. He didn't like taking orders. He'd already annoyed previous employers by insistently pointing out the central problems in their entire product design. "I don't defer," he said with a shrug. He lectures other programmers on the importance of working manically, constantly striving to improve things.

"I tell people, you always take pride in your code. You should always be refactoring it, it should *look* like you've been working on it, when people see it," he said. A single flabbily written function would convey something other than total commitment to the craft. "I'm a firm believer in the broken-windows theory. You find a bug, you hunt it down and kill it." Indeed, when Cohen was working hard, he hated anything that took him out of the flow, even eating. While making himself a sandwich in his kitchen, back when I visited him for *Wired*, he complained that it was taking too long. "Sometimes I wish there were just some way to *install energy* in your body, like the Terminator putting a battery in his chest," he said.

The idea for BitTorrent came when he was talking to Andrew

Loewenstern, a coder friend who liked to record and share online video of performances by bands that were permissive about bootlegging, like Phish. In 2002, it was brutally hard for an everyday person to share a massive, one-hour-long video with someone else. Most home DSL connections were "asymmetric": You could download at a decent pace, but only upload data in a trickle. Because it was so frustrating to upload a big file, almost no one did it; their upload stream was idle nearly all the time. As Cohen pondered this, he had a flash of engineering insight. All that unused upload capacity? It was "excess capacity." If you could figure out how to utilize it, you would unlock a huge amount of uploading power.

The trick was getting lots of people to use their upload capacity together, in cooperation. And that's how Cohen devised the central concept of BitTorrent. It would break the huge file—like last night's episode of *Saturday Night Live*—into small pieces, and share them among several people online, "peers." When someone wanted to download that episode, they'd collect the little pieces from all those peers. Sure, each individual peer had a tricklingly slow upload speed, but if 30 of them all streamed their piece at the same time, the entire file would arrive at the downloader's computer in a brisk flow.

It sounds simple, but devising a protocol like that is a deep challenge. It was also a very new challenge. To get a website up and running, it's pretty easy; you're using the HTTP protocol (Hypertext Transfer Protocol, which was released in the early '90s and is well known and works reliably). But BitTorrent couldn't use any existing protocol like that. In effect, Cohen would need to create a new central element of the internet.

Now *this*—this was a challenge. He quit his job and lived off credit cards for two years while he worked, alone, on BitTorrent. When he released it, at first it had a clunky text-only interface that nobody but other coders could use. Cohen built a slightly more user-friendly visual one, and within a year word had spread, and 40 million people were using it. Cohen had married his girlfriend Jenna, a former system administrator with a five-year-old daughter, Riley; they relocated to Bellevue, Washington, where Cohen could focus monomaniacally on improving BitTorrent while living off donations from grateful users.

Many coders write "on the page," as it were, working line by line, function by function, testing bits and slowly building up the program. Cohen back in 2004 wasn't doing this. He'd wander around all day long lost in his head, pondering the weird timing issues of far-flung BitTorrent peers, modeling it in his mind. Then, after hours of this Romantic idyll, the code would come bursting out—and he'd sit down, composing so perfectly that it appeared to have been dictated. (He compared himself, matter-of-factly, to Mozart in *Amadeus.*) Jenna confirmed it when she wandered into the kitchen while breastfeeding their three-month-old son. "Bram will just pace around the house all day long, back and forth, in and out of the kitchen. Then he'll suddenly go to his computer and the code just comes flowing out. And you can see by the lines that it's clean," she says. "It's clean code!"

She pats Cohen affectionately on his head. "My sweet little autistic nerd boy."

Cohen does, in fact, have Asperger syndrome. I was initially surprised to hear it. Although he was obviously a bit bombastic and prone to launching into minilectures on just about any subject that came up, I liked him a lot and found him witty and frequently charming. He seemed quite good at picking up on the emotional cues of those around him. But it was, he said, the result of years of practice, training himself to act "neurotypical." "I learned how to do it," he said, one night as we drove to a local bar for a drink. He was raised in New York by parents who were intellectuals: a mother who was an elementary-school reading teacher; a father who ran a socialist newspaper. (It was among the first US publications to be designed entirely digitally, Cohen recalls, which is one reason he grew up with computers around the house.) Growing up and into early adulthood, Cohen realized there was something odd about himself; he felt awkward, at sea, when interacting with other kids. Even his body movements seemed strangely stiff. It was the early '90s, before Asperger's was a well-understood condition, so his parents never suspected it or had him diagnosed; they just figured Cohen was, well, nerdy. When Cohen was in his early twenties, he read an article about the behavior of people with Asperger's, though, and felt a jolt of recognition. So he decided to hack his behavior much in the way he'd probe and tweak a computer system. He read books on Asperger's, and as he walked around the city, he closely studied how neurotypical people interacted, gathering reams of what was, in essence, test data.

"I intentionally studied interactions and eye contact in particular," he told me. "I'm really big on duration of eye contact." Indeed, he'd even debugged the Asperger's therapy recommendations; though all the books highlighted the importance of eye contact, they rarely mentioned duration. "So people who have been trained with that often get that really wrong because there's really important subtleties to eye contact duration," he noted.

With Cohen, it was hard to separate the question of on-spectrum behavior from his rock-solid confidence in his own skill, his acerbic humor, his joy in pricking conventional behavior. "He'll come up to me and say, 'I just beat Riley eight or nine times in the Care Bears game,'" his wife, Jenna, says, laughing, when I visited him back in 2004. "And I'm like, 'Great, do you want a medal?' But Bram really teaches her good sportsmanship, how to win and how to lose with grace."

Cohen smiles as he listens to the story. He notes that Riley came home the other day from school after running a race and said the teachers told her there were no losers. "So I told her, of course there are winners and losers. You're just not allowed to *say* who won."

... ...

In the early 1960s, companies began hiring more and more programmers, as firms bought room-sized computers to crunch numbers, run payroll, or produce business projections. Thousands of programmers began working in everyday companies. And their bosses noticed: There was something kind of weird about these folks.

Office workers of the period were known for accepting conformity with a shrug—slapping on the gray flannel suit and playing their part in the hierarchy. They were the "ones of our middle class who have left home, spiritually as well as physically, to take the vows of organization life," as William H. Whyte wrote in *The Organization Man.* They believed in subsuming themselves to a larger whole; they followed orders dutifully; they were, in essence, collectivists. The people who gravitated to coder jobs weren't. They were odder, more idiosyncratic, off-putting.

"When we come right down to it," noted the psychologists Dallis Perry and William Cannon in a 1966 paper, "most people know what it means to be an accountant, a physician, or an engineer; but when it comes to computer programming, most people either have never heard of it, or have only the vaguest impressions of what this relatively new occupation amounts to." To learn more about this new tribe, Cannon and Perry took 1,378 male programmers and gave them a vocational assessment designed to suss out their interests and passions.

When they analyzed the results, three things jumped out. One was that programmers were avid problem solvers, "crazy about finding the answers to problems and solving all sorts of puzzles, including all forms of mathematical and mechanical activities." This probably wasn't terribly surprising to managers of the time. In fact, since there weren't yet many universities formally teaching coding, they used logic and pattern-recognition tests to figure out whether any newbies ought to be a coder. As Nathan Ensmenger notes in *The Computer Boys Take Over*, they were already selecting for puzzle lovers, and were beginning to shift toward hiring men over women. One ad for IBM around that time—entitled "Are YOU the man to command electronic giants?"—asked, "Do you have an orderly mind that enjoys such games as chess, bridge or anagrams . . . ?"

The second finding, though, was that coders liked to learn new things—and dreaded repetitive work. "They show a liking for research activities," wrote the psychologists, "and a tendency to prefer varied and even risky activities, while at the same time avoiding routine and regimentation."

The third finding spoke most directly to American managers' sense of unease that these coders were chilly isolates, the un–Organization Men. "They don't like people," Perry and Cannon concluded bluntly. "They dislike activities involving close personal interaction; they generally are more interested in things than in people." This was true for female programmers as well as men. When Perry and Cannon repeated the assessment on 293 women in coding jobs, they found the coders differed from noncoder women in two major ways: The women programmers had "a strong interest in all forms of mathematics and a lack of interest in people—especially in activities involving responsibility for helping people."

By the late '60s, the idea that programmers were a prickly bunch was cemented in the American managerial class. Richard Brandon, an industry analyst, thought it was risky for US firms to be so actively courting and hiring such hard-to-wrangle people. Many coders were young and coming out of the already antiauthoritarian counterculture of the '60s. When you put these kids in charge of important machines that their managers didn't understand, it was a recipe for insolence—or, as Brandon noted, employees who were "excessively independent."

The average programmer, Brandon continued, was "often egocentric, slightly neurotic, and he borders upon a limited schizophrenia. The incidence of beards, sandals, and other symptoms of rugged individualism or nonconformity are notably greater among this demographic group."

Report after report piled up, testifying to the social reticence and arrogance of programmers. "Too frequently these people, while exhibiting excellent technical skills, are non-professional in every other aspect of their work," declared a fretful 1971 report. Another set of studies in the mid-'70s included one by the psychologist P. H. Barnes, which found that coders were "quiet, reserved, independent, confident, introverted, logical, and analytical" and that they still really didn't want to integrate with other folks: "The computer became a more desirable companion."

In 1976, the computer scientist Joseph Weizenbaum proclaimed that programmers were becoming defective in empathy, because they spent too much time in a tight loop with the machine. Weizenbaum had worked at MIT since the '60s, and wandered into the AI lab, where he'd beheld the unkempt and often-fragrant original generation of hackers. In a famous passage from his book *Computer Power and Human Reason*, he described the lab as a world of hopeless addicts, cut off from humanity:

> . . . bright young men of disheveled appearance, often with sunken glowing eyes, can be seen sitting at computer consoles, their arms tensed and waiting to fire their fingers, already poised to strike, at the buttons and keys on which their attention seems to be riveted as a gambler's on the rolling dice. When not so transfixed, they often sit at tables strewn with computer printouts over which they pore like possessed students of a cabbalistic text. They work until they nearly drop,

> twenty, thirty hours at a time. Their food, if they arrange it, is brought to them: coffee, Cokes, sandwiches. If possible, they sleep on cots near the printouts. Their rumpled clothes, their unwashed and unshaven faces, and their uncombed hair all testify that they are oblivious to their bodies and to the world in which they move. These are computer bums, compulsive programmers . . .

Aloof, introverted, acerbic, refusing to follow orders: The nerdy reputation of coders hasn't changed much since it congealed back in the '70s. Many programmers I know today shrug at the idea that they're difficult (*Eh, true enough*); some regard it as a badge of honor or, possibly, a way of transforming their lazy refusal to be polite into noble badassery. But plenty of coders, too, are deeply annoyed at the stereotype.

"We are unfairly tarnished by this idea that programmers can't communicate, are always awkward," complained my friend Hilary Mason, when I visited her at Fast Forward Labs, her machine-learning firm in New York. She thinks the reputation comes partly from their being comfortable and fluent with machines that intimidate and mystify most of the rest of the population. "If I had to characterize the programmers I know, I'd say there's a certain confidence that comes with being infused with technology. It's that confidence in actually understanding what this device in our hands is doing." Mason is a pioneering data scientist and as committed a nerd as they come; when I first met her years earlier, she enthusiastically told me how she had "replaced myself with a bunch of small shell scripts"—she'd written dozens of short little programs to reply to dull, rote emails (students of hers asking "Will this be on the exam?") because she'd rather save her time for more important stuff. But she's also a connector who's founded or helped start all manner of organizations designed to help bootstrap newbies to tech, including a Brooklyn "hackerspace" and hackNY, which runs hackathons for students. In a very data-scientist fashion, she rebels at the idea that a single archetype can hold true across an ever-larger cohort of coders worldwide. The population has grown so huge that you can't generalize across the entire field anymore when it comes to personality.

She's right that the population of programmers has exploded. Back when Perry and Cannon first started studying this odd new caste, there were barely 100,000 professional programmers in America. In the '70s and '80s, the field was arguably much more homogenous; the early women pioneers were gone, and the scene was much more one of heavily male shoe gazers. The teenagers who were cranking out BASIC on early home computers were brute-forcing their skills out of obsession, which meant the whole scene selected for deep nerdery. Few were getting into coding to make money, because it wasn't yet clear that millions could be made. "I actually had no idea how valuable it was going to become," David Bill, a 46-year-old veteran of coding, told me over dinner one night in San Francisco. To be into computers back then, you were "a very logical person, in fact so much so that you didn't fit in super well with your fellow humans in high school or whatever, because you are more logical and rational—and in fact you found it frustrating to deal with people who were not super rational, right? And watching their lack of logical inferences break down, and things not quite adding up."

By the '90s, the explosion of high-octane start-ups led to another reason for coder surliness: the sheerly brutal pace of work. When Netscape was rushing to get its browser to market in 1994, its coders developed the culture of frantic speed and overwork that left no room for niceties.

"In the early days, we were kind of horrible to each other," recalls

Jamie Zawinski, who was 25 when he worked on the Netscape browser. "We were very awful. People would just lean over the cubicle wall and say, 'This sucks, your code is shit, what are you doing?' And you go grumble, grumble, grumble, grumble. *I fixed it!*" The team of nearly all young, just-out-of-college men developed a proud sense of avoiding all bullshit. Rather like being in the military, worrying about feelings would just slow you down. For Zawinski, the stress and pace of work wracked his hands and wrists with such terrible repetitive-strain injuries that he wound up spending years trying everything from acupuncture to wrist braces to ease the pain. He has a grudging admiration for the culture. "At some level, it's nice to have coworkers who don't have to beat around the bush," he notes. "But in hindsight it might've been better to treat each other better."

When Rob Spectre started coding professionally in the '90s and early '00s he found this self-mythologizing culture—fuck your feelings, we're just here to speak truthfully about whether the code is good or not—was still the default setting. "I remember going on IRC," he says, referring to the text-chat mode popular among coders, "and asking for help with this PHP problem. And the first thing that someone responded to me with was 'Well have you tried *not being an idiot*? Have you tried that? Have you tried not being a *moron*?' There was a part, when I was learning how to program, that was super adversarial. It was really mean. Everybody was super, super, super grumpy. No one wanted to teach you anything."

What's changed since then is the sheer number of coders. The industry in the US alone ballooned in size by the late '90s and early '00s, growing to over 4 million programmers today. It brought many more types into the field.

This was particularly true in the growth of "front-end" design. Front-end design is the world of HTML, JavaScript, and CSS—the code that arranges the way things look and behave in a browser, how the logic flows inside the part of the app that you see. Because it involves the layout side of coding—almost comparable to the design of gorgeous magazine layouts in the world of print magazines—it tends to attract a different phylum of coder. The back-end code is code that a user never sees: the databases that store all the entries in your blog or the server software that sends pages to your browser; it's the plumbing of the web. The ranks of the back end are filled with the programmers who get aroused talking about data structures or arguing about which languages are built for maximum speed. These are the more isolated nerds, the ones who enjoy the intellectual mouth-feel of algorithms for sorting arrays, arguing in sprawling threads online about how to shave milliseconds off a binary-tree search or why bubble-sorts are so dreadful.

Front-end engineers are certainly nerds; the coding required to make apps function in browsers or phone displays is frequently mind-bendingly complex, and many of those working in it today formally studied computer science. Yet it's also a field that—as I found while interviewing coders—was home to a surprising number of those who taught themselves, and who arrived via a cultural side door. Often when they were young, they wanted to make something fun or weird online: Maybe they crafted websites for their friends' bands, online anime tribute pages, or sprawling confessional emo diaries. They obsessed over how to make things look gorgeous, or aggro, or new; they learned how to write a simple HTML page, then realized they could change the way all their pages looked by changing the style in their CSS and then realized, hey, if they learned JavaScript they could make random-insult-generators for their friends. These engineers got into coding not just from the thrill of commanding a machine but because it let them make something compelling and useful, for others to see and use.

... ...

That's how Sarah Drasner shifted into coding. A graduate of art school in the early 2000s, she got a job as an illustrator at the Field Museum of Natural History in Chicago. "I was doing scientific illustrations of snakes and lizards and things like that, for encyclopedias," she said. The museum still needed hand illustrators, because cameras weren't good at taking pictures of specimens; during a close-up they could focus only on the top or bottom of an insect, meaning part of it was always blurry; so illustration was the only way to go. But a few years after she started, the museum bought a new type of camera that could capture every layer of a specimen at once. They told Drasner they didn't need her illustrator skills anymore. "Do you know how to program websites?" they asked.

She lied, saying she did, then fled home to quickly start boning up on how HTML worked. She found it easy to pick up—learning how to make pages attractive triggered all the same aesthetic muscles she used in art. After a few years, she left to teach art at a college on an island in Greece but kept on making "silly websites" in her spare time. When she tired of the low pay, she returned to the US, where her web skills won her a job at an online design firm. They were brutal taskmasters: "If you coded something one week and it took you 12 minutes, and the next week it took you 14 minutes, they'd ask you why," she says. "Work was crazy. Like, I had to clock out to go to the *bathroom*. But I ended up getting better really fast." In front-end design, working speedily—and constantly learning new techniques—is important, because front ends can change suddenly. Back-end code evolves more slowly; once you've built the database for, say, your company's payroll system, it can be difficult or disastrous to change it, so you generally don't. But a client might ask for you to suddenly reengineer the front end of your site to make it perform in a new way, such as including new forms of navigation, or optimizing the code base so it loads and performs more speedily. Or it might be that Google has updated the Chrome browser in some way that suddenly requires all the front-end code to be tweaked, lest the website suddenly stop displaying correctly. Either way, front-end designers often finish a grueling redesign of a site only to find they're being asked to start planning a whole *new* redesign. "It's kind of like the Golden Gate Bridge, where you have to repaint it as soon as you're done," she jokes.

But mostly, what makes front-end coders a breed apart is they have to think deeply about their users—what they're seeing when they look at the page, what the user will or won't understand. They ponder the psychology of attention: How should you design a page so it draws the eye to the right place? Back-end coders need to worry about reliability, about making bits move around speedily. Front-end coders have to do that too, while also guiding the eye and the hand.

Drasner loved the artistry and the gnarly logical challenges, pondering how to delight and guide someone's attention with code. She also discovered that the psychology of visual artists overlapped a lot with that of programmers. The need for precision, for obsessive detail; even the work rhythms were similar. She'd come home from work and do a second shift, from 9:00 p.m. until 1:00 a.m., when—in a pool of isolation on her sofa in the dark—she could finally immerse herself, and the code would flow. She'd always been good, even disturbingly so, at shutting out the world while working: "I have a hard time putting things down. My old boss used to joke that, okay, if you need anything done that takes 30 hours, give it to Sarah. Because I have like a reverse ADD. I can't focus on anything *unless* it takes a really, really long time."

Drasner made her name as a world expert in SVG graphics. That's a way of drawing pictures in the browser by specifying individual shapes—a circle of this size in the corner, a rectangle of that size below it—which can be animated with code: "It's drawing with math." She wrote a best-selling book on the technique and left her day job to travel the world giving talks and workshops on SVG graphics. When I first met her, I'd wandered into a demo she was doing for a handful of rapt programmers at a conference, where she showed off reams of code on an overhead projector, her hair dyed brilliant red.

Drasner seems, in many ways, the orthogonal opposite of the grumpy, irritable nerd. She spends her days teaching; she tweets rallying encouragement of other coders. ("OMG AWESOME"; "This is amazing"; "It's so humbling to see what everyone is making. What a time to be alive, seriously. And I'm not even that drunk, either.") When I visited with her former colleagues at the real-estate website Trulia, they gushed. ("We *looovved* Sarah," one told me when I visited.)

Still, for all the energy she puts out, Drasner is by her own estimation an introvert; she needs isolation to replenish her energy and prefers to work in a cloistered fashion. When Uber tried to recruit her, she visited their main office only to be horrified at the site of the engineers seated at long picnic-style benches, out in the open. "I can't work here!" she told them. "I can't imagine writing even a *line* of code here. Like, there's *no way.* This place is crawling with people."

When Drasner gets deep into a programming challenge, the rest of the world drops away. At night, her husband—also a software engineer—often needs to force her to put away her laptop so she'll stop sharpening and improving her code and go to sleep. One day in the summer of 2015, when they were still dating, he'd suggested at breakfast that they go for a walk later to the Golden Gate Park. She'd agreed, but then got swept up trying to fix a bug in a piece of convoluted SVG code. The time came to leave; her boyfriend came over to get her. *Just five more minutes*, she told him. He came back after five minutes: *No, no, no, I just have to finish this one thing*, she said. This went on for multiple cycles.

He eventually managed to wrest her away, which turned out to be a good thing, because when they finally got to the park he pulled out a ring to give to her. He was trying to propose; she'd been so hard-core on her coding she could barely pause.

Back in 1985, Jean Hollands published *The Silicon Syndrome,* a guide to relationships between computer engineers and the nonengineers who love them. Hollands was a psychologist who'd turned herself into a corporate coach. The book was based on interviews with scores of married couples (primarily, given the era, male engineers paired with women) in the valley. "I recognize the Silicon Syndrome," Hollands wrote, "whenever the wife says, 'But he's always in front of the computer.' Or, 'He's more interested in printed circuit boards than he is in me.'" Her main argument was that engineer/nonengineer relationships arose from a sort of high-tech Mars/Venus disjunct—"a crippling lack of communication caused by differing styles of thinking and feeling. He speaks Chinese; she speaks French. They do not know how to translate each other. They don't even know how to sell themselves or their ideas to their partners." There's a section subtitled "There Is a Heart Inside the Tin Man."

It all struck me as far too glib and stereotypical. After all, I know plenty of coders who are happily married or otherwise paired up. But as I started interviewing more of their partners, many agreed that while things weren't quite "Tin Man" status, dating programmers meant adapting to a 24/7 engineering mind-set.

Jennifer 8. Lee, a 42-year-old investor and vice chair of Unicode Emoji Subcommittee San Francisco, estimates that about half the people she's dated have been programmers. She's drawn to them; this is due, she thinks, to her family background. "My dad is a socially awkward, introverted engineer," she tells me over dinner. "So, total Electra complex. I like socially awkward, quantitative engineers, sue me!"

Lee studied a bit of programming herself, but had no interest in doing it for a living; she likes working with others, and couldn't imagine sitting in front of a screen alone for so many hours. She became a technology reporter for the *New York Times*; I was long an admirer of her sharply observed pieces on Silicon Valley.

"What I like about these guys," she explains, "is they're very reliable." That's the upside to people who build rules-governed systems. "They would be good husbands and good fathers and I was willing to sacrifice some of the warm fuzzy stuff." They were skilled at managing anything in life requiring organization, systematization. She'd go on a trip with a coder; he'd pack items in the SUV trunk like Tetris bricks, ultraoptimizing each inch of space. (Better yet, whatever items they'd need most frequently would be carefully laid on the top layer for easy access, precisely the way you "cache" regularly accessed data for faster access. "They take great joy in that.")

They weren't emotionless; quite the contrary. When she dated Craig Silverstein, Google's third engineer, he was better than her at listening empathetically to her girlfriends' tales of romantic woe for hours. Silverstein broke up with her, in fact, when he realized he most wanted a happy family and wasn't sure Lee was the right one for him. Though when she asked why he was ending things, he enumerated a list of reasons—"they were ranked, too," she says, and laughs. Number four: She "spent too much time on my devices." (This was true, she admits.)

When I spoke to Lee, she had not long ago broken up with her latest programmer boyfriend. He was dreamily attractive, with long dark hair and eyes, "but very Aspy. That's a bad combination! He's very Vulcan. And he takes pride in that. I think some people would be bothered if some people called them Vulcan, but he actually takes pride in the fact that he's Vulcan, he actually kind of enjoys that he's very Spock-like." They broke up after less than a year, when she'd reached her mid '30s, wanted to start a family, and wanted to know if he—eight years her junior—was ready. He wasn't sure. She understood; it's awfully hard to figure out, as a guy in your late 20s, if you've met The One, because you've only had a few long-term relationships. There's not much to compare it to.

It's a common problem, and her boyfriend realized it was a version of a classic dilemma in game theory, the "Secretary Problem." In the Secretary Problem, you're told that you have to hire a secretary, and you start interviewing candidates. Once you decide on the one you like, you hire that person; but you stop interviewing, so you'll never know if the *next* person you might have interviewed would be better. The Secretary Problem has a famous solution, but it arrives at the "best" hire only 37 percent of the time, which is perhaps an acceptable risk for hiring an assistant but is troubling to a math-minded coder who prefers greater certainty about committing to a lifelong partner. On the other hand, the downside of *not* picking a secretary—or partner—is making the opposite mistake, and dithering for long enough that you wind up with a suboptimal pick; or perhaps never meeting any new suitable partners, ever. This is why most people, faced with a romantic decision, go with their gut; it is a usefully somewhat less agonizing way to resolve the problem than relying on raw logic. From an engineering perspective, romantic choices are awfully hard to optimize. After running the numbers, her boyfriend decided to break up with her.

"I don't have enough data," he said.

Coders themselves will frequently cop to the problems their mental style can introduce to relationships. Dozens sheepishly told me of arguments caused by them taking a methodically logical approach to fuzzy-edged emotional issues. Scott Hanselman, a coder for Microsoft and popular blogger and podcaster on software and other subjects, has been married for twenty years to Mo, a nurse who isn't terribly interested in technical issues. He calls it a "mixed marriage—a geek and a normal."

"I'm a 'World Famous Coder,'" he says, "and my wife could give a shit." When they first started dating, he, too, treated each of her bad moods as a technical snafu to be untangled. "She comes home and wants to tell me about her day," he adds. "And I don't care about her day—I care about *her*. So I want to fix her day! I'm like, 'Those interactions with the nurses were hard because of X, Y, and Z!' And she's like, 'Shut up, I don't want you to fix it. I want you to shut up and listen.' And I'm like, 'Okay, I'll shut up and listen for twenty minutes a day.'" (Twenty minutes seems to be the quantum of listening time most of these couples has settled on.)

Hanselman himself has pretty high EQ. He's done stand-up comedy for years and talks in a rapid-fire patter of wry, self-deprecating jokes. In tech circles he's known for energetically advocating to get more underrepresented techies into programming. Yet he still bridles at moments when he thinks his wife is ignoring basic logic and rationality. Scott and Mo once argued hotly for months over the fastest driving route between her sister's house and his; she preferred taking a route that didn't have stoplights and insisted it was faster, which drove him absolutely bonkers since he says he'd charted it on a GPS and proved—with a "preponderance of evidence"—that it was 17 miles longer. Months later, he was still baffled at how, for her, "the logic stuff doesn't work."

In social situations, though, his engineer's mind gets him in trouble because he frequently blurts out precisely what he thinks, witless of the social damage it might cause. When they visited the house of Mo's brother in South Africa, her brother had just repainted the kitchen yellow. "Yellow? I don't like *that*," Hanselman announced bluntly (as Mo recalled it). To her, the comment seemed bluntly rude; in a podcast they did together, they talked about it.

Mo sighed. "You were coming at it," she told Scott, "from a point of, *Hey, I'm giving you honest feedback!* But it really landed like, 'Wow, okay, so here's a rude guy who just flew in all the way from the United States.'" It's a side effect, she's come to realize, of the fact that Scott spends his time all day long in an environment where being utterly blunt is valued. "To the extent that you're spending 8, 9, sometimes 10 hours, 12 hours a day working with engineers, people who think like you," she said to him, "when you shift from that, you almost need to decompress a little bit before you're fit for the rest of society."

He admits it's true. He works in a home office, surrounded by his wide-screen monitor and (when I Skyped with him) a jumble of hardware projects. When he finishes work, he'll immediately start critiquing whatever Mo's doing. "Is this what we're having?" he'll ask, if she's cooking dinner. "Is that how you cook that? Is there enough oil there?" It's as if he's still wrangling with his nerdy colleagues. "It's like, I've immediately gone into *code review* mode of the rice and meat," he admits.

Mo has noticed there's a mental transition-time that comes with coding. The work is so mentally all-absorbing and requires such exactitude that it's hard to step away. The code still haunts the mind.

He agrees. It's like coming up from a long dive underwater: You have to do it slowly. "It's an engineer-technologist version of the bends."

This comparison to deep-sea diving is revealing, because many programmers have told me the same thing. "Most of the time, we're not snippy or weird or overly logical," they say. It's just that they turn into terrifying robots during the actual *act* of coding.

This is because of the peculiarly taxing mental demands of the discipline. Debugging a piece of code is more than just staring at those few lines and trying to figure out why they're wrong. No, it often requires thinking about the enormous hairball of the entire system: how those few lines interact with dozens or hundreds of other modules of code—each one passing bits of data back and forth. You start with one function; see what other pieces of code it talks to; figure out which pieces of code *those* functions talk to. Slowly, slowly, you can begin to build up a mental picture of the many-nested interrelations. After a long effort, you finally load the system's Byzantine logic into your mind's eye; you can see how the Rube Goldberg mechanisms work, how *this* ball knocks over *that* domino that causes *this* switch to flick, turning on *that* light. Or to use a slightly different metaphor, you've soared high enough in the air that you can see the entire city arrayed before you at night. You can see the Matrix. And it is at this point that programmers can actually get serious work done, because they understand the implications of changing any tiny part of the mechanism. Coders typically told me it takes many minutes to get into that state, but for really gnarly coding problems it can take hours.

Once you're there, though, it's glorious. Doing coding when the program is arrayed in your wetware is deeply pleasant. It is a classic "flow" state identified by the psychologist Mihály Csíkszentmihályi—where one is "completely involved in an activity for its own sake. The ego falls away. Time flies. Every action, movement, and thought follows inevitably from the previous one, like playing jazz. Your whole being is involved, and you're using your skills to the utmost."

But it's also an amazingly fragile state. The slightest interruption, and that carefully assembled understanding can vanish in a puff. Thus, the one thing that drives coders into a blind fury is someone pinging them when they're in the zone. They're trying desperately to maintain an ethereal crystal lattice in their minds, and now somebody is asking them, *Hey, did you get that email I sent you?*, and poof, it's gone.

"My wife and kids have learned that when I'm in my room coding, you really can't interrupt me, or else you get this completely different father and husband," as one developer for a social network told me. He's colder, sharper, and usually deeply distressed that the trance has been broken. Many coders who work from home described the same problem. "I often become very black and white, very logical," said another one who's worked in machine vision. Interrupt him midwork and he tries to compress the interaction to a binary variable: *Can you give me a yes/no question?* The rest of the time, he thinks, he's a perfectly nice guy. But the act of coding transforms him. Interrupt him at your peril.

And so while they're working, programmers tend to do everything they can to wall off other people, because other people are, in essence, nothing *but* distractions. They slap on noise-canceling headphones; some save their hard work for the wee hours of the night, when calls and messaging and even political-news tickers fall silent.

This desperation to stay in a "flow zone" isn't unique to programming, of course. Any discipline that requires deep-end immersion in an all-encompassing mental state produces the same hermit behavior and pissy irritation at the slightest social demand: lawyers working on briefs; surgeons planning a surgery; artists of all stripes. Novelists need to build a world in their head and hold it there, so they, too, have historically pursued isolation with the intensity of mad desert prophets. To avoid the internet, Jonathan Franzen removed the wi-fi card from his computer and superglued his Ethernet port shut. Stephen King drew the shades on the windows, lest a glimpse of the outside world break his concentration.

Coders, in other words, have an artistic temperament, something that unsettled their managers back in the '60s and '70s. The managers expected engineers to act in engineer-y, logical ways, and the programmers certainly did. But they also behaved like Coleridge, trying in a frantic burst to get "Kubla Khan" down on paper before a knock on the door could break their creative trance. Indeed, coder workflows are deeply Romantic, in the original sense of the word. If you brought a bunch of coders back to meet the nineteenth-century Mary Shelley or Lord Bryon, they'd warmly agree that one's best work is done in a remote garret on a rocky promontory. (In fact, the description of the enchanted poet in "Kubla Khan" reads pretty much like a description of the coding trance: "And all should cry, Beware! Beware! / His flashing eyes, his floating hair!") My friend Elizabeth Churchill, a social scientist and engineer who currently works as director of user experience for Google, has gotten used to the thousand-yard stare of the coder who's at a meeting, but in body only, while still mentally in combat on the astral plane. "You talk to these guys who are really deep in coding, and they've got this really wild look," she says. "And you have to get them to take a deep breath and get them back. And then a day or two later you get them back to reality, and they can talk about the architecture of what they're working on."

Indeed, many programmers even gravitate toward tools that preserve "flow" at the level of tiny physical movements. For example, many coders loathe using the mouse, since it requires them to move their hands away from the keyboard. That extra movement galls; it's like asking a concert pianist to occasionally reach up and twiddle a knob on her piano lid. So a certain breed of coder gravitates toward Vim, a text editor in which nearly everything—cutting, pasting, paging around—can be done with your fingers on the main letter keys. You don't even need to reach down to the "arrow" keys to move around a document; Vim regards that as too wasteful a motion, too likely to knock you out of the zone by forcing you to glance down at your keyboard. Saron Yitbarek, a coder friend of mine, was required to learn Vim during her first programming apprenticeship. The elder coders all used Vim; "real coders use Vim," they more or less told her. So she forced herself. At first it was ungainly and awkward; she kept on wanting to use her usual mouse-or-arrow-key movements to navigate. But after three weeks of painstakingly reprogramming her habits, it clicked—and suddenly she felt a cyborgic level of oneness with the machine, a sort of "I know kung fu" moment reminiscent of *The Matrix*. "You're one with the keyboard," she marvels. "There's no moment when I have to break away from my train of thought."

Coders are white-collar workers, but the need for deep immersion and concentration puts them at sharp odds with most white-collar rhythms. Paul Graham refers to it as the collision between the "maker's schedule" and the "manager's schedule." The work of managers, he points out, is composed almost entirely of meetings. Managers' jobs are to make sure things are going well, so they block out their day in one-hour increments, meeting with a different employee every hour. So they think nothing of asking a coder to come in at 1:00 p.m. and have a check-in meeting. But for the coders, that meeting destroys any chance of getting into a long flow state.

"I find one meeting can sometimes affect a whole day," Graham wrote. "A meeting commonly blows at least half a day by breaking up a morning or afternoon. But in addition there's sometimes a cascading effect. If I know the afternoon is going to be broken up, I'm slightly less likely to start something ambitious in the morning. I know this may sound oversensitive, but if you're a maker, think of your own case. Don't your spirits rise at the thought of having an entire day free to work, with no appointments at all? Well, that means your spirits are correspondingly depressed when you don't. And ambitious projects are by definition close to the limits of your capacity. A small decrease in morale is enough to kill them off." Calling a sudden, unplanned meeting is even worse: If you've interrupted a coder in the midst of what they thought was supposed to be a long stretch of concentration time, you essentially guarantee you're going to face a remarkably peeved and machine-minded person.

Nobody is more like a stereotypical "coder" than one you've just interrupted.

... ...

Are coders more *depressed* than other people?

Plenty of people asked me that as I wrote this book. Indeed, plenty of coders themselves asked me that. They knew of lots of ones who were depressed or struggled with anxiety or mania. Many had stories of coder friends who had been so tragically beset with mental-health issues that they'd committed suicide. (Some identified strongly with the *Mr. Robot* TV series lead hacker character, Elliot, who suffers from depression and deep social anxiety and self-medicates with morphine.) Was coding, they wondered, somehow tied up with mental-health problems? Maybe programming was more likely to attract people who wrestle with the noonday demon—much as does, mythically and historically, writing and poetry and the other arts. Or maybe the act of coding, all those isolated hours and stress, exacerbated the problems of those who'd otherwise be on the borderline.

These are questions Greg Baugues ponders a lot.

He's a coder well known in the industry for talking openly about his struggles with mental health. The son of a midwestern pastor, Baugues taught himself coding as a teenager, then went on to earn a computer science degree. But it was in college that he began to find himself emotionally unable to cope. Technically, the work ought to have been easy for him; he was already pretty well self-taught. But he'd procrastinate for weeks, leaving assignments so late that he'd collapse under a sense of futility. When a friend showed up looking for him, he hid under a crumpled blanket in his bed. Indeed, he spent a lot of time in bed, sleeping up to 16 hours a day.

"The best times of my day were when I was unconscious, when I didn't have to deal with reality," Baugues tells me, when we met for lunch at a café, having discovered we lived in the same neighborhood; he's a tall, soft-spoken guy with a light orange beard. "I just tried to stay in that as much as possible." *Greg is one of the smartest kids I know,* a friend of his said, *but he's also the laziest.*

Sometimes, though, he was precisely the opposite of lazy: He was manic with creative energy. Ideas for software products would spill out of him in a Niagaran flood, and he'd stay up for days, frantically coding, wondering how he'd deal with all the sudden wealth once he launched these rad new apps. But he'd never launch them, really, because after a day or two his energy would crash, and he'd retreat to his bed again. After five long years at college, he finally dropped out and began bouncing around coding jobs. He'd perform well enough for a few months, but then begin procrastinating again; he'd anger roommates by never paying the rent or utility bills, until the electricity was shut off.

Finally, around age 25, he decided he had deeper mental-health issues than he was willing to admit. While googling "chronic procrastination" he read the symptoms of ADHD, and they certainly sounded familiar. He got himself tested. "You are definitely ADHD," the therapist told him. "Like you're off the charts." She also diagnosed him with type 2 bipolar disorder.

He started taking meds for ADHD. Barely fifteen minutes after his first dose, he was astounded at how much better it made him feel. *Holy shit*, he thought, *is this what normal people feel like?* He initially resisted getting treatment for bipolar disorder, since he didn't want to admit he might have that problem. But once he treated it, his life settled into much more manageable grooves. He still struggles with depression, but the everyday snafus of life don't knock him off his stride as easily as they did before. He was hired by Twilio, a firm that makes it easier for coders to add phone and SMS services to their apps; his job is as an "evangelist," or someone who goes to conferences and hackathons to teach how to use Twilio's products. He also started talking at conferences about his struggle with mental illness in tech.

That's how he discovered he wasn't alone. After his talks, coders would come up to him afterward, wide-eyed with their own stories of depression, bipolar disorder, and other troubles. It became such a common experience that he began to guess that the rate of mental illness among developers is higher than that of the general population.

The industry, he argues, rewards many of the maladaptive behaviors that come with mental-health problems. "Social isolation, irregular sleep patterns? Thoughts of grandiosity—thinking that you can change the world, or that the rules don't apply to you?" Baugues says. "If you're an adolescent or a young adult experiencing these symptoms, and you stumble across coding, it will probably feel a lot like coming home to you. I mean we, as developers, accept the socially isolated and the socially awkward in ways that others don't. We will accommodate your irregular sleep patterns. It's not uncommon for coders to come in whenever they want, work till two in the morning, work as long as you get the work done." Indeed, this sort of behavior is now thoroughly romanticized in pop culture and in business culture. Coders in TV and movies are creatures of the night, uplit by the monitors in isolation; business stories of start-ups fetishize the programmer who pulls 72-hour-long keyboard fests. The industry not only selects for it but actively glamorizes it. In a field like that, who's going to admit they have a problem?

"There's an identity that developers have built up around being the smartest person in the room, around being a high intellectual performer," Baugues notes. "Saying 'I have a mental illness,' by definition, is saying 'My brain does not work properly.' So what—are you going to go to your employer and tell them, 'Hey, this thing you're paying me for doesn't work right?' " It doesn't help that the meds can dull the sort of spiky mental activity programmers rely on, so some coders simply don't want to take them. They worry it'll shut down their creativity, the racing moments of insight, the hours of hyperfocus. Baugues thinks that may be true even of him, that his medication dulls some of his highs. But the trade-off, he figures, is worth it in spades. "Even if the meds decreased the velocity at which my brain came up with new ideas," he says, "it has, by an order of magnitude, increased my ability to execute on ideas. Because now I can ship."

Baugues is lucky, he knows. He's got an employer who's supportive. But much of the rest of the industry is, mostly, happy to churn through bright, young people, grinding them up in remarkably unhealthy workflows. Sixty-hour-week death-marches at work, for weeks at a time? Staying indoors? Nearly cultlike isolation away from your loved ones? If you had a history of mental illness, any doctor would tell you to stay far away from these environments; they're all well-known triggers. Frankly, even if you weren't buffeted by a biochemical issue, that sort of work style can grind healthy people to pieces.

We left the café and wandered over to Baugues's house. It was a bright spring day—"Sunlight! See, this is what I'm *supposed* to get more of," he cracked. Inside his apartment, his dog clambered up his legs, and Baugues showed me a witty little project he'd recently done to show off Twilio. It was a big red button that, when his dog pushed it, triggered a camera to take a picture and text it to Baugues. "A dog selfie!" He laughed. It had been a hit at a recent conference, when he'd showed it off onstage. Offstage, later, he had separate conversations with coders who quietly sought him out to talk about their own mental storms.

Raised by a Methodist pastor, Baugues is still deeply observant himself; his house has scattered testaments to his various passions, including one shelf that holds C. S. Lewis's Narnia books next to a Raspberry Pi minicomputer. "When religious evangelists are doing their job right, the focus is on serving the community," he says. He's found a dual calling.

< Chapter 5 >

The Cult of Efficiency

It wasn't until she got to Japan that Shelley apprehended just how crazy her new boyfriend could be.

She'd met Jason Ho through some mutual work friends and hit it off quickly with the slender, tall kid who had a sly smile. He was a computer programmer in San Francisco who ran his own company, and he had decided to take a vacation break in Japan for four weeks. Did she want to come? Shelley, a petite woman with thick horn-rimmed glasses and a wry laugh, was a bit apprehensive. Spending that much time together could spark a nice relationship—but familiarity could also breed contempt. "I'm worried about how we're gonna be sleeping in the same room," she recalled. But she decided to take the gamble, and they bought tickets.

Ho, as it turned out, had a very strict and peculiar itinerary planned for the trip. He's particularly fond of ramen-noodle dishes, and his goal on this jaunt, he told her, was to visit as many Tokyo ramen shops as possible. To make sure they crammed in as many as they could, he'd created some custom code. *Huh?* she thought. He explained: First, he'd assembled a list of top Tokyo noodle places and plotted them on a Google Map. Then he'd written code that drew the optimum pathway that connected all the shops, so they could travel in the most efficient route between them. It was, he said, a "pretty traditional" algorithmic challenge, of the sort you learn in college, and he used tricks like this all the time to optimize the way he lived his life. He whipped out his phone to show her the map. He told her he was planning on keeping careful notes about the quality of each meal, too.

Oh wow, she thought, impressed. *This guy is kind of nuts.*

As it turns out, Ho was also witty, well-read, and funny, so the trip was a success; they drank beer ringside at a Sumo wrestling match, toured local architecture, visited a petting zoo, and soon they were dating.

Ho, she learned, enjoyed automating lots of things. As a kid growing up in Macon, Georgia, he'd learned to program on a Texas Instruments TI-89 calculator. One day while leafing through the instruction manual, he discovered that the calculator contained a form of the BASIC programming language; better yet, you could draw pictures, pixel by pixel, and make video games to share on your classmates' TI calculators. It was his "Hello, World!" moment, and he was electrified. He spent months patiently re-creating the classic Nintendo Game Boy game *The Legend of Zelda*. It required some clever hacking: The TI could only display pixels that were dark black, but *Zelda*'s graphics were gray. So Ho figured out that if he flickered the pixel on and off rapidly, several times a second, the black and white would blur into gray. Soon he was teaching himself Java to make computer games, and he formed a Java game-building club at school. For college, he went to Georgia Tech, where he studied computer science. He thought the curriculum was too dry; he enjoyed studying abstract algorithmic concepts, but mostly what fascinated him was using computers to get rid of repetitive labor.

"Anytime I have to repeat something over and over," he told me, "I get bored." In his final year of college, he decided that the whole structure of the college system was weirdly inefficient: You had kids taking basically the same classes at different institutions, often with the same lessons, running into the same problems with their work. But these far-flung students didn't have any easy way to talk together. He put together Qaboom.com, a question-answering site that tried to cluster students across the US by common subjects. Though a few investors in Silicon Valley liked it, it never took off because Ho couldn't crack the cultural piece: how to get students to post good-quality questions and answers. As a coder, he was concerned with making the site so elegantly designed and robust that it would scale. But since nobody was posting anything, a million people would never show up, he realized. Content mattered, and he didn't really know how to get that snowball rolling. He shut the site down and, his degree program nearly over, interviewed at a few firms like Google. But he sunk into a funk. He didn't want to work for someone else. Considered as a question of value creation, being an employee was a terrible proposition, he felt. Sure, you earned a check. But most of the value of your labor was captured by the founders. He had the skills to build something, soup to nuts— these "magical powers." He just needed to find something that needed that gift.

A few weeks later, he stumbled into one. He'd gone home to visit his family in Macon, where his father ran a pediatric office. While there, Ho accompanied his father on an errand to Staples to buy two time clocks for his offices, the type where an employee punches their card and the machine stamps it with the time they started and stopped working. Each clock cost $300. Ho was astounded: Had time-clock technology not changed since the era of the Flintstones? *Do people really put these pieces of paper into this machine and manually add it up?* he wondered. *I can't believe this is still a thing.* He realized he could quickly cobble together a website that performed the same task, but better: Employees could check in via their phones, and the site would total up the hours automatically. "Don't buy this time clock," he told his father. "I'm going to code you one."

Three days later, he'd created it. His father's offices began using it, and, to Ho's delight, they loved it. It was, in fact, remarkably more efficient than a paper time clock, saving his father hours he'd spent manually toting up figures and eliminating mistakes that might have previously cost employees missing pay. This, he realized, was a product: Unlike Qaboom, it solved a problem people actually faced. He spiffed up the website, gave it a name—Clockspot—and four months later, amazingly, a paying customer, a local law firm, signed on. When their first payment came through, he was coding with a friend in the library of Georgia Tech and nearly jumped out of his seat: *Holy shit, someone paid me $18.95 to use this software*, he told his friend. A few months later, Ho was doing $10,000 a month in business, with customers ranging from cleaning companies to home-health-care-aide firms to the city of Birmingham. After two years of nonstop work, he'd debugged the code and had it running so smoothly his company ran mostly on autopilot; he was making a healthy income with only one full-time employee, himself, and a part-time customer-service agent he'd hired to work from home in Florida.

When I met Ho and Shelley in San Francisco for dinner at—where else—a ramen-noodle restaurant, Ho was working only a few hours a week. He spent quite a bit of time traveling; once he'd even managed a perilous Clockspot outage while on the base camp of Mount Everest. "He says he works twenty hours a month, but I don't think I've seen him work that much," she said.

She had also, in the two years they'd been together, discovered that Ho's obsessive habit of optimizing everything could leak into almost every part of his life. When he decided to buy a house, he didn't want to sit around going house by house and pondering whether to buy it. So he wrote a little piece of software into which he could dump the information for scores of San Francisco homes—like their locations, prices, and neighborhood statistics—and it would calculate its probable long-term value. (The program recommended a top pick; he duly bought it, and currently lives in it.) Because he hates shopping, he bought dozens of pairs of the same T-shirt and khakis—which, as he notes, also removes any decision-making time when dressing in the morning. A few years ago, tired of being out of shape, he decided to take up bodybuilding, since it seemed like a particularly demented optimization challenge. He began whipping out a food-scale at restaurants to weigh his portions, and devising ways to fit exercise into nearly any part of his day. If he passed a thick metal crosswalk bar, he'd use it to do pull-ups; if he passed a dumpster, he'd lift it up on one edge.

"He tracked, like, every single thing he ate in this massive spreadsheet," Shelley tells me. Ho sheepishly shows it to me on his phone; it's a sprawling beast that meticulously plots out every ingredient in his workout meals, which brought in 3,500 calories per day. (Two ounces of "waxy maize" = 210 calories, apparently.) But it worked: After two years of training, he placed second in a local amateur Californian bodybuilding competition, up onstage flexing his shoulders in "the crab." He flips through his phone to find pictures of himself from the period. "I was down to about seven percent body fat," he says, and you can tell: Lightly oiled down and posing in his underwear before a sunny window, he looks like a Greek statue. He shrugs. It felt good to look so ripped, he says, but mostly he'd gotten his body to that point just to see whether this sort of crazy torquing would even be possible.

He didn't meet any other programmers in his new weight-lifting community, but the other bodybuilders all had a hacker-like approach to their bodies.

Ho shows me another chart he wrote. It's a life guide, of sorts. He decided he only wants to spend time on the things where every ounce of his effort is most likely to produce maximum results. So he charted out sixteen life activities in rows, ranging from "entrepreneurship" and "coding" to "guitar," "StarCraft," "shopping," and "spending time with friends and family." Then, in columns, he plotted out various criteria—like whether the activity can be self-taught ("autotelic"), whether it "can be mastered," or whether it "impacts multiple areas of life." For "coding" and "entrepreneurship," Ho ticked off *yes* for every quality. When he came to the social realm of "spending time with friends and family," he noted that *yes*, this skill "impacts multiple areas of life." For "can be mastered," he wrote *maybe*.

Programmers are obsessed with efficiency. It is the one thing I've encountered in essentially every coder I've met. Coders might be wildly diverse in other ways—politically, socially, culturally, what have you. But nearly every one found deep, almost soulful pleasure in taking something inefficient and ratcheting it up a notch. Removing the friction from a system is an aesthetic joy; their eyes blaze when they talk about making something run faster, or how they eliminated some bothersome human effort from a process.

In part, this lust for efficiency isn't unique to software developers. Engineers and inventors of all stripes have long been obsessed with it. Ho's chart seemed amiably nuts, but it reminded me of the original American hacker and engineer, Benjamin Franklin. Franklin, too, was an engineering genius obsessed with optimizing everyday acts. His vision was terrible both near and far, so he carried two pairs of glasses—one for distance, one for reading—and eventually became sick of swapping one for the other. So he cut one semicircle of the distance lens, glued it to one semicircle of the reading lens, and invented bifocal glasses. To make it easier to warm colonial houses, he invented a stove that more efficiently stored heat in its metal housing, producing, as he boasted, a "great Saving of Wood to the Inhabitants."

Franklin wasn't just interested in optimizing the physical world, though. He also thought his moral and ethical qualities should be tracked and tweaked for maximum improvement, too. So he wrote down thirteen "virtues," including "Industry. Lose no time; be always employ'd in something useful; cut off all unnecessary actions," and "Frugality. Make no expense but to do good to others or yourself; i.e., waste nothing." Then he printed them on a chart, with the virtues in rows and the days of the week in columns, so he could track daily how often he practiced good behavior, with a satisfying Puritan tick of the pencil. It was, as *The Baffler* once put it while describing the present-day Franklin Planner system, "quintessentially American—simultaneously wholesome and insane." But it was also quintessentially an engineer's way of approaching the world. That combination, that shining-eyed ardor for efficiency and order that manages to be admirable while also completely bonkers, is a heat signature of the engineering mind-set.

At the turn of the last century, during industrialism's first bloom, engineers regarded the automation of everyday tasks as a moral good. The inventor was humanity's "redeemer from despairing drudgery and burdensome labor," to quote the engineer Charles Hermany from 1904. And nearly any form of mechanic bridles at any sort of waste and friction in a system. I once had a mechanical-engineer neighbor who repaired motorcycles; if he heard the tiniest unexpected noise coming from the engine, he'd tear it apart to hunt down and eliminate its source. ("Noise," he intoned soberly, "is wasted energy.") Frederick

Winslow Taylor—the inventor of "Taylorism"—inveighed against the "awkward, inefficient or ill-directed movements of men," arguing that workers' movements ought to be carefully prescribed to ensure maximum output. His colleague Frank Gilbreth obsessed over wasted movements in everything from bricklaying to vest buttoning, while his engineer wife designed kitchens as such that the number of steps in making a strawberry shortcake was reduced "from 281 to 45," as *Better Homes Manual* enthused.

But computers, in many ways, inspire dreams of efficiency greater than any tool that came before.

That's because they're remarkably good at automating repetitive tasks. Write a script once, set it running, and the computer will tirelessly execute it until it dies or the power runs out. What's more, computers are strong in precisely the ways that humans are weak. Give us a repetitive task, and our mind tends to wander, so we gradually perform it more and more irregularly. Ask us to do something at a precise time or interval, and we space out and forget to do it. The ability to remember to do something in the future is what cognitive psychologists call "prospective memory," and humans are dreadful at it, which is precisely why we've relied for centuries on lists and calendars to remind us to do things. In contrast, computers are clock driven and superb at doing the same thing at the same time, day in and day out. Back in the '70s when he was cocreating the UNIX operating system, the legendary coder Ken Thompson created "cron," a scheduling command: You can tell the computer to run a program or accomplish a task at a particular time in the future, over and over again. Thompson reportedly called it *cron* in honor of the Greek word for "time." Setting up a repeated task is known as a "cron job," and coders often run dozens of these on their machines, tiny daemons that execute tasks every day so the coder doesn't have to think about them. Computers are useful slaves to time.

The upshot is that most coders arrive at the same logic, which we could summarize thusly: (a) Doing things repeatedly or at the same time every day is boring, and I'm terrible at it. By contrast, (b) slavishly and meticulously doing the same task again and again is easy for the deathless machine sitting on my desk. Thus, (c) I am going to automate every single thing I possibly can.

Many programmers have this aha moment in their teenage years, when they first discover that life is full of blindingly dull repetitive tasks. Even in leisure activities! Lance Ivy, one of the two developers who built the first iteration of Kickstarter, had his efficiency epiphany while a teenager and playing an online adventure game. To make your character more powerful, "you'd have to repeat these mundane tasks": You could play an instrument over and over or have your character read books. *It's just grinding*, he thought to himself. *Why do I have to type these commands each time?* So he wrote a script that took his on-screen character and had it repeatedly practice its "hiding" skill by finding medicinal herbs and hiding them in a cave, ad infinitum. (It worked a bit too well. "I left it running one night," he tells me, "and came back the next day and discovered that my character had been put in a padded room, basically. They tried talking to me after I was away from the keyboard and I obviously wasn't there," so they thought he'd gone nuts.)

School is another swamp of repetition. Many kids, for example, find it annoying when the teacher forces them to "show their work" while doing long division or multiplication. So when Brad Fitzpatrick—the inventor of the pioneering blogging site LiveJournal—was in high school, he simply wrote a program to autosolve them: Type in a problem, and it would calculate the solution, showing each step along the way. "We would get these problems to come out like ten per page or something, and I would type it into the computer and then just reproduce the problems in scribbles," as he later said. "I did the same thing in chemistry to find the orbitals of electrons." Was it cheating? Sort of. But being forced to write a program that could actually, say, break down long division into small steps, Fitzpatrick had to deeply ponder and comprehend the process. He didn't object to the task of understanding long division; he just wanted to get rid of the drudge work of having to do it over and over. What gets more ambiguous is when you automate the mental work of other students. Fred Benenson, a coder friend of mine, wrote a similar program in college to do calculus computations on his Texas Instruments calculator, and once his pals got wind of it, they asked him for copies, so he shared. "But then they were just using a program to basically avoid learning things in the first place," he says. So Fred got an early exposure to the ethical complexities of software: By making something easy to do, you can change the mental habits of other people. Humans are inherently pretty lazy; as Nicholas Carr notes in *The Glass Cage*, when someone offers us the ability to take a shortcut, we take it. We only discover later that we may have traded off an ingrained skill—or, in the case of calculus, never learned it in the first place. But it's incredibly hard to resist because we're constantly given new tools from programmers who've figured out how to remove friction from daily life.

Larry Wall, the famous coder and linguist who created the Perl programming language, deeply intuited this coderly aversion to repetition. In his book on Perl, he and his coauthors wrote that one of the key virtues of a programmer is "laziness." It's not that you're too lazy for coding. It's that you're too lazy to do routine things, so it inspires you to automate them. This, as they noted, is

> the quality that makes you go to great effort to reduce overall energy expenditure. It makes you write labor-saving programs

> that other people will find useful, and then document what you wrote so you don't have to answer so many questions about it.

Eventually, of course, that orientation has a way of bleeding into your everyday life. It becomes hard to turn off, like X-ray vision. "Most engineers I know go through life seeing inefficiencies everywhere," as Christina, a coder in San Francisco, once told me. "Inefficiencies boarding your planes, whatever. You just get sick of shit being broken. 'Hi, this isn't good, and I'm going to fix it!' " She'll even find herself wishing people navigated the sidewalks and street crossings in a more optimal fashion. "It's a fundamental kind of dissatisfaction with the way things are currently running," she says. Jeannette Wing, a professor of computer science—currently running the Data Science Institute of Columbia University—popularized the phrase "computational thinking." It's the art of seeing the invisible systems in the world around you, the rule sets and design decisions that govern how we live. And it often leads her, too, to notice subpar organization around her.

"Whenever I get to a lunch buffet," she says, "I get annoyed when the stations of a buffet are not lined up properly. So when they put the forks and the knives with the thing wrapped around it as the first thing? I find that annoying! Because you have to hold your plate while you hold this thing of knives and forks, too. The cutlery should be at the end! So this buffet is really this linear sequence of stations, and you want it to be sensible. You don't want a lot of latency." It's a classic bit of optimization thinking, something that certainly you don't need to be a coder to appreciate, but which comes to programmers effortlessly, even unstoppably. (Indeed, there's something about the demands of food preparation, delivery, and cleanup that seem to inspire particular annoyance among hackers—which probably anyone who hates housework could identify with. The programmer Steve Phillips once told me he yearned to robotically automate dish drying, when as a teenager he was helping his mother clean up after dinner. "I grab one off the stack, I dry it, I sit it down, I grab one off the stack, I dry it, I sit it down. It's like—this should be a for loop. This is *pissing me off.*")

And of course, the zeal for optimization can reach into social and emotional relations, too. After all, dealing with the fellow humans around you—saying hi to them, asking how their day is going, listening to partners complain about their day—requires one-on-one attention. It's *constitutionally* inefficient; the whole point behind taking emotional care of someone else is to slow down and attend to them. For the set of coders who are already terrible at emotional work—or who regard it as beneath them, or who are simply baffled by it—the instinct is: Well, let's automate our everyday emotional work, too. It can produce a set of efficiency ploys that can be unsettling, if morbidly fascinating in their ingenuity.

Several were on full display in a Quora thread where coders talked about automating everyday life. "I got tired of hearing 'You never message me' from friends and family," as one wrote, so he created a script that would randomly send a text to one of them, created using a Mad Libs–style mash-up. (A text would begin with this gambit—"Good morning/afternoon/evening, Hey {name}, I've been meaning to call you"—and then append one option from a list of greetings: [I hope all has been well., I will be home later next month love you., let's talk sometime next week when are you free?].) Another programmer translated the tale of a Russian coder who'd written a script to automatically send a "late at work" message to his partner if he were still there at 9:00 p.m. (with a randomly generated reason affixed) as well as a program that turned on the latte machine in his company's kitchen and commanded it to brew a "midsized half-caf latte," to be ready in 41 seconds.

("The timing is exactly how long it takes to walk to the machine from the dude's desk," his colleague marveled.) At a hackathon in San Francisco, a middle-aged coder excitedly showed me an app he'd created that would send romantic messages, culled from online quote-databases, to a partner. "So when you don't have time to think about her"—and, yep, he assumed the emotionally needy partner is a "her"—"this app can take care of it for you," he enthused. He seemed only very faintly aware of how nuts it seemed.

Linguists and psychologists have long documented the value of *phatic* communications—the various emotion gambits we humans use in everyday life to make those around us feel at ease or listened to: *How's it going? Crazy weather, eh? What're you up to tonight?* The more I talked to coders, the more stories I heard of programmer colleagues who found that stuff as irritating as grit in the otherwise frictionless gears of their daily lives. One night while I dined with Chris Thorpe, a coder who was a veteran of over a half-dozen tech firms, he recalled a former colleague who fit that bill: "This was a person, this incredibly talented engineer, who was someone who knew the right way to do things. He was very upset with me that we told jokes in all our meetings, because we were wasting time. 'Why are we spending five minutes having fun with twenty people in the office? This is *work* time. Why are we telling these jokes? Everybody is laughing—but, you know, you're wasting all this valuable time.'" This guy would begin rattling off the math: "'Five minutes times twenty, that's like, you know, you've *wasted an hour and a half of person-time on these jokes.*'"

Konrad Zuse, the inventor of the first digital computer, once argued that people wielding computers would be affected deeply by that exposure. "The danger that computers will become like humans," he noted, "is not as great as the danger that humans will become like computers."

... ...

I began to notice this zeal for optimization myself. The more I dabbled in coding, the more I began to notice, and be deeply annoyed by, moments of inefficiency in my daily affairs.

While writing a chapter in this book, for example, I'd find myself consulting an online thesaurus over and over again. (Feel free to judge me.) It was useful, but so slow to load that each time I did a search it took two or three seconds to load the results. Eventually I began to wonder, *hmmm*, maybe I could just code my own command-line thesaurus? Then I wouldn't have to sit drumming my fingers.

So I started hunting around for an online thesaurus that I could ping directly for synonyms. I quickly found one at Big Huge Labs, an online firm that had offered a thesaurus API, and bonus—you could do 1,000 synonym searches a day for free. After a morning of tinkering with Python, I'd created a bare-bones command-line app: Type in a word, and it would spit out a list of its synonyms and antonyms. It was green text on black, unadorned, and crude. But damn, it was fast. Because it wasn't barnacled with user-tracking cruft, the results would spill down my screen milliseconds after I hit Enter.

Granted, the amount of time this little script saved me was . . . not terribly consequential. Assuming I search for synonyms twice an hour on average during my writing work, and (generously) assuming this command-line app saved me a rollicking two seconds per search, that added up to only one hour less waiting in a whole year. I wasn't exactly carving out new vacation time here.

Nonetheless! Software performance is timed in milliseconds, and so I was saving up to 2,000 milliseconds per query, which sounds and feels a lot more impressive than "two seconds." Plus, the burn of velocity warmed my soul. Each single time I searched for a synonym, the zippy speed of the results produced a surge of pleasure every bit as satisfying as the very first time I got the script working. I was applying the drug of efficiency to my veins, and my God, it felt good. (As many software firms have discovered, milliseconds really do matter, not just to machines but to the psychology of we impatient humans. We want the digital world around us to react as crisply; Google found that a mere 100-to-400-millisecond delay in returning search results produced a small but regular decrease in how many searches people type.)

Over the next few months, I began to crave that feeling in every other corner of life. I wrote cron jobs to clean up the desktop on my computer, because it'd get cluttered with weird downloads and I was too lazy to put them in the right place. I'd download the text of YouTube video autotranslations and, sick of cleaning them up by hand, wrote a script to automate it. My son complained of having to sit by his computer waiting for his teacher to post his elementary school homework (she'd do it at erratically different times, and he'd sit there refreshing the page by hand), so I wrote him a bot that would check every few minutes and text us when the homework was posted. One day I found myself annoyed that I couldn't simultaneously clean the house while reading through the *New York Times*. (Other people listen to NPR while they clean, but dammit, I just wanted hard, info-rich news and nothing but.) So I spent an evening cobbling together a text-to-speech web app: I could now pick a few dozen *Times* stories, queue them up, and have a gnarly robotized voice—I preferred the "male Irish" synthesized bot—read them out loud while I swept and did dishes.

Not all these experiments in efficiency worked. Indeed, some backfired. I occasionally discovered it was harder than I expected to automate a task, and in a bit of bleak coding irony, I was spending more time writing the software than it would have taken to simply do the task by hand. (One evening I discovered a transcriber had delivered several transcripts without any identifying marks of who had said what. *No problem*, I figured, *I'll just write a quick Python script that crawls through the text and appends "Question" and "Answer," alternately, to every other paragraph*. But I made a few idiotic mistakes, discovered the data I was trying to clean up wasn't as regularly formatted as I thought it was—and so a script that should have taken about maybe fifteen minutes to bang out wound up eating up over an hour. Winner.)

I finally got it working, but it was a spectacularly dumb waste of time. I'd spent over an hour trying to automate a task that would have only taken me a few minutes to do manually! So much for making life more efficient.

On the other hand . . . the very act of automating the task was, well, fun. It was certainly more fun than doing things by hand. Automating it required me to engage in mental chess, in puzzle solving. Even the head-slapping moments (when I discovered my noob mistakes) contained a burst of eureka joy. Sure, I could have knocked off the task myself manually in a few minutes. But those would have been dull, boring minutes! While I may have wasted an hour of my time noodling with the algorithm, it was an intellectually engaging, electric hour.

As it turns out, this experience of mine is a common syndrome among programmers. The more I asked around, the more stories I heard of people spending more time writing a program to automate an action than it would take to simply do the damn thing itself. Over drinks at a tech conference, one project manager told me about a coder he'd employed who was famous for going down optimization ratholes.

"He was supposed to be working on a database migration," the manager said, "which required a bunch of pain-in-the-ass stuff up front, some scut work of cleaning up data by hand." It probably should have taken the coder a half day or so; it was a onetime affair, and once he was done he'd never have to do it again. But he hated the idea of wasting a morning doing dull, repetitive, by-hand work. So he decided to automate it and plunged into the deep end of optimization madness. The program manager kept on walking by the coder's cubicle and seeing the guy diligently at work, head down, headphones on, cranking away. But when the manager checked in two weeks later ("my fault, bad management, I should have been checking in every couple of days"), he found the migration was still totally incomplete. The coder hadn't finished it. He was still meticulously crafting, tweaking, and perfecting a tool to automate that first step, the data cleanup. He'd blown half the month trying to make a tool just to avoid doing three hours of drudgery. "We were now totally behind schedule, but he was all like, hey, I have this awesome tool now!" the manager said, and sighed.

"Totally useless. But he was super proud of it."

The coderish fixation with optimization isn't really capricious, though. A lot of it stems from the raw dictates of the machine—where efficiency often matters profoundly.

Software can be written carelessly and inefficiently. A sloppily coded process might, for example, require far more RAM than is necessary. An algorithm to sort data might have been written to pick the slowest, lousiest way to do so. To put it in metaphoric terms, imagine you had a team of ten people and had to unload 100 boxes from a van into a house. The simplest but slowest algorithm is to tell everyone to grab a box from the van, carry it inside, and then repeat until the van is empty. That'd require 100 individual trips. A much more efficient algorithm would be to have the ten people line up between the van and the house, and pass the boxes along, bucket-brigade style.

Code has precisely the same dynamics. A coder who's new and clueless (like me) can usually figure out the most obvious, simple way to write a function. I'll get ten people walking in and out of the house. But the more-experienced, talented programmers will intuit the faster way to organize the routines. They can write an algorithm that will be faster—hey, we saved 20 milliseconds per function call—than an inefficient one. Writing the slower code might not matter when only a few people are using your app. But if suddenly thousands or millions of people start pounding away on your server? In that situation, unoptimized code can be ruinous.

That's what Lance Ivy found out when Kickstarter got popular. In 2012, three years after it was launched, Kickstarter started to get its first "million-dollar" campaigns. Among the first was a campaign by veteran video-game designer Tim Schafer to create *Broken Age*, his latest title. Schafer initially wanted to raise $400,000, a huge sum for Kickstarter at the time. But Schafer's fan base rallied around the cause, and within 24 hours they had come close to raising a full $1 million. To capture the excitement of the moment, Schafer's company streamed footage from their office, showing the staff as they followed the increasing Kickstarter pledges. In a cycle of hype, that made fans more excited—so they kept on loading and reloading Schafer's Kickstarter page. Everyone wanted to see the instant the campaign tipped over from $999,999 to $1,000,000.

And that, as it turns out, was a problem for Kickstarter's existing design. "They're all streaming it, saying 'let's all celebrate this moment, this is Kickstarter history,'" Ivy says. But he hadn't yet optimized Kickstarter to deal with such crushing amounts of activity. His development team was pretty small; they'd been focused on all sorts of other design challenges. Neither Ivy nor the other engineers had ever predicted this sort of user behavior would emerge—"people so excited they're sitting there, refreshing the page over and over," Ivy says, laughing ruefully. So he and his teams hadn't bothered to torque how a main campaign page loads. They simply didn't imagine it would ever be an issue. Now it was a problem, and a brutal one. With so many people eagerly pounding away on the same page, they were inadvertently staging a "denial of service attack," a flood of traffic that brings a server to a halt. "We went down," Ivy says. "We went down multiple times."

Nobody minded, thank God. On the contrary, back in those days "breaking Kickstarter" was considered to be a mark of pride, a proof of your campaign's viral success. But Ivy knew fans wouldn't always be so forgiving, and that meant optimizing Kickstarter so it wouldn't burn through so many resources when a campaign heated up. One key trick: They wrote code that auto-updated a campaign's pledge amount in real time, so that rabid fans could sit and watch it tick upward without constantly refreshing the page. So they rewrote and rewrote and rewrote the code base, "refactoring" it, as it's called, to slowly transform it from a mass of rapidly iterated good-enough-for-now code into something that was increasingly streamlined.

"Refactoring" is, in a way, like editing. When you're writing a letter or a speech or an article, often the first draft is a bit baggy and fuzzy. You're getting the ideas out onto the page, but they're not crisply stated. So you repeat yourself; you beat around the bush; your thinking is fuzzy. But that's okay, because your goal isn't to be perfect; it's to get those first words onto the page, so you can reshape and hone them later. Editing sharpens and hones those flabby ideas and language. It pares down your pointless and windy riffs, so that the prose tightens and tempers. This is why editing a piece of writing—lingering over it, molding it—can make it shorter, not longer. "The present letter is a very long one, simply because I had no leisure to make it shorter," as Blaise Pascal once apologized. (As a mathematician, he no doubt understood the value of economy of expression, since mathematicians love a simple, elegant proof.) Or as Shakespeare put it in an appropriately pithy koan: "Brevity is the soul of wit."

So it is with programming. Generally speaking, the tighter the code, the better. That's partly because the fewer lines of code that exist, the less likely one is to introduce a bug. A huge pile of "spaghetti code" is a bog in which bugs can easily hide. When the code is shorter and tighter, it's easier to see precisely what it's supposed to be doing and not doing. It's a forest of spare birch trees; you can see right through it.

Bjarne Stroustrup, the designer of the C++ programming language, has taught many young students. He told me that, much as first-year political science students lard their essays with jargon, the newbie programmers tend to write much longer, more convoluted code.

"The average solution that the students come up with was between two and three times larger than my solution," he says. "I simply don't write as much code as they did. My code is more reliable, usually faster, better error checking. But it's half to a third of the size of what they write." Why the difference? It's experience, he says. Inexperienced programmers dive in and start devising their code on the screen; they mistake the length of the program, the act of writing lines, with productivity. But the more experienced programmer ponders the problem they're trying to solve, the task the code is supposed to do. When they write, it's guided by years of pattern recognition: They've seen this sort of array-sorting problem, and know the most elegant way to address it.

And they know the converse is true, too. When a particular function—a subroutine in a program—becomes long and labyrinthine, odds are high it could be made more efficient. Often the act of fixing it, refactoring it, involves bits of fractal meta-efficiency: You find places where your code repeats itself, and you remove the duplication. One of the oldest dictums in coding is to be DRY—"Don't repeat yourself."

Occasionally the high-tech press will report on the mammoth size of a software base; Google's various services in 2015, for example, consisted of fully 2 billion lines of code. It conveys the idea that size is value—that the bigger the line count, the more impressive the result; and thus a valuable coder is one who adds more, sitting and typing in an ecstatic trance. But in reality, a truly useful programmer is one who makes a code base smaller, elegantly removing duplication, the junk DNA of an app. After three years at Facebook, the engineer Jinghao Yan checked all his contributions to the company's code base and found that the math was negative. "I've added 391,973 lines to and removed 509,793 lines from the main repository," he wrote. "So if I coded 1000 hours a year, that's about 39 net lines removed per hour!"

Code is, in this way, oddly reminiscent of poetry. Poetry is a form where the power often comes from compression. "In a well-crafted poem, every single word has meaning and purpose," writes the coder and entrepreneur Matt Ward. "[The] entire piece is meticulously crafted. A poet can spend hours struggling for just the right word, or set aside a poem for days before coming back to it for a fresh perspective." Early modernist poetry in English in particular fetishized compression; among the first famous "modern" poems, inspired by the age-old concision of haiku, was Ezra Pound's "In the Station of the Metro":

> The apparition of these faces in the crowd;
> Petals on a wet, black bough.

"In just two lines and fourteen simple words," Ward notes, "Pound paints a striking image, ripe with meaning and begging to be devoured by scholars and critics. Now, that's efficiency." Pound turned out to be deft not only at composing terse verse but also at refactoring others' poetry. When T. S. Eliot showed Pound the first draft of his epic poem *The Waste Land*, Pound took out his pen and savagely trimmed it—chopping it from almost 1,000 lines down to 434 and transforming a baggy work into a tight clutch of imagery. (In thanks, Eliot dedicated the poem to Pound as "Il Miglior Fabbro," Italian for "the better craftsman.")

When programmers talk about the efficiency of code, the language is aesthetics, and how software appeals to the senses. Code that's efficient and well organized is "clean," "beautiful," "elegant"—metaphors of visual art, the delights of the eye. But when code is a slow, wretchedly designed mess that's hard to maintain, programmers talk about stench. The metaphor becomes olfactory, the swill of gases that rise from rot and decay.

Programmers have likely spoken of pungency ("This code stinks") for decades. But the language became institutionalized in the late '90s when two eminent coders—Martin Fowler and Kent Beck—were writing a book on refactoring and trying to describe the ontological mark of terrible code. Beck, possibly "under the influence of the odors of his newborn daughter at the time," suggested "smells," and so their chapter on the problem was called "Bad Smells in Code." Ever since then, the usage has become common; type "smells" into a coder forum like Stack Overflow, and behold the throngs of programmers bemoaning the reek wafting off their work and begging for help in sanitizing it. (Even academic computer science researchers now routinely use this language. One paper called "When and Why Your Code Starts to Smell Bad" studied why malodorous code gets written; it found, among other things, that overwork and tight deadlines are to blame. "Developers with high workloads and release pressure," they noted, "are more prone to introducing *smell instances*." The emphasis is mine.)

This shift from the eye to the nose strikes me as funny, and revealing. Because the goodness of good code is regarded as something visible, it suggests the beauty of structure, something before which you stand in awe: a ticking watch work, a suspension bridge, a Modigliani. The point is to admire the artifice, the *made*-ness of the thing. But by associating bad code with odor, suddenly we're in the world of organic matter, of rotting fish and stale milk, of suspicion that something is wrong *somewhere*, hidden where we can't see it. Or the world of morality: Researchers are increasingly finding that we associate terrible smells with ideas or politics that offend us.

The programmer Bryan Cantrill notes that coders vastly prefer to figure out the cleanest fix for a bug, some solution that will neatly deal with every possible situation the software might face. When they can't do that—maybe they don't have the time, or maybe the bug simply defeats them—they'll reluctantly use an inelegant solution: a sprawling bunch of lines that deal with the four specific fail-conditions, the strange edge cases they're aware of, and they'll just cross their fingers and pray there aren't *more* edge cases out there lurking. Technically, they'll have solved their problem. But they'll hate themselves so much that they'll often add a comment next to their solution, castigating themselves—and calling their code "gross," "disgusting," or "vile." For programmers, inefficient code is an offense to the soul.

A while ago I sat down for dinner, which was going to be only one thing: a bottle of Soylent.

I uncorked the beverage and poured it into a tall, 12-ounce glass. A vaguely goopy white, Soylent looks pretty much like "a glassful of pancake batter," as a coder friend of mine joked, adding that "it tastes like it, too." He was right, as I discovered as I slugged back the drink. But I finished the entire meal in five minutes, which is the point: Soylent is the ultimately optimized meal.

Soylent was invented in 2013 by a 25-year-old Rob Rhinehart, a programmer whose start-up—founded with two young collaborators—had gotten $170,000 from Y Combinator to create inexpensive, newfangled cell phone towers. Alas, their tech was a bust. But Rhinehart had, in his spare time, been pondering a new problem of everyday life: food.

Food, Rhinehart decided, was an enormous waste of time. He figured he spent about two hours a day on meals. "Typically I would cook eggs for breakfast, eat out for lunch, and cook a quesadilla, pasta, or a burger for dinner," he calculated on his blog. "For every meal at home I would then have to clean and dry the dishes. This does not include trips to the grocery store." The problem, he decided, was that food preparation itself was an inefficient way to resupply his body. It'd be a lot easier if he just assembled the necessary nutrients into a liquid. He began researching the problem online, tinkering in his kitchen, and a few weeks later had produced the first batch of Soylent. He began subsisting on a diet of nothing but his beverage, and over the next month, somewhat to his roommates' surprise, he didn't die. Indeed, he seemed to thrive, reporting he'd found "my skin is clearer, my teeth whiter, my hair thicker, and my dandruff gone."

What's more, he had abstracted away all that time spent on food. "Now I spend about five minutes in the evening preparing for the next day, and every meal takes a few seconds," he noted. "I love order of magnitude improvements, and I certainly don't miss doing dishes. In fact I could get rid of the kitchen entirely, no fridge sucking down power, no constant cleaning or worrying about pests, and more living space. I just need a water source." He wasn't forswearing regular food entirely. "As any Instagram user knows, food is a big part of life. Food can be art, comfort, science, celebration, romance, or a reason to meet with friends," he added. "Most of the time it's just a hassle, though."

Many coders thrilled to Soylent. Rhinehart's initial post was rapidly passed around hacker sites, and his crowdfunding campaign to turn Soylent into a product reached its goal in two hours. Several startup programmers told me they kept their pantries stocked with it. It was perfect for the nonstop workload endemic in Silicon Valley, the 14-hour-long jags of coding; if you don't want to break your flow state, inhaling a bottle of goo at your laptop "literally requires less physical movements than even ordering and eating a pizza," as one said to me. And it's not just programmers buying Soylent; the drink has won fans among oil-rig workers, construction-site managers, commuters, and whoever wants to eradicate time spent on eating.

Soylent is, in many ways, a perfect metonym of the software engineers' obsession for efficiency. It's what you get when that worldview—speed everything up, remove friction everywhere—becomes second nature, a twitch instinct. Each moment of life becomes a target for Taylorization, a world of nails to be hammered into submission. Everything should move as rapidly as possible: communicating, working, shopping. Indeed, Amazon's success is predicated on rendering life into a carousel of instant gratification, as the computer engineer and director Ruhi Sarikaya notes: "Friction is any variable that impedes your progress toward a goal, whether it's purchasing a product or navigating traffic to make your nine a.m. meeting on time. Amazon is obsessively focused on reducing or eliminating friction—think one-click ordering, Amazon Prime, or Amazon Go."

This fetish for efficiency is what has driven the delirious explosion of "on demand" services. By the mid-2010s, Silicon Valley entrepreneurs started flooding the market with apps designed to optimize nearly every fiddly task of everyday life—offering to do work so you didn't. There was Washio (a service that dispatched laundry "ninjas" to pick up your dirty clothes), Handy (on-demand apartment neatening), Instacart (to pick up items from the local grocery store), and the endless phalanxes of TaskRabbits (to do basically anything else for you).

In terms of pure demography, there's a deep narcissism at work here. The blizzard of "do stuff for me" apps is what you get when you populate a tech hub—San Francisco—with a plurality of young men just out of college, and give them the tools of optimization and geysers of money for start-ups. The odds are high the "problem" they'll decide needs most urgently to be solved is the re-creation of the conveniences of dorm and home-life—where everyone prepared their meals, cleaned up after them, and ferried them around in vehicles. As my friend Clara Jeffery, the editor in chief of *Mother Jones* noted in a tweet, "So many Silicon Valley startups are about dudes wanting to replicate mom." It's also a symptom of how coding has evolved.

To be sure, some of these services are popular outside the coterie of coddled recent grads. Many people—the ones wealthy enough to pay for it, certainly—are happy to pay to have the pain-in-the-ass parts of life abstracted away. Uber's enormous success was predicated on reshaping the hailing of a car, an act that was historically plagued by a teeth-gnashing inefficiency: Taxis didn't know where customers were, and customers didn't know where taxis were.

But optimization inevitably produces side effects; there are winners and losers. Uber flooded the streets in cities worldwide with cars, which was terrific for riders—but less so for drivers, many of whom began to find it harder and harder to piece together a steady living, given the frenetic competition on the streets. (In New York City alone, in 2018 there were only 13,578 traditional taxis, but the number of ride-hail drives had exploded to 80,000.) Certainly, drivers who were only doing it for spare money were thrilled to have a way to quickly pick up some extra pocket money; Uber and Lyft made it possible to do driving as piecework. But it was bad news for anyone looking to drive as a reliably steady gig, a job that historically has been one of the easier-to-acquire forms of work for immigrants in big cities. "What Uber and Lyft have done is come into the industry and wreck it," as the Nigerian cabdriver Nnamdi Uwazie told NBC. By 2017, several cabdrivers had committed suicide and blamed the ride-hail firms for destabilizing their work so massively that it wasn't possible to rely on driving for a predictable income.

Indeed, even the programmers themselves can be surprised, and disenchanted, by how their zeal for optimization can produce unexpected and unsettling freaky side effects.

That's what happened to some of the early engineers of Facebook's Like button.

Back in 2007, several Facebook employees wanted to speed up the metabolism of the service. One designer, Leah Pearlman, suspected that people weren't posting as often as they might because it took too much work to think of a clever statement responding to a friend's update. Or if they did leave a comment—but didn't really have much to say—they'd leave a single word, like *Yay!*, and those single-word comments were cluttering up the look of Facebook. Pearlman envisioned there being something much more stripped down and efficient, some one-click way of letting users affirm a friend's post. It might unlock a ton of positivity, Pearlman figured. ("How much *more validation* might I be missing out on?" she once joked in a speech.) Another of her colleagues at Facebook, Justin Rosenstein—a coder who'd come to Facebook after helping to prototype Gchat for Google—was coming to a similar conclusion. "What if we could make it so that the effort required was so low" to approve of a post that people only needed a single gesture? (Or, as he told *The Ringer*: "There was an opportunity there to use design to make the path of least resistance to engage in certain kinds of interaction.")

Soon their idea got traction among Facebook employees. The team decided to make an "Awesome" button that would display next to every post: Click it, and you'd provide an upbeat vote of confidence, neatly, cleanly, and quickly. Better yet, it would be great for Facebook's analytics: The News Feed team could watch to see on what types of posts a user clicked on, the better to figure out what types of posts to promote in that user's News Feed. (Are you clicking on baby photos? Let's show you more of those!) The advertising people loved it, too, because it would give them ever more data on what type of content each user most loved, the better to customize ads.

After a long night of hacking, Pearlman and Rosenstein and a group of Facebook employees patched together a crude prototype. By six in the morning, they could click Awesome on a few internal, experimental accounts. Zuckerberg was initially dubious, but the concept quickly gained popularity among the firm's employees. Eventually, design teams switched the name from "Awesome" to "Like" and designed the fat blue thumbs-up that was to become one of the world's most famous computer icons. It launched on February 9, 2009, and soon became one of the service's most-used features. It's been clicked well over a trillion times by now.

In one way, Like achieved precisely what Rosenstein and Pearlman had hoped. By making it enormously faster and more efficient to show approval or support on Facebook, it unlocked a flow of positivity.

But it also, as people began to notice, produced some disturbingly addictive behavior.

That's because Facebook displays a numeric total of how many Likes each post has amassed. Users became obsessed with staring at that number and tracking precisely how much approval they were getting. They'd post a picture, then maniacally refresh the page to watch the Like total rise. It became another self-interruption that attention researchers track—we'd be working on an email or talking to a friend and then suddenly flip over to Facebook to nervously check our quantified popularity.

"It's our generation's crack cocaine," as the tech entrepreneur

Rameet Chawla wrote. "People are addicted. We experience withdrawals. We are so driven by this drug, getting just one hit elicits truly peculiar reactions." Psychologists probably could have predicted this behavior: Back in the '70s, the social psychologist Donald Campbell pointed out that if you use a single measurement to reward people, they'll do everything they can to goose that number higher. (It's now known as "Campbell's Law.") And so the Like button also arguably changed what people posted about on Facebook. Facebook users began putting up updates and photos designed specifically to trigger ever-higher Like counts, material psychologically juked to elicit reaction from onlookers: hot takes, controversial over-the-top statements, explosive emotionality, clickbait.

Even more discomfitingly, Like gave Facebook a powerful new way to track users. Once Like became popular, Facebook published cut-and-paste code so any website could slap a Like button on any post. Soon, there were Like buttons on news sites all over the web. But this code doubled as a snooping device. If you were a Facebook user and logged into the site, any time you visited a page with a Like button on it, the button would tell Facebook where you'd been, even if you never clicked it. In effect, the galloping popularity of Like gave Facebook a neat way to build a mammoth dossier of every user's surfing habits, across the entire web.

A few years later Rosenstein and Pearlman left Facebook; he, to found the workplace-productivity service Asana, and she to found several companies and pursue a new career as an artist. As time went on, they began to regret the side effects of the Like button.

"I find myself getting addicted—yes, in some cases, by the very things I've built," as Rosenstein told *The Verge*. He found himself increasingly pecked to death by social-media notices ("bright dings of pseudo-pleasure," as he called them) and was alarmed to learn that our app use is so compulsive that we touch our phones up to 2,617 times a day. For her part, Pearlman was unnerved by people's quest for Like-brokered approval. She watched "Nosedive," an episode of the dystopic sci-fi TV show *Black Mirror*, in which a young woman lives in a world where everyone rates everyone else on their daily interactions; this produces a world of bleak artificiality, as she posts faux exuberant photos (complete with expressions she practices in the mirror) and showers compliments on cashiers, lest they ding her with a low rating.

It haunted Pearlman. Indeed, she'd found that checking Facebook had become such an emotional chore that she installed software to block the News Feed. (The browser plug-in replaces News Feed with a random quote about self-control, such as philosopher Mortimer J. Adler's koan "True freedom is impossible without a mind made free by discipline.") "I check and I feel bad," as Pearlman said. "Whether there's a notification or not, it doesn't really feel that good. Whatever we're hoping to see, it never quite meets that bar." Rosenstein, too, decided he needed to decouple from the distractions of social media. As he told the *Guardian*, he blocked sites like Reddit, stopped using Snapchat, and even set up parental controls to keep himself from downloading new apps on his phone.

His assistant has the password. It turns out there's some use, after all, in techniques for slowing things down.

< Chapter 6 >

10X, Rock Stars, and the Myth of Meritocracy

In the annals of coding, Max Levchin is considered something of a colossus.

Levchin is the programmer who almost single-handedly willed PayPal into existence, in a frenzy of programming. He'd grown up in the Ukraine in the Soviet '80s, where he first discovered the fun of mucking around with programmable computers. When his family fled the region—they were Jewish, a persecuted minority—they arrived in Chicago in 1991. They were broke, but they respected their nerd son's passions and scrounged enough to get him a computer. When he studied computer science at the University of Illinois, he learned about Marc Andreessen, the graduate who just a few years earlier had done insanely well, cocreating the Netscape browser and becoming an overnight millionaire. Levchin was an intense introvert, but the romance of start-ups entranced him. While a student, he founded not just one but three companies, and when the last one sold for $100,000, he packed his worldly possessions—mostly a pile of electronics—into a truck and drove with some friends to Silicon Valley.

He crashed with a friend, and one day, wandered into a lecture on political freedom by Peter Thiel at the Stanford campus. Back then, Thiel was a lawyer and former Wall Streeter who'd studied philosophy and was a fervid libertarian. Levchin enjoyed the talk and met Thiel later to pitch his latest software ideas. He'd been writing ingenious encryption that ran on the weak chips of handheld PDAs like the PalmPilot, the hot tech of the day. They hit on the idea of creating software to let people transfer money digitally—from one PalmPilot to another, or even online: PayPal.

At school, Levchin had developed a reputation for being a ferocious font of code, working in relentless, multi-day-long jags. So it was with the origins of PayPal: He'd already written scads of crypto emulators, so he was able to crank out the prototype of PayPal's initial PalmPilot software by himself. When PayPal secured $4.5 million in financing, for a PR stunt he and Thiel decided to have the investors beam $3 million of it in digital money from one PalmPilot to another. They pondered doing a fake mock-up, one that just pretended to send the money. But Levchin was "disgusted" with the idea, as he later recalled in the book *Founders at Work*. "What if it crashes?" he thought. "I'll have to go and commit ritual suicide to avoid any sort of embarrassment."

Instead, he and the two other programmers they'd hired by then hunkered down and coded almost nonstop for five straight days. In *Founders*, Levchin recalled not sleeping at all during that period: "It was just this insane marathon." They finished their work barely one hour before the demo was set to begin. At a local restaurant called Buck's, with reporters crowded around a pair of PalmPilots, the $3 million was successfully transferred from one device to another. Levchin ordered an omelet for breakfast but fell asleep before he could eat it. "The next thing I remember," he said, "I woke up, and I was on the side of my own omelet, and there was no one at Buck's. Everyone was gone. They just let me sleep."

As PayPal took off, he kept up those empyrean feats of coderly productivity. One example was his fight against fraud: As PayPal grew, criminals began to use PayPal to steal money, about $10 million a month and growing. A panicked Levchin calculated that at its current rate of increase, fraud would in short order consume the company's profits and strangle it. *The world is going to end any minute now*, he thought. So he and his engineers threw themselves at that task, with Levchin heading again into sleepless-hacker mode. He'd go for days without bathing, as the journalist Sarah Lacy noted in her book *Once You're Lucky, Twice You're Good*; when he became so fragrant it was a workplace issue, he'd take a quick shower at the PayPal offices and put on a new company-logo'd T-shirt. ("At any given time, Max's workload could be measured by the growing pile of these used shirts at his almost-unlived-in Palo Alto apartment," Lacy wrote drily.)

Still, his work was quite inventive. During these work jags, Levchin came up with cunning anti-fraud ideas; he cocreated one of the first commercially used CAPTCHA tests, stretching and skewing a piece of text so that a human could read it but a bot couldn't. He also devised a tool that would help PayPal's human fraud agents parse huge amounts of transactions quickly; when he showed it off to them, one wept in joy. Levchin's anti-fraud work arguably kept PayPal from being destroyed by theft the way its many competitors were ruined.

Propelled by these epic feats—clever software, endurance, nutty work hours—the PayPal crew developed a proud sense of having won the gold ring in the meritocracy of Silicon Valley. While most other high-tech firms were collapsing in the dot-com bust of the early '00s, PayPal was soaring. It successfully went public, then was bought by eBay for $1.5 billion, making its founders remarkably wealthy. PayPal, the vice president Keith Rabois proclaimed, was a "perfect validation of merit." As he told the journalist Emily Chang: "None of us had any connection to anyone important in Silicon Valley. . . . We went from complete misfits to the establishment in five years. We were literal nobodies. People wouldn't talk to us."

Or as David Sacks, a former management consultant who joined PayPal as its COO, later told the *New York Times*, "If meritocracy exists anywhere on earth, it is in Silicon Valley." When you're remarkable enough, you win, as the story goes.

Rock Star. Ninja. Genius. In the world of tech, there's a long-standing cultural mythos around the software developer who is not merely good but possessed of an Olympian brilliance and productivity.

Pop culture long ago glommed on to this idea, such that depictions of coding invariably fixate on lone geniuses who summon digital palaces into being, in Levchinesque bouts of mania. The spectacle of the sleepless night of hacking is, at this point, practically a Jungian archetype. Early on in *The Social Network*, the Mark Zuckerberg character engages in an evening of frenzied keyboard-clattering, producing his app that let Harvard students vote on the relative appearance of female schoolmates. In the TV show *Silicon Valley*, the night before the start-up is about to go down in flames onstage at the TechCrunch conference, the coder-founder Richard has an epiphany and—again, in a single night—rewrites his entire compression algorithm, nearly doubling its performance and trouncing his competition. The hacker Cameron Howe of the TV show *Halt and Catch Fire*, as a favor to her friend's firm, creates what is essentially the Google PageRank algorithm. It's so artful that the firm's resident head of software wincingly admits he can't even understand how it works; she's that good.

This belief in the unicorn programmer isn't just a piece of pop culture. Indeed, in the real world of software, it's so well known that the concept has a name: the "10X" coder. As the moniker suggests, it describes a programmer who is provably better, multiple times so, than the average code monkey.

The concept can be traced to 1966, when three researchers at System Development Corporation issued a white paper with a mundane-sounding title: "Exploratory Experimental Studies Comparing Online and Offline Programming Performance." Originally, the trio—led by Hal Sackman—had set out to answer a fairly mundane, technical question about coding: Was it better to do it on paper or while sitting typing at a computer? Back in the '60s, having your own computer was pretty rare. Most programming was done in "off-line" mode. For example, you'd write your code by hand, get it punched onto punch cards, then give the cards to the computer operator. Then you sat back and waited, sometimes hours, for your batch of cards to be processed. But emerging by the mid-'60s were new online systems that let a coder control a machine interactively, in real time. Now you could sit at a keyboard, type your code, and see the results immediately.

As you might imagine, most coders far preferred online systems. It was considerably more fun to have the machine respond to you instantly; they felt it made them much more productive. But online systems were enormously more expensive, and companies weren't sure it was worth the expense. So Sackman examined the results of two bake-off tests. Groups of programmers were split into two conditions, one that had to write and debug a piece of code using the old-school offline style, and the other that was allowed to use the new online mode. He measured various aspects of their performance, including how long it took them to write the program, how long it took to debug it, how long the code was, and how fast it ran. When he crunched the numbers, voilà: Online programming won out. Those who were able to interact fluidly with the machine did "significantly better" than those working in the offline mode.

But Sackman also discovered something else that he didn't expect. There were "striking individual differences" in the abilities of experienced coders. The good ones seemed to be much better than the average ones—and wildly, crazily better than the bad ones.

For example, when asked to write and debug code that would navigate a maze, the fastest programmer was about 25 times faster than the slowest one. The same thing with writing and debugging an algebra problem: The fastest one was 16 times faster than the slowest while writing the code, and 28 times faster while debugging. What's more, the software of the good ones ran more speedily; the top-ranked maze code ran 13 times faster than the slowest. "To paraphrase a nursery rhyme," Sackman and his colleagues wrote,

> When a programmer is good,
> He is very, very good,
> But when he is bad,
> He is horrid.

Their takeaway? Coders differed in skill, "typically by an order of magnitude."

Now, Sackman's study had manifold problems, as many critics have noted. Among the biggest deficits, it tested a tiny sample of people, which makes the results impossible to extrapolate to the world at large, and subject to mere chance.

But those criticisms never took hold. There was something poetically *affirming* about the idea of a 10X difference, for many in the computer industry. It resonated with anecdotal evidence, an emerging folk wisdom of people who worked alongside computers and observed the hackers who coded them. Even back in the '60s and '70s, managers were claiming that—for reasons they couldn't quite explain—a minority of their coders could build software and debug problems with supernatural speed and grace, as if they observed the machine via a third eye their peers did not possess. (Even one of Sackman's harshest critics—who otherwise claimed the study was "quite unreliable"—noted that there was an "extreme shortage of high performers" in coding. "Programming differs little from other creative work in this regard.") Soon, other researchers were proffering data sets that seemed to document the existence of an elite virtuoso class. The software manager Bill Curtis published a paper showing a test he'd done with 54 coders, in which the best outperformed the worst by factors of 8X to 13X. A headline in the computer magazine *Infosystems* posed the field of programming as a star system of rare heroes looming over their mediocre colleagues: "The Mongolian Hordes versus Superprogrammer."

The legend arguably cemented into place in 1975, when Fred Brooks published *The Mythical Man-Month*, a book about the dark art of managing software projects. Brooks noted Sackman's findings approvingly. "Programming managers have long recognized wide productivity variations between good programmers and poor ones," he noted. "But the actual measured magnitudes have astounded all of us." The upshot, he argued, is that one could hypothetically construct the world's best coding team by simply paring down to nothing but the most awesome performers. If a team had 200 coders and only 25 were true stars, "fire the 175 troops" and leave the 25 rock stars alone to get the work done.

Tech firms, Brooks noted, were treating coding as if it were a form of manual labor, where you can speed something up by simply adding more workers. You need wheat to be picked twice as quickly? Easy: Add twice as many wheat pickers. But coding is an insight-based form of labor, more akin to writing a poem. Merely adding more people doesn't help, because the solution to any problem is liable to come not from the sweat of many brows but from the lightning-strike aha moment of a single insightful individual. The upshot is that coding was literally the opposite of ditch digging, organizationally. Throwing more coders at an intractable problem can make things *worse* because it increases the amount of meetings and communication overhead. Brooks formulated this into a pithy koan: *Adding manpower to a late software project makes it later.*

It wasn't long before coders worldwide were confidently citing this concept to explain why it was so crucial to hire not just adequate programmers, not just good ones, but absolute titans. It gave justification to the talent wars that broke out in the '90s up to today, with highly oxygenated sums and campus-like perks paid to lure top students leaving top schools or to acquire the talent from fizzy start-ups. The very phrase "10X" appealed to the high-tech world's nerdy self-image of logic, data, and precision. Other industries might have star systems, but in software they had a *number*—an actual measurement of how much a star eclipsed a mere mortal. "A great lathe operator commands several times the wage of an average lathe operator," as Bill Gates once said, "but a great writer of software code is worth ten thousand times the price of an average software writer."

In the upper echelons of the tech industry, the idea of coding as strong meritocracy is powerful. When I ask venture capitalists and founders whether 10Xers really exist, many immediately say: Oh yes. *Hell* yes.

"I think it's probably 1000X," as Marc Andreessen, that cofounder of Netscape who so impressed Levchin, tells me. "If you make a list of the great software built in the last fifty years, you'd find that in virtually every case, it's one or two people. It's almost never a team of three hundred. It's at most a team of one or two."

He's right: The list of small-person or one-person innovators is long. The first version of Photoshop was created by two brothers; the version of BASIC that launched Microsoft in 1975 was hacked together in weeks by a young Bill Gates, his former schoolmate Paul Allen, and a Harvard freshman Monte Davidoff. An early and influential blogging tool, Live-Journal, was written by Brad Fitzpatrick. The breakthrough search algorithm that led to Google was a product of two students, Larry Page and Sergey Brin; YouTube was a trio of coworkers; Snapchat a trio (or, the level of the code, one person, Bobby Murphy). BitTorrent was entirely a creation of Bram Cohen, and Bitcoin was reputedly the work of a lone coder, the pseudonymous "Satoshi Nakamoto." John Carmack created the 3-D-graphics engines that helped usher in the multi-billion-dollar industry of first-person shooter video games.

The reason so few people can have such an outsize impact, Andreessen argues, is that when you're creating a weird new prototype of an app, the mental castle building is most efficiently done inside one or two isolated brains. The 10X productivity comes from being in the zone and staying there and from having a remarkable ability to visualize a complex architecture. "If they're physically capable of staying awake, they can get really far," he says. "The limits are awake time. It takes you two hours to get the whole thing loaded into your head, and then you get like 10 or 12 or 14 hours where you can function at that level." The 10Xers he has known also tend to be "systems thinkers," insatiably curious about every part of the technology stack, from the way currents flow in computer processors to the latency of touchscreen button presses. "It's some combination of curiosity, drive, and the need to understand. They find it intolerable if they don't understand some part of how the system works."

This is why 10Xers love start-ups. It's where, starting from a fresh page, they can peacock and show off what they can do. In contrast, many big and established firms require a slower pace. They've got huge legacy systems, code going back years—maybe decades—and customers who rely on things working. The job is less "moving fast and creating" than patiently and gently fixing bugs in systems that otherwise more or less work. Andreessen remembers being at IBM as an intern in the early '90s, where the company had a firm standard: "Ten lines of software code a day. Ten lines written, tested, debugged, documented—no more, and no less. Less than ten lines a day, and you were being lazy. More than ten lines a day, and you were being reckless."

Many veteran coders agree that 10Xers not only exist but can be the key to productivity. Joel Spolsky, the cofounder of Fog Creek Software, once wrote in a blog post about how when he was at Juno, an online service and free email provider, they had a bug checker named Jill McFarlane. She "found three times as many bugs as all four other testers, combined. I'm not exaggerating, I actually measured this. She was more than twelve times more productive than the average tester. When she quit, I sent an email to the CEO saying 'I'd rather have Jill on Mondays and Tuesdays than the rest of the QA team put together.' "

As Sam Altman, the head of the famous start-up accelerator Y Combinator, suggests, the idea of singular geniuses in code shouldn't be surprising; every insight-based field has standout individuals. "It doesn't seem that controversial in other fields," he tells me. "There are 10X physicists, and we give them Nobel Prizes and that's fine. There are 10X writers and we give them number one on the *New York Times* Best Sellers list." Some computer science teachers, it would appear, not only agree but believe 10X talent is visible even when kids are first learning to code. A small study by computer science professor Clayton Lewis found that 77 percent of faculty in a computer science department disagreed with the statement, "Nearly everyone is capable of succeeding in the computer science curriculum if they work at it." The way they see it, you either have the gift, or you don't.

One day in 2017 I visited Dropbox to meet with its founder, Drew Houston. He is what some might easily call a 10X coder himself. He started programming as a young child, attended MIT, and once in his spare time wrote a bot that played a sharp game of online poker. After graduating from MIT, he was so annoyed at leaving his USB drives behind he decided to hand roll his own system for automatically syncing files from a computer to a server. That turned into a prototype for Dropbox; he realized others might want to use it; and after taking it through Y Combinator, Houston turned it into a company that was worth about $10 billion in late 2018.

Sitting on a purple sofa in Dropbox's music room—the company has an entire music studio, complete with a drum kit, guitars, and amps, for employees to use for relaxing—Houston argued that part of his success was built on finding and hiring 10Xers. Some of their aptitude is built on experience—they've put in their 10,000 hours, have seen every rare error message, and over time have honed techniques for getting out of jams. "That's trainable," he argues, nurture over nature. But there's some nature stuff in there, too, he says: a passion, an intensity and love of the craft that seems characterological. "They're kind of a force that you unleash in the general direction of a problem," he says.

Houston introduced me to an employee he regarded as one of his highly productive coders: Ben Newhouse, a 28-year-old who was at the time an engineering lead for the firm. (He later left to head back into entrepreneurship.) Newhouse had—like Houston—created some valuable code while still a student: As a 21-year-old Stanford undergraduate, he created one of the first augmented-reality apps for the iPhone. He was interning at Yelp when he realized he could use the compass and GPS sensors inside the iPhone to make the screen respond to the world around it. After a spree of coding, fueled by a case of Red Bull and—you could probably see this coming—a sleepless night, he'd built a feature that let you hold up your iPhone and see Yelp reviews for nearby businesses floating in the air around you.

At Dropbox, he fixated on an interesting problem that had begun to plague the firm. People often use Dropbox to back up their entire hard drives, so over years of use, customers might have 300 or 400 gigs of photos and movies and documents in their Dropbox account. Then one day they'd decide to upgrade their old, clunky laptop—and they'd opt for an ultralight model like a MacBook Air. But these new light computers only have tiny hard drives, holding perhaps 128 gigs. Now the Dropbox users had a problem. They can't reverse-sync all 400 gigs of files they have stored in the Dropbox cloud back down to their new, now-tiny little laptop drive. Effectively, it ruins their use of Dropbox. They now have to sit around thinking, *Okay, which files should I manually transfer back to my tiny laptop? I can't fit 'em all.* It wrecks the set-and-forget convenience that Dropbox prides itself on providing. The company's coders had long pondered how to deal with that problem. But they'd concluded it was just too damn hard to solve.

But the problem bugged Newhouse. *There must be a more elegant solution to this*, he thought.

A few months later, he got the chance to tackle it. Dropbox held one of its regular internal hackathons, where "we nail the doors shut and just spend the week trying out new ideas," as Houston told me. Newhouse was thinking about the backup problem when he made a useful, stray mental connection: antivirus software. When you go to open a file, your antivirus software uses "minifilters" to quickly check its contents. He could borrow this technique and graft it onto Dropbox. It would make your tiny MacBook Air appear as though it had all your Dropbox files on it—but it would just grab the file from the cloud the instant you asked for it, let you edit it, and store it back in the cloud. But it'd happen so quickly you wouldn't feel any painful lag.

This wouldn't be easy to do. It'd require code that works deep in the "kernel" of your computer's operating system—one of its more central parts—and tinkering at the level of the kernel is like a form of neurosurgery. Change how Dropbox works without caution, and you could give millions of its users a digital stroke, destroying some of their backed-up info. "The kernel is a much more intense and dangerous place to work," Newhouse says, "because if you fuck up, everything goes boom." That's why for years, the engineers thought it was best not to mess around with something so drastic. "A lot of people thought it was impossible," Jamie Turner, another Dropbox engineer, told me.

Working from home, Newhouse hunched over his laptop, coded like mad, and by the end of the week had a demo of his concept. Impressed, Houston gave Newhouse a six-person team to make it a reality. When Newhouse quietly deployed the new feature in-house, Turner, his colleague, didn't realize it was turned on yet; he discovered it when he set up Dropbox on his wife's new MacBook, expecting it would take about 48 hours to sync the massive number of files. Instead, the sync took only a few minutes. *How the hell is that done yet?* he wondered, when he suddenly saw a "Smart Sync enabled" message. "Well, I'll be damned." When the feature went public as Smart Sync, it was one of Dropbox's more significant new upgrades in years.

This type of thing, Houston told me, is precisely the punch-above-your-weight creativity he tries to hire. Get one of them, and you can launch ideas that ten others won't.

"No matter how long I sat in a room and tried to compose a symphony, I couldn't," he says. "You can have ten or a hundred designers," he concludes, "and you won't have one Jony Ive."

It's not hard to understand why so many coders love the idea that programming is a world of pure willpower, raw talent, and 10X meritocracy.

On the sheer level of everyday coding, it can certainly *feel* true. One cannot bullshit the computer, or bluster through a failed code test. "You can't argue with a root shell," as programmer Meredith L.

Patterson wrote in a 2014 essay, adding: "Code is no respecter of persons. Your code makes you great, not the other way around." Let the noncoders hand wave and plead and persuade; real programmers respect only running code, dammit. As Mark Zuckerberg wrote in an open letter when Facebook went public: "Instead of debating for days whether a new idea is possible or what the best way to build something is, hackers would rather just prototype something and see what works. There's a hacker mantra that you'll hear a lot around Facebook offices: 'Code wins arguments.' . . . Hacker culture is also extremely open and meritocratic. Hackers believe that the best idea and implementation should always win—not the person who is best at lobbying for an idea or the person who manages the most people."

As another metric of merit, some coders pointed out to me that programming is a rare form of engineering in which totally self-taught individuals can be accepted by highly credentialed peers. "To me, the most fascinating thing about computer science is that it's the only STEM field I'm aware of where there's this combination of people who are super highly accredited working side by side with people that just taught themselves everything," says Johanna Brewer, who, indeed, taught herself programming beginning in middle school, then earned a PhD in information and computer science and founded several companies.

There can also be, though, a fair amount of self-flattery underpinning the belief in meritocracy, as several coders added.

Sometimes the desire to believe in meritocracy is a way of coping with teenage nerd damage. For anyone who grew up as an awkward introvert, feeling at sea in the ultrasocial pecking orders of high school or early corporate life, the idea that programming is a neutrally objective world is deeply appealing. Cynthia Lee, who has a PhD in high-performance computing and teaches at Stanford, remembers working as a coder at start-ups in the '90s and early '00s alongside colleagues who'd all felt shunned or misunderstood as youths. Finally, they exulted, a realm where those damn poseurs don't automatically win!

"There was a lot of suspicion of people who were going to come into our tech space with a suit on or looking a little too sharp," she says. "Because they were sort of the enemy. In the 1980s movies about high school, they were the popular kids, and we were making the utopia of the nerds."

Tracy Chou, a programmer known for ferociously productive stints at Quora and Pinterest—"an absolute rock star," as Pinterest cofounder Ben Silbermann once gushed to me—has seen similar dynamics. "I think a lot of the people who have been successful in software have not been the people who would classically be successful in other business pursuits," she says. "And to be successful there, they want to really *own* that success." She also points out that programming's often-obtuse nature—to noninitiates and sometimes even to other colleagues—makes mystical claims of extreme merit easier to pull off. "There's something about the fact that code is incomprehensible to most people or hidden from most people," she adds. "And even when it's not hidden, it's most often incomprehensible. And so it's easier to hide behind a charade of, 'oh, it's meritocratic, and if you understood, you would know.'"

The world of open source software—in which code is released online for anyone to examine and tweak—is often particularly seen as a meritocracy, because it's all about competing to have your (often volunteer) contribution accepted into a project.

Consider the most well-known success story of open source, the operating system GNU/Linux—often referred to simply as "Linux." Like Windows or MacOS, it's an operating system that runs a computer; unlike those, it's free to use, and anyone can download and inspect its roughly 25 million lines of code. Linux was initially started in 1991 when the Finnish university student Linus Torvalds decided to create his own kernel for an operating system just for fun. It wouldn't be "big and professional," as he wrote when he first announced it online. Soon Torvalds had a simple kernel running, and he put the source code online for any other hackers to look at.

Then the snowball began to roll. Pretty soon programmers worldwide were writing him to suggest new features to add to Linux, offering snippets of code, or posting bug fixes. Torvalds adopted the suggestions he liked (he "pulled" in their code, to use the jargon). Linux gradually grew more and more features, all contributed by strangers worldwide. Eventually there were hundreds, and then thousands, of contributors. To make it easier for so many disparate coders to all tinker with a single code base without their changes getting mangled up together, Torvalds wrote "Git," a piece of software that is now also widely used by coders. Git lets you pull in someone else's contribution or quickly revert back to an older copy of the code if a new change you've made has screwed things up.

As some of its fans argue, open source becomes a sort of market competition of merit—a race to see whose idea can be judged so good that other coders will agree and go, *Okay, sure, let's accept that code into our project.* Open source thus can, to many involved, feel like a pure distillation of merit. With Linux, Torvalds became the "benevolent dictator," accepting only the contributions to the Linux code base that struck him as truly useful and excellent. The barrier to contribution is, in theory anyway, quite low: Just download a copy of the Linux source code, make any changes to it—which will show up in the "tree" structure of the code, if you're using Git to manage your changes—and then send a request to the core Linux contributors, saying, *Hey, check out my contribution.* If they like it, it'll go in, and thus become used by millions of firms worldwide. It's a little more complicated than that in practice, but this is a microcosm of how most open source projects tend to work.

"You have your own fucking tree and you can do whatever you want to your tree," Torvalds told me, when I visited him in Portland in 2016. "You do your crazy thing, and if it turns out you're right—and it turns out your crazy thing wasn't crazy after all—you can now publish your tree and say *Look what I did.* And if it's really good, everybody else can just pull in."

Indeed, by the time I met up with him, Torvalds was writing very little code himself. He just adjudicated, a Solomon of software: He'd sit at his small office in his Portland house, littered with cables and gear (including some scuba-diving stuff; he's a fan of the sport and programs his own specialized diving software), and spend his day viewing the latest submitted code. To get to him, a contribution generally had to pass through a gauntlet of Linux's maintainers, a group of central contributors who'd proved their mettle to Torvalds and to each other, volunteering for hours to write Linux code and critique and assess that of others. The inner sanctum has a heft and power to it. Linux is now so widely relied upon by the computing world that many tech firms, including Intel, Red Hat, and Samsung, pay employees to be full- or part-time contributors to Linux. Being a core Linux contributor is regarded as a pretty auspicious item on a coder's CV.

Contributing to a popular open source project is usually a career-enhancing move, which is why so many coders try to get involved in one. It's also why so many release their weekend just-for-fun side-projects as open source code on sites like GitHub. There's the pleasure of letting others see your work, the joy of discovering that some weird tool you crafted for yourself is also useful for others; plus, you can learn a lot from looking at other people's open code and seeing how they built things. Coders also told me they felt an ecological sense of obligation: They open source their own code—and contribute to other people's projects—as a way of giving back to a scene that has helped them out. Virtually every coder I've met uses tons of open source code at work; million-dollar businesses are routinely built on top of open source code. Open source thus contains a curious combination of motivations—an Adam Smithian competition to impress, blended inextricably with a communitarian ethos that might make Karl Marx smile. And through it all is the ideal that, well, the code doesn't lie: If it's good, your fellow programmers will see it and accept it.

"Bullshitting, in general," Torvalds says, "is very much frowned upon."

That's a seductive idea. But the reality of a world run by superhero talent gets messy quickly—and can be considerably less productive than it might appear, as Jonathan Solórzano-Hamilton discovered.

Solórzano-Hamilton is a software architect who worked with a self-professed rock star programmer. "Rick"—as Solórzano-Hamilton pseudonymously called him, while recounting the story in a blog post—was known throughout the firm for his ability to solve anyone's problem: Ask him, and he'd sketch out a quick solution on his in-office whiteboard. He was the head architect, designing projects, and also the top programmer, cranking out the code itself. He often stepped in with a lifesaving fix.

It appeared that sense of being indispensable grew on Rick, worked its way into his psyche, and turned sour. He started regarding himself as the coding superstar of the firm, the 10Xer towering over all the mere mortals. Convinced that his skills were crucial to everything, he took over more and more tasks, more and more pieces of code.

But despite Rick's work, the project was blowing past its deadlines. If a project is big enough, there's no way a single person can do it all, no matter how talented they might be. The project ran fully one year late, and the managers realized it wasn't likely to be ready for two more. Rick was monomaniacally trying to be the single hero; worse, it seems his managers indulged his self-mythologizing.

"Rick was churning out code faster than ever. He was working seven-day weeks, twelve hours a day," as Solórzano-Hamilton wrote. "Everyone knew only Rick could pull the team out of this mess. Everyone held their breath and waited for Rick to invent the miracle cure that would mend this crippled project." Meanwhile, overburdened with work, he was getting grumpier and isolating himself from others.

Solórzano-Hamilton was asked to help and see if the project could be saved. A meeting with Rick did not go well. "You will never be able to understand any of what I've created," he raged. "I am Albert fucking Einstein and you are all monkeys scrabbling in the dirt."

When Solórzano-Hamilton looked at Rick's code, though, he realized it was so idiosyncratic and undocumented that nobody else would be able to maintain it. They spoke to Rick and told him they wanted to build a new product from scratch, with everyone collaborating on it. Rick angrily dismissed that, too. As things worsened, Rick wouldn't take time off, he'd revert code written by others, and belittled colleagues.

Eventually they fired him. And then, behold, things improved: His teammates set about building a new, dramatically simpler product. By the time they were done, the replacement product was less than 20 percent as big and complex as the previous project. That meant it'd be much easier for new employers to read it, grasp it, and maintain it. They wouldn't *need* superheroes. Better yet, the teammates got it done in just over six months. "There were no Ricks left on the team. We didn't have any mad geniuses building everything from scratch. But our productivity was never higher," Solórzano-Hamilton wrote.

The company had run smack into the noxious downside of the worship of coder merit: It can create "brilliant jerks," a caste of programmers who come to believe the myth of their own irreplaceability. You wind up with aggro blustering types who not only drive other talented folks away but often wind up producing work that—ironically—isn't even very useful, because it's so locked up inside their own heads. Sure, their talent might be quite real, but who cares about that when their cult of personality wrecks the enterprise?

Quite a few coders I spoke to had horror stories of working alongside talented but entitled assholes. One Y Combinator firm hired a Russian coder whose work was superb but who'd snark "I hate it here" if you asked him how things were going. *Why do you hate it?* "Because everyone's work is shit," he'd reply. "He was a total diva," the head of programming said with a sigh. As Bonnie Eisenman, a programmer for Twitter who's an expert in React—a code library increasingly used to make apps—puts it, "the whole myth of the rock star coder creates dysfunction."

Brilliant jerks aren't even necessarily worth it, because while they might be useful for solving a hard, short-term problem, the wreckage they cause to morale can be hard to repair. Other talented employees flee, unwilling to wrangle with the jerk. "I've met people who are the most brilliant programmers, but they make things that never see the light of day because nobody else can work with these people," Grady Booch, a veteran IBM coder, tells me.

Even when the 10Xers like Rick are productive, writing that much software—that quickly—tends to produce what's known, in a lovely phrase, as "technical debt": a bunch of wreckage produced by moving too quickly. A fast-cranking coder will almost always use shortcuts and employ patched-together solutions that will require, in the years to come, careful and patient cleanup by later colleagues. "The 10X engineer is not actually 10 times more productive than everybody else," says my friend the developer Max Whitney. "A 10X engineer is out there—I'm quoting somebody off the internet—generating 10 times more work for everyone. So, they're like the tip of the iceberg, making something flashy and beautiful, and leaving so much technical debt everywhere."

Part of the reason coders love building start-ups, as Andreessen notes, is they get to move quickly. But even here, the early heroes can create messy code bases that only function well enough to acquire early users, and soon need cleanup by more patient programmers who bring order to the chaos.

When Tracy Chou was hired at Pinterest, she headed up a heavy rewriting of its back end. While poking around in the code base, she discovered something weird: Whenever a user searched for a term, the server would run the query twice. *What's up with that?* she wondered. Chou eventually discovered that the code for making a query had been accidentally cut and pasted twice. Someone in the early days of Pinterest, it seems, had been working too quickly. By cutting that single mistaken line of code, Chou doubled the efficiency of Pinterest search. Oftentimes the real 10Xing isn't in writing the code but rather in fixing someone else's blunders.

Perhaps the most insidious problem with the idea of 10xing is that it mythologizes a sort of behavior that almost no one can get away with but young white dudes.

After all, recall those stories of unwashed, unshowered, wild-haired keyboard jockeys, so fragrant their colleagues have to drag them to the showers: "Can you *imagine* a woman getting away with that?" as Sue Gardner, who ran the Wikimedia Foundation for almost seven years, tells me.

Jacob Kaplan-Moss, a well-known Python developer, tells the story of watching a student at a conference present her work predicting the seasonal flooding of the Kansas River. She'd used a welter of languages and tools to do so, from Python to PostgreSQL to GeoDjango. Impressed, he approached her afterward to see if she wanted to interview for a job at his company. She demurred, saying she "was not really a programmer."

Kaplan-Moss was stunned. But this is, he argues, precisely the problem with venerating rock star coding. In purely objective terms, the student was remarkable, assembling her own custom system for analyzing satellite data. But she'd internalized the industry's trope that the only "real" coders were the slovenly, wild-haired types cybernetically merged with the keyboard. "Programming is something you *are* in this myth, not something you *do*," he concluded.

In a sign of the times, the meritocratic ideal in Silicon Valley has come under more scrutiny of late. Consider the case of PayPal: It was, quite genuinely, an example of smart and relentless work by people like Levchin and his gang of sharp and intense developers, designers, and marketers. After all, plenty of other payment systems failed. PayPal plugged on. The early hires wound up rich and influential; the company minted what's known as the "PayPal Mafia," a group of millionaires who became even wealthier and more influential as they invested in the next generation of tech firms, like Facebook and Uber.

But as Emily Chang, the author of *Brotopia*, points out, PayPal—like many of those early start-ups—was hardly the pure meritocracy many of its founders claimed it to be. For one, the company didn't actually hire people based purely on how talented they were. Quite the contrary: Levchin and Thiel picked people who were like them, to "make their early staff as personally similar as possible," as Thiel wrote in his book *Zero to One*. They worked their school and friendship networks to find young geeks suspicious of government. "We were all the same kind of nerd," Thiel writes. "We all loved science fiction: *Cryptonomicon* was required reading, and we preferred the capitalist *Star Wars* to the communist *Star Trek*. Most important, we were all obsessed with creating a digital currency that would be controlled by individuals instead of governments. For the company to work, it didn't matter what people looked like or which country they came from, but we needed every new hire to be equally obsessed." In principle, sure, those PayPal hires could have come from any country or any walk of life. In practice, they cast a surprisingly small net, nabbing a group that was primarily young, white, male, and highly educated. You could call their hiring strategy a deeply pragmatic way to quickly assemble a team, one that had excellent group cohesion. But it's hardly a blind, talent-is-the-only-thing-that-matters form of meritocracy.

The idea of tech being a pure meritocracy crumbles even further if you look at the background of who, precisely, creates start-ups. One study by a pair of business professors found that one of the biggest common threads uniting entrepreneurs is that they come from well-off families. This makes sense: It's easier to embrace risk when you have a safety net. And when it comes to "who gets investment money," things narrow even further: Research by Reuters found that just under 80 percent of the Silicon Valley founders who got Series A funding from the top five VC firms had already worked at well-connected tech firms or attended Stanford, Harvard, or MIT. This looks less like a system for rewarding scrappy, wild-eyed, unconnected idealists with a brilliant concept and more like what the economist Robert Frank would call a "winner take all" dynamic. Early success leads to a cascade of good fortune downstream. And that good fortune is quickly spun into a mythology: We deserved it.

It's often hard for wealthy businesspeople of any stripe to admit the role that fortune played in their success. It's probably even harder if you're a techie who pulled insane hours and solved countless hair-pulling bugs on a server that truly didn't care about anything other than your code being executable. You worked unbelievably hard; you enjoyed remarkable good fortune. Both things are true, but you focus on the first. That's particularly true for someone hired early, in a stroke of good fortune, at a company that later skyrockets to success.

"If you made fifty million dollars because you were early in Google, a lot of people are like, 'Yeah, that's because I was good—and all those other engineers who didn't make that money, that's just because they weren't that good,'" Josh Levy, a longtime veteran of several Silicon Valley firms, tells me. "It's really hard for people to see that the economic value can be random."

As Chang points out, PayPal enjoyed enormous luck; it could have been easily destroyed, save for a few massive strokes of chance. If it'd tried to get funding a few months later than it did, it would have slammed into the "dot-com bust," when investors suddenly shut their purse strings. But when fate swings one's way, self-mythologizing comes easily, particularly when—frankly—journalists like myself come looking for stories of "lone geniuses." "What was an improbable bonanza at the hands of the flailing half-blind becomes the inevitable coup of the assured visionary," Antonio García Martínez, a former Facebook ad-tech employee, jokes in his book *Chaos Monkeys.* "The world crowns you a genius, and you start acting like one."

Even the world of open source becomes rather less of a meritocracy the closer one looks. After all, it's predicated on a coder having tons of free time to crank away on volunteer labor. This works remarkably well for time-rich young people but less so for anyone with lots of coding talent but more real-world responsibilities. "Meritocracies say 'your GitHub is your résumé,' then they act surprised that their candidate pool doesn't include a lot of single moms without time to hack on hobby projects," as Johnathan Nightingale, a former general manager of the open source browser Firefox, writes. Firefox, he notes, was like that, too. The company that ran Firefox, Mozilla, had a female founder, "and still our 'meritocracy' was full of people who looked an awful lot like me." A survey done by GitHub found that 95 percent of its respondents identified as men; 3 percent identified as female and 1 percent as nonbinary. Though solid numbers are hard to come by, other surveys and estimates of the amount of women involved in open source projects appear to find that it's around 10 percent or lower.

More directly yet, many women had reported incidents of outright harassment or assault at open source conferences. Even online, participating in projects like Linux could require one to withstand verbal storms of derision, particularly if you fell afoul of Torvalds himself, who had long been known to write furious emails to contributors he thought were acting like idiots. ("Please just kill yourself now. The world will be a better place," as he wrote in one sample email; "SHUT THE FUCK UP!" in another.) One analysis of his emails suggested he wasn't any more venomous to women than to men. But even Torvalds decided, at long last, that his behavior was a problem for the Linux community: In September 2018, he temporarily stepped down as benevolent dictator to "get some assistance on how to understand people's emotions and respond appropriately," as he wrote.

One could reply, as some coders do, that the low proportion of women and minorities in projects like this—or at Silicon Valley startups like the early stage PayPal—is also just a function of meritocracy. Maybe women are just inherently less good at the discipline. The short answer to that is no. The longer answer is something I take up in the next chapter, "The ENIAC Girls Vanish."

Coders and techies, certainly in the US, have a reputation for leaning libertarian—which we could define here very loosely as believing that each person is responsible for their own fate, that government regulations generally stifle liberty, and that society is best served by letting the best rise by their own efforts.

Why? It's partly because some iconic tech CEOs espouse libertarian stances so stark they practically come off as cat-stroking Bond villains. Peter Thiel, perhaps the best-known example of this phenomenon, is a man so committed to his loathing of "confiscatory taxes" that he once explored the possibility of "seasteading"—creating a floating city beyond the reach of the weary giants of flesh and steel. He's also proclaimed, "I no longer believe that freedom and democracy are compatible." Then there's Uber's former CEO Travis Kalanick, who once argued on the question-answering section of the site Mahalo that California is a sorry ethical morass of moochers living off the spoils of the affluent. "One of the interesting stats I came across was that 50 percent of all California taxes are paid by 141,000 people (a state with 30mm inhabitants). This hit home as I had recently finished *Atlas Shrugged.* If 141,000 affluent people in CA went 'on strike,' CA would be done for . . . another reason you can't keep increasing taxes to pay for unaccountable gov't programs that offer poor services."

There is, of course, a deep irony—even hilarity—in virtually *anyone* in technology being a libertarian, given that the entire American industry has been built on innovations patiently funded by the government. Without the military buying early microchips en masse, that whole industry would probably have been stillborn. The federal government also paid for foundational original work into tech inventions as crucial as relational databases, cryptography, voice recognition, and, most of all, the Internet Protocols themselves. (If you like today's AI, thank the Canadian government. It spent plenty of taxpayers' dollars helping to support crucial deep-learning research for years at the nation's public universities, back when that style of AI was being pooh-poohed worldwide.) When *R&D* magazine surveyed the top innovations from 1971 to 2006, they found 88 percent had been funded by federal research dollars. None of these fields were being sufficiently funded by the free market. It took the slow-moving, long-term patience of a government to produce the core inventions that make it possible for us to hold a phone and order one of Kalanick's Uber cars.

Nonetheless, the libertarian protestations of a certain set of coders continues apace. In recent years, blockchain technology has been the latest site of tech's anti-government fervor. That ranges from Bitcoin—a currency specifically designed to create money that couldn't be controlled by dough-printing central banks—to Ethereum, a way of creating "smart contracts" that, its adherents hope, would allow commerce so frictionless and decentralized that even lawyers wouldn't be necessary: The instant someone performed the service you'd contracted them to do for you, the digital cash would arrive in their digital wallet. One survey of people in the cryptocurrency community found that fully 27 percent called themselves libertarian, more than double the rate Pew Research Center found in the general population.

At first blush, it's not hard to figure out why there'd be a strong Venn overlap between coders and libertarians. Both inhabit realms where first principles and logic are heavily touted. They're also arenas heavily populated by young guys—who often have little experience with the messy priors and injustices of the real world, and thus are prone to breezily hand wave these aside. In the programming scene of the '80s and '90s, "I think you had a lot of people who were poorly socialized, they liked formal systems, they like solving problems," as the Netscape coder Jamie Zawinski tells me. He thinks the philosophy is naive, but he gets why coders are drawn to it: "Libertarianism is nothing if not coldly logical, right?"

Google's Peter Norvig, who's been programming since the '70s, adds that libertarianism is probably also just a function of how relatively well-off coders are: "I think everybody who has a job as a programmer has a comfortable life and so that means, for some of them, this libertarian approach is attractive because they're saying, 'Let's just get government out of the way. I'd be fine without it. Therefore, everybody else will be fine without it too.'" When discussions of sexism in the industry emerged, posts on Blind—an app for anonymous discussion limited to employees of top tech firms—bemoaned the idea that techies would even deign to discuss this sort of social-justice stuff. "Can we go back to the time when Silicon Valley were about nerds and geeks, that's why I applied [to] Google and came to the US," as one wrote. "I mean this industry used to be a safe place for people like us, why so fking complicated now."

The thing is, this reputation for libertarianism may not be entirely deserved.

Recently, two Stanford researchers and a journalist got interested in this question, and did some intriguing research. As they noted, on a purely partisan level, Silicon Valley's denizens tend to vote—and donate—to Democratic Party candidates. Indeed, in the 2016 presidential election, employees at the biggest tech firms donated a hefty 60 times more money to Hillary Clinton than to Donald Trump. And when Trump won, staff at many companies were plunged into near mourning, as I found when visiting the city that week.

To gather some hard data on political attitudes, the researchers polled almost 700 high-tech founders and CEOs. In one part of the survey, these heads of tech were asked how much they agreed with a short statement of libertarian ideals: "I would like to live in a society where government does nothing except provide national defense and police protection, so that people could be left alone to earn whatever they could."

Interestingly, only 23.5 percent agreed with it. This was much less than the agreement rate for Republicans on average (62.5 percent)—and it was also lower than for Democrats on average (43.8 percent). In other words, the tech founders and executives were, remarkably, less likely than *garden-variety* Democrats to agree with basic libertarian ideas.

The study also found that the tech folks were extremely globalist in their worldview, with 44 percent, more than any other group, agreeing that "trade policy should prioritize the wellbeing of those abroad instead of Americans." And they supported many classically redistributionist tax-and-spend policies: 82 percent supported single-payer health care even if it meant raising taxes, and 75 percent supported spending federal money on programs that benefited only the poor. Nearly all supported same-sex marriage, and 82 percent favored gun control. "In other words," as the researchers concluded in their paper, "technology entrepreneurs are not libertarians." They were, in many ways, just traditional Californian leftish thinkers.

With one big exception: regulating corporate behavior.

The tech entrepreneurs were hotly opposed to any government rules that determined how they did their business. They didn't like any regulations governing how employees get hired or fired; they wanted to see the influence of unions and organized labor diminish. Fully 82 percent thought it was too hard to fire employees and that regulations should be changed to make it easier. They also were far less likely than average citizens to pass ethical judgment on corporate behavior. When given examples of "surge pricing"—including Uber suddenly jacking up the price of rides when demand is high or florists abruptly charging more for flowers on holidays—well over 90 percent said it was fine in both cases. This put them out of step with both Democrats and Republicans, who were much less approving. (Only 43 percent of Democrats and 51 percent of Republicans thought surge pricing was fair for Uber; and 61 percent and 58 percent, respectively, objected to surge pricing for flowers.)

In other words, tech founders and CEOs were a curious political blend: as liberal as Democrats on taxing-and-spending or civil rights, but straight in line with Republicans when it comes to corporate regulation. "It's not a simple picture," Neil Malhotra, a Stanford professor of political economy (and one of the trio who did the research), told me.

If we wanted to synthesize it, we could view tech founders as people who are aware that people can fall through the cracks of a capitalist economy—and even that "disruption" can wreak havoc in the lives of the average Joe and Jane. But tech leaders are also themselves devoutly committed to creating those disruptions: Amazon's online sales leading to the shuttering of local stores, Uber and the on-demand world producing a patchwork of unreliable "gig" labor, automation destroying entire classes of jobs, such as legal assistants. Nonetheless, tech founders fundamentally believe that talent and merit and innovation—as they define it, anyway—shouldn't be restrained a whit. So they square that circle by supporting redistributive policies that attempt to make madcap disruption vaguely survivable. Hence the approval of universal health care—or even universal basic income, something that's a warmly approved talking point at tech conferences. They're happy to share some of the wealth via taxation but want to ensure nothing stops them in their process of *how* they amass their wealth. They believe that, fundamentally, they know what's best for society: Their view is "trust us," as the philosopher and technologist Ian Bogost says. And of course, it's a redistributionist view that leaves their political power intact. Workers getting handouts from a small coterie of stratospherically wealthy 1 percenters are not liable to form any sort of strong counterweight to the power of tech giants.

The vision is, in essence, that of a digital-age version of the robber baron. Back in the late nineteenth century, the Andrew Carnegies of the world were happy to pay for civic goods like public libraries, so long as they controlled the munificence—and so long as workers were clubbed back into place when they tried to organize to improve their lot. In the worldview of the tech elite, they got where they were by sheer effort and smarts; their judgment shouldn't be questioned.

The truth is, some of the most genuinely talented coders I've met—the ones who themselves get *called* 10X—are uneasy with the hero-worship in tech.

When I met with Ben Newhouse, the coder who'd invented the clever Dropbox app by himself, he scoffed at the idea that he was particularly unique. "I'm a product of basically every privilege you could stack on a person," he said drily. Sure, he'd created a useful prototype, he agreed, in the classic nerd-hammering-a-keyboard fashion. But to actually take his prototype and turn it into something fully fleshed out, debugged, and tested, and with an interaction that everyday humans could understand? That took the heads-down work of oodles more Dropbox engineers and designers, over months of patient time. And as many other coders also pointed out to me, programming talent that seems innate or inborn is very often just the product of the proverbial 10,000 hours: You code and code and code and code, gradually getting better, until years on you're at the top of your field.

But there's another point, which Newhouse noted: Most truly useful coding isn't a lone-gunman activity. It's a deeply social team sport.

As an example, consider a mammoth coding project that Dropbox had just completed: the creation of their own personal cloud storage system. For years, they'd stored documents by renting space on Amazon's enormous cloud. But as Dropbox grew huger, using Amazon's cloud began to raise issues of cost, efficiency, and how readily Dropbox could customize its technology. So Houston tapped two coders to spearhead a "moonshot" project to write the code for their own cloud. It would have to sync tens of thousands of hard drives, stitching them together to act as one, without losing a single document. Houston needed two rock stars.

The two he needed were already on staff. One was James Cowling, a tall, voluble Australian whom Houston had met when Cowling was a student at MIT, doing his MA and later a PhD; Cowling specialized in "highly distributed systems" before decamping to Dropbox. The other was Jamie Turner, a bearlike, bearded coder with a dry wit. He'd gone to UCLA to study English but dropped out for a coding job, landing later at Dropbox.

"So Jamie's a college dropout, start-up experience," as Cowling told me.

"I've actually *done* shit, instead of just thought about it!" Turner replied.

At first, the project was precisely the sort of heroic, 10Xing scenario that wouldn't be out of place in a cheesy hacker movie. They set a goal of rewriting the core of the system in six weeks, and even put a clock on the wall to threateningly count down toward the deadline. They'd crank away for 16-hour days, taking breaks around 11:00 p.m. to head to Dropbox's music room to play music. "There was a month and a half period where we didn't sleep basically," Cowling says.

"I forgot my kids' names briefly," Turner adds.

"They forgot *yours*," jokes Jessica McKellar, then a director of engineering at Dropbox. When the team was finally done, McKellar told me, they'd been so immersed for so long they struggled to reenter the normal flow of work: "They didn't know what to do with themselves." The code worked. They'd figured out how to make a cloud.

But that early burst, the keyboard-clattering craziness? That was the quick part. As with Newhouse's invention, the longer, slower part was expanding the project now to include many more Dropbox engineers, who'd now be working intimately with the code, testing it and improving it and making it stable, becoming expert in it. Cowling and

Turner would have to articulate their design principles, why they built things the way they did. If they got "hit by a bus," as they put it, the system would live on. "In the culture of heroics, you get rewarded for heroics," Turner notes. "But ultimately your job should be to eliminate the heroism."

As we talked and ate at dinner, they were, like Newhouse, dismissive of the entire 10X idea. "That concept of 'rock star engineers,' I don't like that phrase," Cowling said, wincing. Lots of people can code, he argues. Lots can make a prototype. But the most important things that software firms do isn't the coding. It's higher up the cognitive ladder. It's deciding what type of system to build: what it should do, what it *shouldn't* do; what your customers need, what they don't; what architecture to use, how to actually make the project happen. "Being a programmer is being a bricklayer, we're all bricklayers, we all write code," Cowling says. "But if you want to build a skyscraper you don't say, 'Quick, find me the world's greatest bricklayer!' You say, 'Find me an architect and people who can build a team.' "

The next day, Cowling and I wandered over to his desk, where they'd rolled by a huge whiteboard, filled with flowcharts depicting the cloud system's various parts. He and Dropbox's coders and designers and project managers all spend an enormous amount of time huddled around that whiteboard, trying to decide: How does one part of Dropbox interact with the others? How should the Dropbox iPhone app request files from the cloud? How should those servers squirt back the reply?

"People think coders code, but we spent a lot of the time in meetings," Cowling says, laughing. He waved theatrically at the whiteboard with a green dry-erase marker. "That's what we do most of the time! Sit around *figuring out* what to do, arguing about it!" Newhouse agreed, noting a great irony of top coders: The better they are at their craft, the more likely they'll be superb at planning and visualizing the big architecture of a massive project, figuring out how to break it into small chunks and motivate team members to do them. And of course, the better you are at doing that, the less code you write yourself. You get into programming because you love to make things; but if you're *really* good at making things, you wind up becoming a manager, helping other people do the fun stuff.

"It's like lawyers," Newhouse joked, "who think they're going to litigate."

A 10X coder can produce amazing things.

But can a 1Xer? Or a 1/100thXer?

Dennis Crowley is a curious test case here, because as he tells me, point-blank: "I'm the worst programmer you've ever met."

Back in the mid-'90s, Crowley was a twentysomething kid fascinated by technology and culture. He loved going to dive bars and soaking up music, and he also dreamed of working in tech. But every time he tried to learn coding, he failed miserably. Attending Syracuse University, he tells me, "I really wanted to learn how to do computer science. I took an entry-level course, and I was so bad at it. I just could not write any code that could be compiled." Assigning variables, getting functions to call each other; none of it made any goddamn sense. He eventually decided not to pursue it. At most he tinkered with simple web pages: "Putting pictures on the internet was about all I could do."

After school, Crowley moved to New York City and got a job at Jupiter Communications, a consultancy where he interviewed tech firms and wrote market research reports. But he still craved the feeling of making something, not just sitting on the sidelines writing about it. He'd spend evenings roaming the city's bars and clubs with friends, and—back when texting was still new—they'd SMS each other all night long, describing where they were and what they were doing. But in those early days of texting, friends were limited in their ability to connect and find one another because phones could only text phones on the same carrier. Crowley noticed that inefficiency and wanted to make a tool that would connect not just his close friends but friends of friends, across all networks. They'd all read *Harry Potter* and knew of the "Marauder's Map," a magical document that showed everyone's location in Hogwarts.

"It was this idea of software as a superpower," he says. "If you're a superhero, you can see through walls and you can see around corners." He imagined the sort of floating social awareness that would emerge when friends could see each other's travels through New York's winding bar scene. You could read the trails of each other's boozy travels, or even connect face-to-face when you checked your phone and realized a workmate was at a club two blocks away, *right now*.

In 1999, he gritted his teeth and borrowed, off a colleague's desk at work, a huge red manual on the Active Server Pages coding language. Over the next two years, with painful trial and error, he learned enough to get some code working: first, a city guide, and then a prototype of his original vision, a rickety server that could ferry alerts from one friend's phone to another. "It wasn't pretty. It barely worked," he adds. But it gave him and his small group of friends a glimpse of what might be possible. The projects also helped him get his next job at Vindigo, a company that made city guides for PalmPilots.

Crowley was still, by most standards, startlingly awful at coding, though. The Vindigo engineers knew that Crowley had made a primitive texting-alert app, and they were intrigued. They offered to train him to be "a real C++ coder," but after several fruitless months they concluded he'd never learn. He got laid off, and around that time, much of the internet industry evaporated in the dot-com bust. With the extra free time, Crowley kept on improving his location-based prototype. He also decided to pursue a graduate degree at MIT's Media

Lab, but they wouldn't let him in; his meager coding skills fell below their minimum requirements. But he also applied to the Interactive Telecommunications Program at New York University, a high-tech program known for taking midcareer people—sometimes artists, sometimes people just looking to make a switch into tech—and giving them just enough technical chops to make vivid, oddball tech projects.

Touring ITP was a welcome shock. The students were all making inventive, quirky creations—a "remote hugging" machine, tiny printers that spat out algorithmically generated poetry, shoes that displayed LED patterns based on the style of your dancing. They were dabblers in code, learning the bare minimum necessary to get their fanciful projects working, but no more. They'd cut and paste a code snippet, slightly tweak it to suit their needs, and hit *compile*. "You don't worry about writing the most elegant code?" he'd ask them, amazed. "I don't know what I'm doing! It just works!" they'd cheerfully reply. *These are my people*, he thought: an entire campus full of people who didn't care about writing ultra-optimized software, but who were making interesting things nonetheless.

He enrolled in ITP in 2002 and joined forces with his fellow student Alex Rainert. They dusted off Crowley's prototype, rewrote it in a fresher language, PHP, and shared it with other ITP students. It still wasn't much to look at, code-wise: "1,000 lines of IF-ELSE statements," as Crowley recalls, a giant hairball of repetitive commands. But it worked reliably, and the duo continued to beaver away at it, until in 2004 they released it for broad public use.

He and Rainert dubbed the service Dodgeball, and within a year, thousands of urban hipsters had discovered its weird joys. The duo gradually added new features, like the ability to list five "crushes," which the crushees could optionally respond to, almost like a proto-Tinder. (Indeed, some people began dating those they'd met on Dodgeball.) Soon pundits were pondering this crazy new high-tech behavior of young people broadcasting their physical whereabouts. After years of grinding away at his idea, Crowley had, in essence, invented the "check-in."

By the fall of 2004, Dodgeball was popular enough that Google executives were intrigued. They invited Crowley and his partner to Google's Times Square offices in New York City. Their first order of business was for senior Google engineers to do a technical interview with Crowley and his partner to assess their skill level; and they'd later review the quality of the Dodgeball code.

The collision of cultures was almost comic: some of the world's best engineers, talking to some of the least capable. One Google interviewer was Orkut Buyukkokten, a Turkish-born coder who'd invented an eponymously named social-networking service owned and run at the time by Google. Crowley recalls being asked to solve some classic Google-interview engineering puzzles, the ones they assume most computer science graduates of Stanford or Harvard would know. *You lose your keys in the Lower East Side*, they asked him. *What algorithm would you write, how would you navigate the streets so that you can find your keys without ever going over the same street twice?*

"I clearly have no idea," Crowley confessed. "I don't know what you're talking about. I've never taken a programming course!"

Eventually the interviewers gave up on the algorithm question and began asking about Dodgeball itself. *How much did it cost to run?* "Nineteen ninety-nine," Crowley said. *Okay, $2,000 a month.* "No—$19.99," Crowley corrected them.

Finally, over the months-long courtship, the Google engineers combed through the Dodgeball code. They were, as Crowley recalls, generally appalled. "They'd look at the PHP code like, 'This is crazy.' And it was not *Beautiful Mind* crazy! It was like, 'You're an idiot' crazy. That's how it felt anyway." Crowley told them he knew it was terrible, inefficient, messily written. "For other programmers, that were taught at schools, it didn't make any sense. But I was like, 'I didn't know any other way to do it. This is the only way I know how to do it.'"

But ultimately the Google engineers had quite a bit of respect for what he and Rainert had built, Crowley says. Sure, they could code circles around Crowley, and there were thousands of them. If these engineers were given a coding challenge, they could knock it out of the park. They could optimize a sorting algorithm so that it executes in 15 milliseconds instead of 150 milliseconds; a 10X improvement from a 10X coder. But Crowley had something equally as powerful, if even more so: a bold, crazy new idea like Dodgeball. He had a unique way of looking at the world, while wandering from bar to bar, noticing the joy that came from friends sharing their travels through a dazzling, mazelike city. You could call him a 1X engineer or worse, but, hey—he invented a new category of daily behavior, the check-in.

Crowley now runs Foursquare, the company he started after Dodgeball. I came by one day to have lunch in Foursquare's cafeteria, where Crowley sat surrounded by his employees, engineers who are all dementedly better than he is. Every once in a while one of them forwards around the old, now-defunct Dodgeball code to the company's internal-messaging platform, and the engineers recoil in horror, gently, at what a crap show it was. He tells them it's a lesson in the value of having a great idea.

It doesn't matter if you're "The Best Engineer in the World," if your prototype does something truly weird, new, and fun. "It's a piece of shit," he says, laughing. "But you're going to use it!"

< Chapter 7 >

The ENIAC Girls Vanish

What makes programming so often inhospitable for women?

This is something Cate Huston has thought about a lot during her 15 years as a coder. She started hacking as a kid in boarding school in Scotland, then went to study computer science at college in Edinburgh, where she quickly discovered, no surprise, it was a boys' club; there were few other women in her computer science class. But it didn't deter her. She craved the weird, head-spinning challenges of coding, like "the time I had to implement my own floating-point math for an app on a Motorola mobile phone during an internship." She dropped out and joined Google in 2011, helping code the mobile and tablet apps for products like Google Docs and Google Plus.

On the surface, Google touted itself as a "nice" workplace, where executives talked bouncily about inclusion and diversity. But when Huston arrived in 2011, she found that many of her mostly male colleagues could be frosty and pedantic, and some were clearly convinced that women weren't suited to coding. One day she asked a coworker about how to mount a hard drive on an Android device to take a photo; he replied by sending her some code "that I wrote two years ago when I was an intern," as he added snippily. (He was also too late; she'd already figured it out.) During "code reviews"—when colleagues appraise each other's code, offering suggestions for improvement—male colleagues would spend hours nitpicking over tiny details in her code or frowning and saying *I would have done it differently*. ("And I'd have to go look into and discover that the way they did it wouldn't have worked," she says.) One colleague was so endlessly antagonistic about her code—and aggressive during code reviews—that her manager, she says, barred the guy from meeting with her alone or emailing her without also copying the manager. "By that point, it happened so often that it seemed normal to me," she adds. "But my manager was, like, no," this was unacceptable. Indeed, she got so used to being harangued by male coders that she worried she was mimicking their behavior. "I wondered if *my* code reviews were usually harsh because I'd learned to do code review that way, by having it done to me by these guys."

Still, the actual programming work remained fun. She got to help program the mobile interface for Google Plus, which back then was a high-profile product for the company. Many of her coworkers were collegial and respectful, and she tried to focus on them. But then comments would come out of the blue that stunned her. "Women should be in the kitchen, not writing code," one coder told another female Googler, a friend of hers. "Women just don't *like* coding, women just like pretty things," others would tell her. And then there was the frenemy head-gaming, quiet undercutting moves. She'd show up at work to discover a colleague had written a piece of code for her ("just wanted to save you time!"); she watched as female colleagues were pulled off a project, claiming, "Look, I know this project isn't really your style."

Was she actually inadequate? By the most objective standards possible, it didn't seem so. She was getting perfectly solid reviews from superiors. But the message from these young-guy colleagues was like a constant hum of background radiation: We don't really think you're suited to be here. When someone pointed out only 15 percent of Google engineers were female, they'd argue that, well, it was probably genetic. Google was a meritocracy; they only hired the best, right? If they weren't hiring women, it must be because women just didn't have enough in-born logic or grit.

And on top of that were actual bursts of flat-out harassment. One coworker called her a "cunt" to one of her friends; and although Google took some action in response to her complaint, in the end, she didn't feel they took it seriously enough. Then there was a tech worker for another firm who started feeling her up on a flight to a conference—"He was treating me like a part of the in-flight entertainment," Huston says. And there was the coder from another Google office who would stalk young female interns so persistently that a female Googler sent a memo to the women in her office warning others about him when he was traveling to visit: "Please stay away from him."

"It was very normal for me to find other women crying in the bathroom—and for *me* to be crying in the bathroom once a week was normal." She even, she notes drily, began looking for new jobs that would allow her to work from home once a day so she could at least do her weekly breakdown in private. "For me, crying at work was so normal that I decided to *optimize for it in my job search*," she says, with mordant wit. "I don't want to cry in the bathroom anymore! I want to cry on my sofa and have access to my Clinique products after to repair things."

After three years, she was heading into her middle career, and she was sick of being undervalued. Huston fled Google and was later hired as the head of mobile development for Automattic, the firm that builds apps for blogging using WordPress, the popular open source blog software. When I spoke to her, she'd been there for a year, managing a staff of 25, and she'd found it a breath of cool air. Her boss, WordPress cofounder Matt Mullenweg, had actively recruited her; women hold many high-up positions in the firm. Plus, WordPress's code is all open source, so the culture is less know-it-all than at Google. And, hey, nobody'd called her a bitch!

"I just love to write code," she says, half brightly, half sadly.

Computer programming is a strange aberration in the world of high-stakes, high-pressure professional work. In the last few decades, the number of women has grown rapidly in many such fields. In 1960, they were only 3 percent of lawyers; by 2013, they were 33 percent. In the same time frame, women went from 7 percent of physicians and surgeons to 36 percent. In many parts of science and technology, it's the same story. Women were 28 percent of biologists, and were fully 53 percent by 2013; they've risen from 8 percent of chemists to 39 percent today.

One big exception? Computer programming. Back in 1960, 27 percent of workers in computing and mathematical professions (they're grouped together in US government stats) were women, and that number rose until 1990, when it reached about 35 percent. But then it reversed trend—and started falling. By 2013, the participation rate for women was back to 26 percent. Things had regressed to worse than the point they were at in 1960. Nearly every other technical field has had increasing numbers of women arrive. Programming is the one field where things went backward, and women were actually chased away. Why?

It's often noted that the first computer programmer ever was a woman: Ada Lovelace. As a young mathematician in Victorian England, she met Charles Babbage, the inventor who was trying to create an Analytical Engine. The Engine was a steampunk precursor to the modern computer: Though designed to be made of metal gears, it could execute loops and store data in memory. More even than Babbage, Lovelace grasped the enormous potential of computers. She understood that because they could modify their own instructions and memory, they could be far more than rote calculators. To prove it, Lovelace wrote what is often regarded as the first computer program in history, an algorithm that the Analytical Engine could use to calculate the Bernoulli sequence of numbers.

True to coder form, it contained a bug! Perhaps even truer to form, though, Lovelace had a clear and bombastic view of her own brilliance. "That *brain* of mine is something more than merely *mortal*; as time will show," she wrote in a letter. When she listed her personal qualities, item 2 was "my immense reasoning faculties." ("No one knows what almost *awful* energy & power lie yet undeveloped in that *wiry* little system of mine," she added in another letter.) Lovelace intuited the enormous power that computer programmers would one day wield. It would be like, she said, being an "*Autocrat*" of information, commanding "the most *harmoniously* disciplined troops—consisting of vast *numbers*, & marching in irresistible power to the sound of *Music*." Alas, Babbage never managed to actually build an Analytical Engine, so Lovelace died of cancer at 36 without ever having seen her code executed.

When the age of truly electronic computers began in the 1940s, women were again at the center of the action, as Janet Abbate recounts in *Recoding Gender*. Men, certainly, were central to this new machine age—but at this point, they felt all the rugged glory and heroism lay in making the hardware. The first programmable digital computer in the US, the ENIAC, was a more than 30-ton behemoth made of 20,000 vacuum tubes and 70,000 resistors. Getting that contraption merely to function? That was the manly engineering task. In contrast, programming it—figuring out how to issue instructions to the machine—seemed menial, even secretarial. Women had long been involved in the scut work of doing calculations. In the years leading up to the ENIAC, many companies used huge electronic tabulating machines they bought from companies like IBM; they were really just glorified adding-and-subtracting machines, but very useful for, say, tallying up payroll. Women frequently worked as the punch-card operators for these machines; they'd punch holes in punch cards that represented, for example, how many hours an employee worked, then feed them into the tabulating machine to add them up. It was noisy, painstaking, and inglorious work.

So when ENIAC came around, programming seemed—to the men in charge—to be similar enough to menial punch-card work that they happily hired women to be the ENIAC's first programmers. Indeed, the first ENIAC programmer team was all-female: Kathleen McNulty, Betty Jennings, Elizabeth Snyder, Marlyn Wescoff, Frances Bilas, and Ruth Lichterman, known later as the "ENIAC Girls." The men who ran ENIAC would figure out what they wanted the program to do; they'd spec out the code, as it were. It was up to the ENIAC women to physically crawl around—and even inside—the machine, hooking up wires that "programmed" the machine to execute the instructions. It was head-scratching, pioneering work; they wound up understanding how ENIAC worked even better than many of the men who'd built it.

"We could diagnose troubles almost down to the individual vacuum tube," as Jennings, one of the women, noted. "Since we knew both the application and the machine, we learned to diagnose troubles as well as, if not better than, the engineer."

They invented groundbreaking ideas in coding. Snyder realized that if you wanted to debug a program that wasn't running correctly, it'd help to have a "break point," a moment when you could stop a program midway through its run. When she pitched the idea to the male engineers running ENIAC, they agreed to implement it; and to this day, coders use break points as a key part of debugging. Indeed, the ENIAC women were particularly adept debuggers, the first coders to discover that software never works right the first time. In 1946, the ENIAC heads wanted to show off the computer to reporters for the first time. They asked Jennings and Snyder to write a program to calculate missile trajectories. After weeks of intense hacking, they had it working, except for one embarrassing bug: The program kept running after the missile had landed. The night before the demo, Synder suddenly intuited the problem. She showed up early the next day and flicked a single switch inside ENIAC, fixing the bug. "Betty could do more logical reasoning while she was asleep than most people can do awake," Jennings later said.

Nonetheless, the women got little credit for their pioneering work. At that famous press conference, the managers of the ENIAC project didn't mention or introduce the women; that's how unimportant coding, and the work of the women, was considered.

After the war, coding jobs shifted from the military to the workplace, and industry desperately needed more programmers—and thus some way to make coding easier than having to onerously write cryptic, number-based "machine code." Here, women again wound up being pioneers. They designed some of the first "compilers." These were programs that would let you create a programming language that more closely resembled actual English writing. A coder could thus write the English-like code, and the compiler would do the hard work of turning it into 1s and 0s for the computer. Grace Hopper was wildly productive in this field, often credited as creating the first compiler, as well as the "FLOW-MATIC" language aimed at nontechnical businesspeople. Later, she worked with a team to create COBOL, the language that became massively used by corporations, and the programmer Jean Sammet from that group remained influential in the language's use for decades. (Her desire, she said, was "to put every person in communication with the computer.") Fran Allen became so expert in optimizing Fortran to run swiftly that, decades later, she became the first female IBM fellow. Women were at the forefront of bringing coding to the masses, by dramatically exploding the number and style of programming languages.

Coding jobs exploded in the '50s and '60s, and for women, this weird new field was quite receptive to them: Since almost nobody knew how to code, in the very early days, men had no special advantage. Indeed, firms were struggling to figure out what type of person would be good at coding. You needed to be logic-minded, good at math, and meticulous, they figured. In this respect, gender stereotypes could work in women's favor. Some executives argued that women's traditional expertise at fastidious pastimes like knitting and weaving imparted precisely this mind-set. (The 1968 book *Your Career in Computers* argued that people who liked "cooking from a cookbook" would make good programmers.) Mostly, firms gave potential coders a simple pattern-recognition test, which many women readily passed. Most hires were then trained on the job, which made the field particularly receptive to neophytes of any gender. ("Know Nothing about Computers? Then We'll Teach You (and pay you while doing so)," as one ad enthused.) Eager to recruit women, IBM even crafted a brochure entitled *My Fair Ladies*. Another ad by the firm English Electric showed a bob-haired woman chewing a pen, noting that "Some of English Electric Leo's best computer programmers are as female as anything."

Even some black women could find a toehold, so hungry was the field for talent. In Toronto, a young black woman named Gwen Braithwaite had married a white man, and they discovered that given the racism of the time, nobody would rent to them. That meant they had to buy a house, which meant she needed a job. After seeing an ad for "data processing" jobs, Braithwaite showed up and convinced the all-white employers to let her take the coding-aptitude test. When she placed in the 99th percentile, the supervisors thought she'd pulled a prank and grilled her with verbal questions. When she passed those with flying colors, they realized she was the real thing and hired her. She became one of the first female coders in Canada, and she led several big projects to computerize insurance companies. "I had it easy," she later told her son. "The computer didn't care that I was a woman or that I was black. Most women had it much harder."

By 1967, there were so many women programming that *Cosmopolitan* magazine commissioned a feature on "The Computer Girls." Illustrated with pictures of beehived women piloting computers that looked like the control deck of the Starship *Enterprise*, the story described this crazy, new Wild West that was paying women $20,000 a year—or over $140,000 in today's money. Coding had quickly become a rare white-collar professional field in which women could thrive. For an ambitious woman in the '60s, the traditional elite professional fields—surgery, law, mechanical engineering—accepted almost no women. Programming was an outlier; the proportion of female coders was fully one in four, a stunningly high figure, given the period. The options for women with math degrees were limited—teaching high school math or doing rote calculations at insurance firms, for example—so "women back then would basically go, 'Well, if I don't do programming, what *else* will I do?'" notes Janet Abbate, a professor in the department of Science, Technology, and Society at Virginia Tech, who has closely studied the era. "At the time back then, the situation was very grim for women's opportunities."

Women even pioneered their own entrepreneurial opportunities, as Abbate's research has documented. One female coder, Elsie Shutt, learned programming while a college student doing summers with the military at the Aberdeen Proving Ground. In 1953, she was hired to code for Raytheon, where the programmer workforce "was about fifty percent men and fifty percent women," she said. "And it really amazed me that these men were programmers, because I thought it was women's work!" She left the job to have children, and she quickly discovered that motherhood killed her chances of getting another programming job; the '50s and '60s may have been welcoming to female coders, but it wanted them childless. No firms were willing to offer part-time work, even to superb coders. So Shutt founded Computations, Inc., a for-hire consultancy that would write code for corporations. Remarkably, Shutt hired stay-at-home mothers as part-time programmers; if they didn't know coding already, she trained them. They'd look after their kids during the day, then code at night, renting time on local computers. "What it turned into," as Shutt told Abatte, "was a feeling of mission—in providing work for women who were talented and did good work and couldn't get part-time jobs." *Businessweek* dubbed Shutt's workforce the "pregnant programmers," running a story that was illustrated by a picture of a baby in a bassinet in a home hallway, with the mother in the background, hard at work on a piece of code. (The article's title: "Mixing Math and Motherhood.")

By the late '60s and '70s, though, the field of coding began to tilt male. As the historian Nathan Ensmenger has documented, as programming became more crucial to firms and coding projects got bigger, they needed to promote coders to management; it didn't feel right to be putting women in such important positions. Culturally, the industry was beginning to professionalize. It was moving away from the "anyone can do this" ethos and demanding advanced degrees and accreditation at a time when women were less likely to have either. And the very image of what a coder looked like was becoming more male in the central-casting eye of employers. The industry was increasingly believing coders *ought* to be introverts who had terrible people skills and unkempt grooming. On top of it all, the economics of coding were becoming more lucrative, too, making it obviously less secretarial and menial than it seemed back in its ENIAC days. As sociologists have long noted, when a field suddenly becomes increasingly well paid and prominent, men who previously spurned it rush in: *We'll take this over now, ladies, thank you very much.*

"One of the big takeaways is that technical skill does not equate to success," says Marie Hicks, a former UNIX administrator who became a historian and studied the same shift taking place in the UK. "They wanted people who were more aligned with management."

The prominent women in coding could see things changing as the '70s wore on. Fran Allen of IBM watched as she became the increasingly rare female coder in the room. "As it became a profession . . . it became an avenue that women were pretty much shut out of—in general," Allen told Abatte. "There were fewer women." There were still pockets of prominent women, particularly in professional societies, and areas like compilers. But "in lots of places, there was a huge glass ceiling." She stuck through to the end, though, and by the early '00s she was still hacking at IBM, helping to create Blue Gene, the computer that would become the basis of Watson, IBM's elite artificial intelligence.

If we wanted to pinpoint a moment when things flipped, we could look at one year in particular: 1984. That's when women started moving out of bachelor's computer science degrees.

In the decade leading up to 1984, both men and women were increasingly interested in programming. Ten years previously, for example, a study found that the number of men and women who expressed an interest in coding as a career was equal; men enrolled more than women did in bachelor's computer science programs (women were only 16.4 percent), but women were certainly intrigued by the prospect of doing it. Women soon began to act on that interest. Their participation in computer science programs rose steadily and rapidly throughout the late '70s and early '80s, such that by the 1983–84 academic year, women were fully 37.1 percent of all student coders. In only one decade, they'd more than doubled their participation rate.

But that was the peak. From 1984 onward, the percentage slid downhill, slumping even more in the '90s, such that by the time 2010 rolled around, it had been cut in half. Only 17.6 percent of the students in computer science programs were women.

What happened? Why was there such a sudden and dramatic inflection point?

One thing that transpired was the shift in when kids could learn to program. As personal computers emerged in the late '70s and early '80s, it altered the population of students who were streaming in to computer science degrees.

In the years leading up to 1984, pretty much every student who showed up on a college campus to learn programming had never touched a computer before. Frankly, they'd probably never even been in the room with a computer before. Computers, in those decades, were rare, expensive machines mostly available only at companies or research labs. So all the students were on equal footing. They were nearly all new to programming. They learned their first "Hello, World!" program at the same time.

But beginning in the late '70s and early '80s, the first generation of personal computers arrived—like the Commodore 64 or the TRS-80. Now, a teenager could mess around with a computer at home, spend weeks and months slowly learning the major concepts of programming in their spare time: for loops, if statements, data structures. So beginning in the mid-80s, some of these showed up for their first class having *already done* a lot of programming. These kids were remarkably well prepared for the introductory work—perhaps even a little jaded about what the first Comp Sci 101 class would offer. As it turns out, the kids who'd had this previous experience were mostly boys, as two academics discovered, when they researched the reasons why women's enrollment was so low.

One of these two researchers was Allan Fisher, then the associate dean of the computer science program at Carnegie Mellon University. CMU had founded a computer science undergraduate program in 1988, and for the first few years, up into the early '90s, Fisher noticed that the proportion of women in the major was consistently below 10 percent. So in 1994, Fisher decided to figure out what was going on and how to attract more women into CS. He partnered with Jane Margolis, a social scientist (and now a senior researcher in the UCLA Graduate School of Education and Information Studies), to embark on an ambitious study. For four years, from 1995 to 1999, she and her team interviewed about one hundred of CMU's computer science undergraduates, male and female, and she and Fisher wrote up their findings in the book *Unlocking the Clubhouse.*

Margolis discovered that these teenagers arriving at CMU with substantial experience were far more likely to be boys. Boys had been given much more exposure to the machines than girls; for example, boys were more than twice as likely to have been given a computer as a gift by their parents. And if parents bought a computer for the whole family to use, the parents most often put it in their son's room, not their daughter's. On top of that, the fathers tended to have almost an "internship" relationship with their sons, working through BASIC manuals with them, figuring out coding and encouraging them, while very rarely doing that with their daughters.

"That was a very important part of our findings," Margolis told me. Nearly every girl at CMU said "it was the father who worked with her brother, and they had to fight their way through to get some attention with their father." The mothers, in contrast, were more absent from the home-computing scene. Girls, even nerdy girls, picked up these messages and adjusted their enthusiasm accordingly. And these were, of course, pretty familiar roles for boys and girls. By the early '80s, parents had for decades or centuries tacitly—and openly—nudged their kids into these categories: Boys do the technical stuff, while girls play with dolls and socialize. It wasn't terribly surprising to Margolis that when a new tech came on the scene, it fit into that groove.

At school, too, the girls got the message that computers were a boy thing. Male nerds would form computer clubs, in part as a relief to finally have their own arena away from the torments of jock culture. The cliques of boys that formed around computers created "a kind of peer support network," as Fisher puts it. But they often wound up, intentionally or not, becoming exclusionary themselves—snubbing, as studies in the '80s found, not just girls who came by, but often the black and Latino kids, too. And the girls also policed their own behavior around computers. Being a girl nerd, they realized, was going to be harder than being a boy nerd. Male nerds were certainly picked on, but being single-mindedly obsessive about something, anything (football, cars, computers) was encouraged in boys. Girls, by contrast, have long been harshly judged—including by their own parents—for seeming too narrowly focused on a single subject. But coding rewarded the obsessive.

This helped explain why CMU's first-year classes were starkly divided—between the sizable amount of men who were already confident in basic programming concepts and the women and minorities who were frequently complete neophytes. A cultural schism had emerged. The girls—and some of the boys who didn't have previous experience—would start doubting their ability. How would they ever catch up?

As Margolis heard from the students (and from faculty, too), it started to feel that if you *hadn't* already been hacking obsessively for years, why in God's name were you there? The stereotype emerged: The only "real programmer" was the one who "had a computer-screen tan from being in front of the monitor all the time," Margolis notes. "The idea was, you just have to love being with a computer all the time, and if you don't do it 24/7, you're not a 'real' programmer." The truth is, many of the hacker men themselves didn't fit this monomaniacal stereotype. They had hobbies; they did social things; they wanted a more rounded life. But there was a double standard: Whereas it was okay for the men to want a rounded existence, women who expressed the same wish were judged for not being "hard-core" enough. The experienced male students would talk openly about how the inexperienced women didn't seem up to scratch and were unsuited for computing. If the women asked a question in class, they'd get snarked at: *How obvious.* By the second year, many of these women, riddled by doubt, began dropping out. The same was true for the few black and Latino students who also arrived on campus without teenage hacking experience. They, Margolis found, were dropping out far more often, too, at a rate of 50 percent (though she cautions that it's hard to generalize from that statistic, because there were so few enrolled to begin with—only four.)

Here's the thing, though: A student's decision to drop out—or, conversely, to stay—was not correlated with raw coding talent, it seems. Many of the women who dropped out were getting perfectly good grades, Margolis discovered. Indeed, some who decided to leave had been top students. And the women who did persist, and made it to the third year of the degree, discovered that by then they'd generally caught up to the teenage hackers. The degree was, in other words, a leveling force. Doing BASIC as a teen might teach you a ton of cool skills, but the pace of learning at college was so much more intense that by the end of the degree, everyone—the total neophytes and the hacker teens alike—eventually wound up graduating with the same amount of knowledge about coding and making software.

This is a counterintuitive finding, of course. We'd assume the teenage hackers would remain eternally ahead. CMU itself, operating under the assumption that self-taught teenage coders would forever outclass neophytes, had actively selected for teenage coders; it was more likely to admit applicants with previous experience into the CS degree. But as Allan Fisher told me, this intuition was wrong. "It turned out that having prior experience is not a great predictor, even of academic success," he says.

One can't solely blame the hacker boys, who'd created a high school culture of programming that prepared them for college, for being sexist or exclusionary. They were kids, pursuing their love and interest in coding, and the adults around them—the ones whom you'd hope might have been better at being less sexist—were validating the boys-only culture. What's more, the early pioneering work of women in programming had by now been generally erased, making it a lot harder for women of the '80s to understand not only that they could belong here but also that women had been, in fact, among the biggest pioneers in the field. Meanwhile, mass culture was hammering home the boys'-club message. Hollywood was energetically producing hit movies (from *Revenge of the Nerds* to *Weird Science* to *Tron* to *WarGames*) in which the computer nerds were nearly always white or, more occasionally, Asian young men. Video games, a major conduit into becoming intrigued with computers, were, as studies at the time found, pitched far more often at boys. The mass public sense of "What does a programmer look like?" was becoming heavily male and white.

It didn't help, amid all this cultural love bombing, that women themselves tend to rate their technical competence lower than men. When Lilly Irani—then a graduate student of Science, Technology, and Society at Stanford University—did a study of the computer science students similar to Margolis's, she took men and women and asked about their level of "confidence in solving problems with computers."

Irani knew that the women and men had essentially identical grades. But *they* didn't know that. And the confidence rates, it turns out, were gendered: Women rated their confidence an average of 7.7 out of 10, and men rated theirs at 8.4 out of 10. More striking was when they were asked to compare themselves to their fellow students: Did they think they were more or less confident than their peers? Women pegged themselves, on average, as being half a point less confident than the other students, while men rated themselves as being "six tenths of a point *more* confident than their peers." Despite being identical in performance, the men were simply more confident, more convinced they *felt* like real coders. The women were precisely the reverse.

Some of the women's sense of inadequacy appeared to be based in their preferences. Some research has found that men more often like software tinkering for the sheer pleasure of tinkering; women, in contrast, more often thought coding would be cool because they wanted to have an impact on the world. So it was at CMU: As Margolis found, most of the women didn't like the idea that they'd just be sitting in a room with a computer for the rest of their lives. This was another factor that drove women away during the first two years of computer science, because that's precisely what students traditionally do during that period: They're heads-down, learning how to write algorithms and work with data structures, without the work being connected to any larger mission. It's the definition of tinkering. To be sure, some of the CMU men also didn't like the idea of ignoring their hobbies and social lives while coding endlessly at abstract tasks. But this "geek myth"—that the only real coders were endlessly glued to their keyboards—was more damaging to the women's sense of belonging, Margolis found.

Nonetheless, the women who managed to stick it out for the later years discovered that things flipped. In the later years of computer science at Stanford and CMU, students moved on to working in teams to build full-featured apps; at Stanford, they also began learning a new language, Java. And here, the comfort level of men and women shifted. Since everyone at Stanford was now struggling to master a new language, it reduced some of the perceived differences between students. Plus, working in teams required new skills in managing projects and collaboration, again putting the students on more equal footing—plus bonding them together, because the success or failure of the project was a common fate. For the women who actually made it to that stage, their confidence suddenly soared. They loved it.

But there were fewer of these survivors reaching this epiphany. At CMU, by this point, about half of the women who'd started out had already left.

There was another force that drove women away from computer science beginning around 1984: the "capacity crisis."

Computer science schools had become victims of their own popularity. Programming was hot. So many students were rushing to enroll that universities ran into a supply problem: They didn't have enough professors to teach that many students. To make matters worse, good professors were getting hired away by the already fat and rapidly ballooning salaries in the private sector. "Industry was paying *so* much more," recalls Eric Roberts, who taught computer science at Wellesley College and Harvard in the '80s, before moving to Stanford. (He's currently at Reed College in Portland.) He remembers making offers to five different people before finding one willing to teach.

How did the schools cope with their limited teaching resources? By chasing students away. Some began imposing "weeder" courses, which students had to pass before they'd be accepted into the computer science major. These were classes with intentionally punishing workloads, and which moved along so quickly they left behind anyone who didn't immediately "get it." Some professors would offer little to no assistance to anyone confused by an assignment; *Figure it out on your own or you don't belong here* was the message. (Many professors overloaded, in any case.) Indeed, some schools demanded perfect scores on first-year pre-CS classes—at Berkeley, you needed a 4.0 average on these—before you'd be allowed into the major. The schools, in other words, created an environment where the people who'd be most likely to get through were the ones who came in already exposed to coding—which was, again, mostly boys, and generally white ones.

"Every time the field has instituted these filters on the front end," Roberts says, "that's had the effect of reducing the participation of women in particular." It's also possible that by weeding out so many kids, the programs wound up actively sorting for kids who were extra competitive and who never suffered from any deflated sense of their worth. Roberts adds, "Women are socialized to like much less the kinds of competitive situations in which there are real losers. And as a result they were quicker to leave the stage."

The capacity crisis hit both men and women, in its first few years. As the schools chased students away, the number of CS degrees overall—for both men and women—declined by over 40 percent. But by the mid '90s, the crisis abated, and computer science programs began to grow again, taking in more students. By now, though, the culture of "who goes into coding" was becoming set. When the crisis eased and more students surged back in, most of them were men. The interest among women had been blunted. It never recovered to where it had been in the late '70s and '80s. And for the women who did show up, they often felt like sore thumbs. In a room of 20 students, perhaps 5 or fewer would be female.

With so few women around, computer science departments became hotbeds of testosterone. In 1991, the computer scientist Ellen Spertus wrote a report on women's experiences in programming classes. She catalogued a landscape of guys who'd snigger about women's presumed inferiority; professors who'd tell female students "you're *far* too pretty" to be studying electrical engineering; when a few CS students at CMU asked men to voluntarily stop using pictures of naked women as their computer desktop wallpaper, the men angrily replied that this was censorship straight out of "the Nazis or the Ayatollah Khomeini." A similar study at MIT produced equally bleak tales. Male students would muse about women's mediocrity: "I really don't think the woman students around here are as good as the men," one said. Behavior in research groups "sometimes approximates that of a locker room," the report found, with men rating how "cute" female students were, in front of them. ("Gee, I don't think it's fair that the only two girls in the group are in the same office. We should share," said one.) "Why do you need a degree for marriage?" others would ask women coders. When women raised their hands in class, they'd typically get ignored by professors and talked over by other students. They'd be told they weren't aggressive enough; though when they actually did behave aggressively, challenging other students or contradicting them, they'd hear comments like, "You sure are bitchy today; must be your period." Some would even grope them, massaging their shoulders or fondling their breasts.

As women fled computer science, the workplace, too, increasingly became a monocrop of mostly white men. According to data from the Bureau of Labor Statistics, in the US in 2017, about 20 percent of workers categorized as "computer programmers," "software developers," or "web developers" were women. If you added in other job categories that involve technical chops—such as database administration or statisticians—it goes up a bit higher, to 25.5 percent. The ratios for black employees were 6 percent and 8.7 percent, respectively, and Latinos were in that range, too (6 percent and 7.3 percent), which is in each case about half, or slightly more than half, their rate of participation in the rest of the private-sector workforce. (Other stats I've found suggest things might be lower; Data USA, for example, pegged the ratio of black coders at 4.7 percent in 2016.) In the more rarefied world of the top Silicon Valley tech firms, the numbers were sometimes skimpier: An analysis by *Recode* suggested that 20 percent of Google's technical employees were women, while only 1 percent were black and 3 percent were Hispanic. Facebook was nearly identical, and Twitter was 15 percent, 2 percent, and 4 percent, respectively.

These were sufficiently small minorities that, many would tell me, arriving for work felt like they were intruding on some weird, cloistered, all-white priesthood. Certainly, every underrepresented coder I spoke to told me stories of individual colleagues or managers who treated them as equals. But nearly every one also described the daily fight of having to prove themselves over and over again to the wide swathe of industry peers who, tacitly or openly, assumed they didn't have serious technical chops, that they *couldn't*.

One coder, Stephanie Hurlburt, was a classically nerdy math-head who'd cut her teeth doing deep work on graphics. "I love C++, the low-level stuff," she tells me. She'd worked for a series of firms, including Unity (which makes a popular game-design tool), and then for Facebook on its Oculus Rift VR headset, cranking mad hours to release their first demo. Hurlburt was accustomed to shrugging off neg hits. There were many: She'd been told, including by many authority figures she admired, that girls weren't wired for math. While working as a coder, if she expressed ignorance of nearly any niggling concept in graphics, some male colleagues would pounce. "I thought you were at a higher math level," one sniffed. At one firm, she was sexually harassed; when she later tried to raise concerns about the company's overall toxic environment with HR, they were—in a pattern that I heard repeatedly from female coders—unwilling to address it.

So Hurlburt eventually founded a start-up called Binomial, programming, along with Rich Geldreich—a friend who shared her love of graphics math—a tool that helps compress the size of "textures" in graphics-heavy software. Working for herself, she figured, would mean no dealing with undermining bosses. But still, she couldn't entirely escape the default assumption that women aren't techies. When she and Geldreich would go into meetings to sell the product, some customers would assume she was just the marketing chick. "I don't know how you got this product off the ground when you only have one programmer!" one customer told Geldreich. Still, owning the company gave Hurlburt some moments of sweet karma. At one of her first coding jobs years earlier, the chief technical officer had told her "I shouldn't waste time working in a field my brain wasn't made for." Now, years later, the staff working under that CTO has been sending her emails asking if they can use her texture-compression code, because they desperately need to speed up their own software. She told me she hadn't figured out if she would sell it to them yet. "People who *aren't* sexist get the best compression!" she laughs.

Women coders get mistaken as PR lightweights. Black and Latino programmers also get mistaken for security or housekeeping. That's one of the things that happened to Erica Baker, a build-release engineer for Slack when I met her in 2017. Baker had grown up as a nerdy African American kid teaching herself QBasic and Hypercard, and eventually wound up working for Google in Atlanta. She faced some racist comments: She had a coworker who said "Does your boyfriend beat you?" she recalls. She eventually moved to Google's Mountain View office, where at one point a temp mistook her for security, and another employee mistook her for an administrative assistant. When she asked to be put in training programs to move up into new technical levels, she never got in, though she "saw white dude after white dude get the chance to do it." Her time at Google wasn't uniformly bad, she notes; several of her fellow Google employees were more supportive and mentored her. (One executive noticed the incident with the temp when it occurred and checked in with her to make sure she was okay.) But eventually, she says, "I had realized that, even though it's the best place for the majority of people at Google, it was not the best place for those of us who were the minority." In 2015, Baker moved to Slack, which had a somewhat higher proportion of minority-group coders than many other valley firms. (Across Slack's worldwide offices, women were 34.3 percent of those in technical jobs; in their US offices, 12.8 percent were from "underrepresented racial and/or ethnic backgrounds"; and 8.3 percent identified as LGBTQ. Later, she moved on to be a senior engineering manager for Patreon.)

Levels of harassment are high, too, for LBGTQ employees in tech. When the Kapor Center for Social Impact surveyed tech workers to find out why they'd voluntarily left a tech job, 24 percent of LBGTQ workers said they'd experienced "public humiliation or embarrassment" at work. They were also the group in the survey that reported the highest rates of being bullied, at 20 percent. (Fully 64 percent said bullying "contributed to their decision to leave.")

Not every minority-group coder I interviewed had trouble being taken seriously. Of the scores of women I spoke to for this book, a small number said they'd experienced essentially no problems in the tech workplace. "I don't find it," noted Yan Zhu, a well-known expert in encryption, who was working as an engineer helping to create the new web browser Brave. "I think that I'm not very sensitive to this, honestly." She certainly knew many tales of discrimination: "I've talked to a lot of women whose experiences don't match mine at all. . . . I don't want to discount theirs."

And while blatant harassment and assault get the most headlines, many coders I spoke to said the bigger problem was the side-eyed doubt, the undermining comment. It was less often a guillotine than death by a thousand tiny cuts. Studies bear this out: One analysis of 248 performance reviews for tech engineers found that women were considerably more likely than men to receive reviews with negative feedback; men were far more likely to get reviews that had "only constructive feedback," with no negative material. In another experiment, the tech recruiting firm Speak with a Geek submitted 5,000 résumés to jobs firms, with identical specs and information. When they put non-gender-specific names on the résumés, fully 54 percent of the women got callbacks; when they included gendered names, only 5 percent of them did.

Another part of what froze women out of many coding jobs is the concern, among start-ups, about "culture fit." A start-up involves a small number of people working often in tight quarters for long hours. This leads, as you might imagine, to founders primarily smiling upon hires who are socially and culturally similar to them. One female coder arrived in Silicon Valley after being scouted by a firm, and the founders took her out to bars three times for long, breezy hangout sessions. They never discussed work. She was baffled. "When are they going to interview me for the job?" she asked a friend, a veteran of the valley. "Oh, those drinking sessions *were* the interview," her friend assured her. "That's what they do. They want to know if you can hang." Many start-ups I interviewed told me they simply hired their close friends from Harvard, Stanford, or MIT.

"It's all this loosey-goosey culture thing," says Sue Gardner. After her stint running the Wikimedia Foundation, she decided to study the reasons so few women were employed as coders, and surveyed over 1,400 women in the field, doing sit-down interviews with scores more. She realized the takeover by guys in the '90s had now turned into a self-perpetuating cycle. Since everyone in charge was mostly a white guy, they preferred hiring people like them; they only recognized talent when it walked and talked as they did. For example, when hiring a coder, most firms relied on whiteboard challenges, in which a prospective employee is asked to write code—often a sorting algorithm—live, on a whiteboard. This type of performance bears almost zero resemblance to what coders actually do in their jobs. But whiteboard questions *do* resemble classroom work at Ivy League schools. So it feels familiar to the guys doing the hiring, many of whom are only a few years out of college, as Gardner notes.

The same cycle governs the all-crucial arena of getting venture-capital funding; they fund people who seem to map onto successful folks they've seen before. Indeed, they even have a term for it: "pattern matching." The venture capitalists, nearly all white men who made their money from the '80s to the early '00s, take a shining to young entrepreneurs who look and behave like younger versions of themselves. In 2008, the leading VC John Doerr said this explicitly, when he described the tech entrepreneurs he most admired. "They all seem to be white, male nerds who've dropped out of Harvard or Stanford and they absolutely have no social life. So when I see that pattern coming in—which was true of Google—it was very easy to decide to invest."

"What I came to realize," Gardner tells me, "is that it's not that women are excluded. It's that practically *everyone* is excluded if you're not a young white man who's single."

This sort of subtle preference—and quiet, understated jabs at women—are more often the barrier to women in coding. But it certainly isn't hard to find stories of flat-out sexism and boiler-room boys'-club behavior, either.

There was the female coder who told me of being assaulted in the offices of a well-known, bold-face tech firm. There's the penchant coders have for including porn in their presentations—like the Ruby on Rails coder who put porn images in his slide deck. There was the Ubuntu lead coder who said in a speech that he was excited about their new software release because "we'll have less trouble explaining to girls what we actually do." There was the guy who, in a seminar on database queries, illustrated how to optimize queries with the example of ranking women by "hotness"—"`WHERE sex='F' AND hotness>0 ORDER BY age LIMIT 10`." There was the leader of a Bitcoin meetup who responded to a female Facebook client-solutions manager who showed up, saying "You don't look like someone who would even know about Bitcoin!," followed by "Women don't usually think in terms of efficiency and effectiveness." (What's more, another attendee groped her while she was there.) Women who talk about these sorts of experiences online face clear threats and harassment; when former Google engineer Kelly Ellis retweeted examples of harassment she got, it just invited even more. ("How much do you charge to spread them?" one replied, in a later-deleted tweet.) And there were tales of tech start-ups that operated as what seemed like thinly veiled frat houses, such as Upload VR—where, as a lawsuit alleged, male employees designated one room with a bed as a "kink room," and a manager once announced he was going into the bathroom to "rub one out" so he could continue to focus while coding.

This frat-like nature of tech start-ups was, in part, the emergence of another trend in the mid-'00s: the arrival of the "brogrammer."

As the portmanteau suggests, the brogrammer—a frattier, bro-ier coder—is a bit of an evolution away from traditional machine-minded men who'd colonized the field. There had been plenty of frat-style environments in the late '90s and early '00s; Facebook's early keg-equipped, graffiti-bedecked office was a good example of that. But the trend accelerated, many techies say, after the 2008 financial crisis. Because many banks contracted, investment banking became a less reliable route for young, cocky, hard-drinking men of the Ivy League to make instant riches. So they moved to the next gold-rush industry—software in Silicon Valley, where investment money flowed at endless silly start-up concepts. (The same trend had appeared, in a smaller format, in the '90s during the dot-com boom, when many Madison Avenue types, from the world of advertising and marketing, glommed on to the start-up scene.)

Eric Roberts remembers seeing a few more of these guys appear in his classroom after the financial collapse. They just wanted to secure a pathway to wealth, and they certainly didn't suffer from lack of confidence. Some were deeply interested in software; others, to his dismay, weren't. "One of the most disturbing things," Roberts tells me, "was that we had a number of students who were majoring in computer science who *hated* the field. But they knew that was where their millions were coming from." As he adds: "It's sort of boiler-room ethic of going and working like mad to become the next billionaire, that we saw so much in the run-up to the crash. And when that fell apart, where did those people go? The only game in town was computing."

Conversely, even as the brogrammers were emerging, many midcareer women were realizing Silicon Valley was never going to give them a break—and were leaving the field. When Sue Gardner surveyed those 1,400 female techies, they told her the same story: In the early years, when they were junior coders, they'd shrugged off the ambient sexism they encountered. They loved programming, were ambitious, and were excited by their jobs. But over time, "they get ground down," Gardner says. As they rose in the ranks, they found few-to-zero mentors willing to take them on. As many as two-thirds either experienced harassment or saw it happen to someone, as she found when examining *The Athena Factor* (another study of women in tech); and one-third reported that their managers were more supportive and friendly to the men they worked alongside. It's often assumed that having kids is the moment when women get sidelined in tech careers, but Gardner found that wasn't often the breaking point for these midcareer women. It was more that they got sick of having equally or less-qualified men around them receive better opportunities and treatment.

Ironically, the higher these women rose, the worse and more blatant the harassment could get. This was particularly true when female programmers decided to found their own firms and had to interact with the world of venture capital, where about 96 percent of the investors are men. Female founders who'd make a pitch for venture capital would get propositioned for sex by investors. Women who actually became venture capitalists themselves discovered the world could be cartoonishly piggish. That's what happened to Ellen Pao: While working as an investor for Kleiner Perkins Caufield & Byers, a tech CEO on a private-jet flight bragged about meeting porn star Jenna Jameson, then asked the members what sort of sex workers—"girls"—they liked. ("Ted said that he preferred white girls—Eastern European, to be specific," Pao later wrote.)

So midcareer women in tech were leaving. Sick of years of that crap and not seeing it getting any better—indeed, seeing it getting worse—they voted with their feet. It wasn't that they'd leave coding entirely. They'd go off and find technical jobs where they'd do programming and oversee coders, but not at software firms. They'd do it in the fields of medicine, in law, in government—anywhere, really, but Silicon Valley.

"What surprised me was that they felt 'I did all that work!' They were *angry*," Gardner tells me. "It wasn't like they needed a helping hand or needed a little extra coaching. They were mad. They were not leaving because they couldn't hack it. They were leaving because they were skilled professionals that had skills that were broadly in demand in the marketplace, and they had other options. So they're, like, 'Fuck it—I'll go somewhere where I'm seen as valuable.' "

The irony is profound. In the early days of coding, women flocked to software because *it* was the place that offered more opportunity, compared to the sexist fields like law. But that situation has now completely inverted itself. Software is the laggard, the boys' club.

... ...

Does it *matter* that the world of tech is mostly white guys? One could, in a purely bloodless way, entertain that question. To which there are two obvious answers: Sure, as a matter of economic opportunity, it certainly does. This is a question about "whether half the human race" is going to get to participate in "a certain kind of work, work that is exciting, lucrative, prestigious, and actually very desirable to many of us," as one female engineer said at a roundtable on gender for Y Combinator.

But it's also a problem for the rest of the world outside coding, too. That's because the monoculture of tech affects the type of products that get made. When you have a homogenous cohort of people making software and hardware, they tend to produce work that works great for *them*—but can be useless, or even a disaster for people in other walks of life.

Consider the engineers at the VR firm Magic Leap, who excitedly developed their augmented-reality headset only to be told by women on staff that it was wildly uncomfortable for many women, since ponytails interfered with the headband, and the device required you to holster a small computer on your belt, which many women don't wear. (The engineers appear to have ignored this input: "None of the proposed changes were made to the design," as a lawsuit alleged.) Or as my friend the comedian and writer Heather Gold has written, consider the UI design Google and Apple introduced in their video chat for groups, where the software is designed by default to take whoever is talking and increase the size of their face on everyone's screen. As she points out, this winds up exacerbating the well-known dynamic of face-to-face meetings: White dudes who bloviate nonstop usually wind up commanding all the attention.

Or consider the case of Twitter, where an originally all-male design team failed for years to appreciate their service's growing problem of abuse on their platform. It was likely quite invisible to them; men experience far fewer threats and less harassment on social networking. If there had been even a few women on the original team who'd been active online in, say, the preexisting early '00s world of blogging, they might have given a heads-up warning: Hey, women online get targeted and harassed, even for pretty innocuous posts! Let's figure out how to get ahead of this problem! ("An opinion, it seems, is the short skirt of the internet," as the social critic and writer Laurie Penny once cracked.) Forewarned thusly, Twitter might have implemented more features to minimize abuse and worked harder to instill a more civil culture among its user-base, as companies like Flickr had previously attempted. But instead it allowed malevolent users to flourish—well unto the point that flat-out neo-Nazis were happily thriving on the social network.

By 2016, the Twitter designers' lax approach to harassment had become an economic liability. Growth had flatlined, but corporate suitors were leery of investing. Disney had decided against buying Twitter, worried that, as the *Boston Globe* put it, "bullying and other uncivil forms of communication on the social media site might soil the company's wholesome family image."

Former Twitter CEO Dick Costolo, after years of denying abuse was a problem, openly recanted. "I wish I could turn back the clock and go back to 2010," he says, "and stop abuse on the platform by creating a very specific bar for how to behave on the platform." That would've been much easier to do if there had been someone around to warn them.

In the summer of 2017, a 28-year-old senior software engineer named James Damore wrote a memo arguing that there was a different reason for the dearth of female coders: It was biological. As anti-feminist theories often do, this argument quickly went viral. Damore worked for Google, a company that—at least publicly—had embraced the notion that gender discrimination was a problem. They'd instigated training sessions to alert staff to "implicit bias," unconscious expressions of bigotry. In meetings, managers discouraged sexist language. None of this stuff had significantly increased Google's ratio of women in technical jobs yet (to do that, you'd have to hire many more women, not just talk about it), but the topic was in the air.

Damore, though, saw these policies as political correctness that avoided the real facts. In a note on the company's internal bulletin board, titled "Google's Ideological Echo Chamber," he argued that the gender gap wasn't necessarily a cultural problem. The real issue was that men were more suited—cognitively, biologically, based on evolutionary adaptation—to be coders. Damore argued that research shows women were more prone to feeling anxious and were less competitive, thus making it harder to thrive in the hard-driving world of Google. The company could, he wrote, adapt more to women by changing its culture to accommodate these traits, such as "allowing those exhibiting cooperative behavior to thrive," or making tech less stressful. But he also insisted that Google's current attempts to increase the hiring of women and minorities had led to reverse discrimination. After reading the memo over a few times, I found his message pretty clear: The overall distribution of men and women in coding was pretty biological.

Damore himself clearly identifies with the modern model of the ultrafocused, highly competitive, socially awkward male coder. When we met near his apartment in Mountain View, he told me that he'd played competitive chess as a child, before becoming fascinated by game theory, evolution, and physics. As an adult, he was diagnosed as a high-functioning autistic, which, he added, explains why he's so intrigued by systems. Damore was fluent in the work of Simon Baron-Cohen, a British clinical psychologist and professor of developmental psychopathology at the University of Cambridge, who argued that because male brains get such a larger bathing in testosterone in vitro, they are hardwired to be systemizers, interested in things, while women are hardwired to be interested in people. Damore himself never formally studied computer science; while working toward his degree in systems biology, he read a book on algorithms, started doing coding challenges, and won a Google code contest. Once the company recruited him, he worked on search infrastructure.

Damore told me he didn't think women at Google were treated more poorly than men. As he saw it, aggro male coders say snippy critical things about their male colleagues, too—and also talk over them, and take credit for their ideas. "That happens to everyone, I feel," he told me. Men just didn't mind, because they were less sensitive, he said. Essentially, from Damore's perspective, programming workplaces were culturally "male," alienating to women, in a way that was at least partly innate. "I'm not denying that there is sexism or anything," he adds. "But that's definitely not the full picture."

Indeed, Damore argued that the act of talking about sexism in the workplace could make things worse for women: "I really think that there's nothing more disempowering than saying the system is rigged against you." I told him that sexist behavior in tech seemed straightforwardly real—there were copious stories and documented examples, including scores I'd heard myself. "I think a lot of this also gets overblown by the media," he responded, and went on to argue that men faced sexism, too, because people presumed they were physically dangerous, and they were overrepresented in highly risky jobs like coal mining or garbage collectors. Listening to Damore was, I realized, like reading the standard arguments on men's rights activists websites: that offenses against men were equal to, or even outweighed, those against women.

If it's true that women are seldom coders because of biology, I asked him, does that also explain why so few US programmers are African American or Latino? Here, he claimed, the issue *wasn't* biology. "I think culture plays a much larger role than genetics in a lot of these things," he mused. An African American child raised in a high-achieving Asian family would be high-achieving, too, he argued. In essence, Damore was claiming the gender gap had roots in biology but the racial gap did not.

Damore's memo was leaked to the public, most likely by his colleagues at Google, and the company swiftly fired him. In the aftermath, Google CEO Sundar Pichai wrote an email to staff explaining that he couldn't employ someone who judged coworkers based on their sex. "To suggest a group of our colleagues have traits that make them less biologically suited to that work," Pichai noted, "is offensive and not OK." More to the point, as former Google engineer Yonatan Zunger noted, any manager would now find it difficult to put Damore on a team. How would a female coder feel having their code reviewed by Damore, knowing he believes their biological makeup was a liability?

Damore is not alone, though, in this argument that biology is a key to all mythologies. Other coders at Google posted in support of him on their internal forums. Nearly every female coder I spoke to had stories of coworkers citing the same research Damore did, dilating on how nature has decreed rigid male and female roles. When Kelly Ellis, that former Google coder, tweeted about sexual harassment she'd experienced, a tech worker emailed her to explain it was all evolution at work—"We have hundreds of thousands of years of evolutionary history to deal with. Men are wired to approach you as an egg donor first."

There are pretty clearly self-serving reasons why some programmers prefer to believe that biology explains why men are employed in coding far more than women. It's a very pat, neat theory. If you accept it, then you can believe nothing needs to change, that merit obtains. It also suggests, despite reams of evidence to the contrary, that white male coders have no social and cultural advantages over others—no extra encouragement, nudges, mentorship—and that their success is purely a matter of talent and work.

"Programmers tend to be extremely self-assured of their own rationality and objectivity," says Cynthia Lee, who should know: She's a programmer who was an early stage hire at several tech start-ups, became an expert in high-performance computing, and now teaches computer science at Stanford. "They have a very big blind spot for biases on their side, because they don't think they *have* blind spots." Lee wrote an essay for *Vox* explaining why so many women in tech were enraged by the memo: because they spend their days responding wearily, over and over, to precisely this sort of biological skepticism. "When men in tech listen to the experiences of women in tech," she wrote, "they can come to understand how this manifesto was throwing a match into dry brush in fire season."

Is it possible that Damore is right—and biology does explain why so few American women are coders?

No. Many scientists who spend their days seriously studying sex differences say biology alone simply isn't that strong a determinant of preference and ability. Certainly, psychologists have long documented lower levels of self-confidence among many professional women and students. But the detectable differences in boy and girl cognition and behavior are too small to explain such divergence in life and career paths; instead, they're like a biological signal that gets heavily amplified by cultural feedback into life-changing decisions and preference. The portion attributable to pure genetics becomes a force that is small and scientifically murky—particularly when compared to the manifold, well-documented examples of phosphorescent sexism radiating in the halls of tech workplaces.

Indeed, if women's biology made them temperamentally unsuited to coding—and uninterested in it—it's difficult to explain why they were so prominent in the early years of American programming. After all, the coding back then was, if anything, harder than today's programming. It was a head-bustingly new field, in which you had to do math in binary and hexidecimal formats, and there were no helpful internet forums offering assistance with your fiendish bug. It was just your brain in a jar, solving hellish problems.

What's more, if women were so biologically neurotic that they couldn't endure the competitiveness of coding, then the ratio of women-to-men in programming ought to be similar around the world.

But it isn't. In India, over 40 percent of the students studying computer science are women. And this is despite its being even harder to be a female coder there; India has such rigid gender roles that female college students often have an 8:00 p.m. curfew, meaning they can't even stay to work late in the computer lab, as the social scientist Roli Varma found when she studied them. The women had one big cultural advantage over their US peers, though, Varma found: They were far more likely to be encouraged by their parents to go into the field. The Indian coders reported much more equal treatment when girls; only 12 percent said their male peers had been given more exposure to computing, compared to the majority of girls in the US who watched their fathers shower their brothers with nerd attention. (Indeed, the women regarded coding as a safer job because it was an indoors profession, exposing them to less of India's rampant street-level sexual harassment.) It was, in other words, considered normal in India that girls would code. The picture is similar in Malaysia, where in 2001—precisely when US women's participating in computer science had slid into a trough—women were 52 percent of CS majors and 39 percent of the PhDs at University of Malaya, a large public university in Kuala Lumpur. These international comparisons are the sort of A/B test of which

Silicon Valley is so fond. The Occam's razor explanation for why US women aren't much in tech seems pretty clear: It's not biology. It's culture.

"Computing being very masculine—that is not a universal image," Varma tells me. And of course, if you wanted to show the fragility of biological and evolutionary explanations, it's worth noting that they could be flipped on their head: You could argue that women's evolutionary history ought to make them *better* at coding. All those centuries of being "gatherers," attentive homemakers, and all that knitting and weaving: Hey, wouldn't that imbue them with the fastidiousness and precision crucial for coding? Indeed, in the early days of the '50s and '60s, as we've seen, this is precisely how executives argued for women's suitability for programming. (One computer-firm head in 1968 claimed that "girls sometimes make better programmers than boys" because "an intelligent girl who has the patience to do embroidery has just the right mentality to do the job.") Once you have your conclusion in mind—*Women are biologically natural coders! No, they're not!*—you can easily craft an evolutionary story (and a definition of what "programming" entails) to get you there, which illustrates precisely why it's difficult to use biology and evolution to explain complex society-level modern phenomena.

If anything, the nature-nurture debate shows the power of the stories society tells itself about talent. People used to say that women were natural coders—they were the winners of that meritocracy. Then in the '80s a different tale emerged.

The very word *meritocracy* has an ironic provenance. When the British politician Michael Young wrote his novel *The Rise of the Meritocracy* in 1958, he intended it as a biting satire—of governmental attempts to determine an individual's value based on an IQ test. The term was later adopted as an unironic, literal concept, something that deeply disappointed Young.

It turns out that believing your field is purely meritocratic has lousy side effects. As the cognitive scientist Sarah-Jane Leslie found in a fascinating study, women and African Americans fare less well in domains where people believe that "raw, innate talent is the main requirement for success." Stats show they're well represented in fields such as molecular biology and neuroscience, where it's understood that hard effort, teamwork, and rigor carry the day; but in fields that pride themselves on requiring raw, isolated genius—like math or philosophy—they're not. Other experiments verify this. Two scientists from MIT and Indiana University took MBA students who had management experience and gave them two hypothetical companies to manage: one where the "core company values" included being meritocratic and one that didn't emphasize that value. Then the students were given written job evaluations for a hypothetical male employee and a female employee, who had identical and positive evaluations. When the MBA students were told the company was "meritocratic," they gave the men 12 percent higher bonuses. Why? Because, the scholars suspect, when you prime someone with the idea of meritocracy, it seems to give them permission to act on their internal biases—evoking, it would seem, the cultural idea of the man as more likely to be the lone genius.

Is it possible to reverse the historic shift of women out of coding?

Back in the '90s, CMU's Allan Fisher decided to try. Impressed by Jane Margolis's findings, he and the faculty instituted several changes designed to break the confidence-destroying loop of neophytes who'd glance around their first classroom, worry they'd never catch up with their confident teen-hacker peers, and begin the slow drift toward leaving. One strategy was to create classes that clustered students by experience: The kids who'd been hacking since youth would start on one track; the newcomers to coding would have a slightly different curriculum, allowing them more time to catch up. CMU also offered extra tutoring to all students, which was particularly useful for the newcomers to programming. If Fisher could get them to stay through the first and second years, he knew, these students would catch up to their peers. They also changed the early courses so they showed how code could affect the real world—so a new student's view of programming wouldn't be just an endless vista of algorithms disconnected from any practical use. He wanted students to get a glimpse, earlier on, of what it's like to make software that works its way into people's lives. Back in the '90s, before social media and the mainstream web, the impact code could have on daily life wasn't so easy to see.

The faculty, too, shifted their culture. For years they had tacitly endorsed the idea that the kids who came in already knowing code were, in an essential way, born to it. "CMU rewarded the obsessive hacker," Fisher notes. But the faculty now knew this wasn't true; they'd been confusing *previous experience* with *raw aptitude*, as Fisher puts it. They still wanted to encourage those obsessive teenage hackers, but they now understood that the neophytes were just as likely to bloom rapidly into remarkable talents, and were just as important to encourage; and students pick up on faculty's expectations and respond to them. "We had to broaden how faculty see what a successful student looks like," he notes. Meanwhile, led by professor Lenore Blum, CMU began building more community participation and support groups for women studying CS, events that made them more visible to their fellow students. Admissions, too, changed; it no longer gave so much preference to students who'd been teen coders.

There was no silver bullet, no single policy that improved things, Fisher notes. "There's really a virtuous cycle," Fisher adds. "If you make the program accommodate people with less experience, then people with less experience come in"; then faculty become more used to seeing how neophyte coders evolve into accomplished ones, and how to teach that type.

CMU's efforts succeeded, dramatically. Only a few years after these changes, the percentage of women entering CMU's computer science degree boomed, rising from 7 percent to 42 percent; and graduation rates for women rose to almost match those of the men. The school vaulted over the national average, and indeed improved on the historical record. When word got out of CMU's remarkable success, other schools began using approaches similar to Fisher's. At Harvey Mudd College in 2006, the school created an intro-to-programming course aimed specifically at complete newbies, and they rebranded their Intro to Java as Creative Problem Solving in Science and Engineering Using Computational Approaches—which, as the president Maria Klawe tells me, "is actually a better description of what you're actually doing when you're coding." These changes weren't always smooth. Mudd has a famously intense curriculum, and the neophyte students drawn into the program have found it hard going. But it, too, eventually worked: By 2018, 54 percent of Harvey Mudd's grads in the computer science major were women, Klawe notes.

There's also been a broader cultural shift, too. In the last few years, women's interest in coding has begun rapidly rising across the entire US. In 2012, the percentage of female undergraduates "who plan to major in computer science" began to rise—for the first time in 25 years, since the collapse in the mid-'80s.

What's behind this? Possibly, women are being influenced by the growing national conversation about how they stopped coding. There's also been a boomlet of institutions training and encouraging underrepresented techies to enter the field, like Black Girls Code and CodeNewbie. And on purely economic terms, coding's appeal has become something akin to law in the '80s and '90s. It's seen as a bastion of good-paying and engaging work.

Perhaps most subtly, the reputation of coding as a profession is changing. Software truly has eaten the world. So anyone who wants to help other people is realizing that programming, as social critic Douglas Rushkoff puts it, is a "high leverage point." In an age when Instagram and Snapchat and iPhones are part of the warp and woof of daily life, potential coders worry less that the job will be isolated, antisocial, and distant from reality. People who used to want to do something "creative" or artsy stayed away from code, including many women. Research shows this is no longer true.

"Computer science today is attracting a more creative woman," notes Linda Sax, an education professor at UCLA, who has pored over decades of demographical data detailing which students, and what genders, opt in or out of STEM fields. This shift is abetted by the fact that it's much easier to learn programming without doing a full degree, using free online code schools, relatively cheaper "boot camps," or even meetup groups for neophytes—all of which emerged only in the last ten years. "This whole thing that let you imagine building an app in a weekend, that was not there back in the '80s and '90s," says Irani, the academic who studied at the Stanford computer science program. "Coding is, on its face, a more social thing now."

There's even been some pushback to the frat-house-style culture of many Silicon Valley firms. Consider Uber: In February 2017, the programmer Susan Fowler single-handedly shone a light into its culture, with a blog post describing "one very, very strange year" of working there. The company had a famously hard-driving, macho culture. Fowler enjoyed the insane pace of work and her talented colleagues; in her few spare hours, she authored a best-selling book on programming "microservices." But she quickly discovered that harassment ran wild at Uber. Not long after Fowler had arrived, her manager propositioned her, via chat, for sex. When she told HR and top management, they said her manager was a "high performer" and this was his first offense, so "they wouldn't feel comfortable giving him anything other than a warning and a stern talking-to." That didn't seem true, though. Fowler later encountered other Uber women who'd previously reported the same man for the same behavior. When Fowler continued to complain about how female employees were treated, HR officials struck back in a creepy fashion—demanding the personal email addresses the women used, implying the women were part of a conspiracy, and telling Fowler that "people of certain genders and ethnic backgrounds" may not be suited for coding jobs. About a week after *that*, Fowler fled Uber for a new job. She wrote her story up on the blog post, but, hey: Another blog post complaining about sexism in the valley? What impact could that have?

A huge one, it turns out. Uber's board had already been hit with years of bad press about its employees' behavior. Employees had used a "God View" internal mapping tool to help ex-boyfriends stalk their ex-girlfriends; a senior executive had threatened to send private investigators after a female journalist. Then CEO Travis Kalanick had crowed about how he'd gotten so much sexual attention from Uber that he calls it "Boob-er." Fowler's story—so carefully documented, with so much official malfeasance by HR—seemed like a particularly humiliating piece of news, and it was the straw that broke the camel's back. In a few short months, Kalanick had been forced out. And almost as if a cracking of the ice had begun, over the next few months, female coders and entrepreneurs told stories of prominent investors harassing and propositioning them, including Dave McClure, founder of 500 Startups, and venture capitalists Chris Sacca and Justin Caldbeck.

These shifts are only a small beginning. Most coders and experts I spoke to could easily, if wearily, generate a long to-do list of ways the industry could begin to break up its monoculture. They could hire less on "culture fit." They could recruit from more schools outside the Ivy League. They could kill the "algorithm challenge" whiteboard interview. "They have created their interviews to select for one specific type of candidate, which is, in the case of Google, for example, a Stanford grad. I mean, they design their whole company around it," to the point where it looks like a Stanford grad school, notes Erica Baker. And perhaps most of all, tech firms—and venture-capitalist titans—would need to actually punish and fire employees who harass others.

Sue Gardner keenly wants a wider range of people going into tech. But she also worries it might be *unethical* of her to encourage the young women she meets to go into the field. She knows they'll arrive excited, thrive early on, but—unless things change—gradually get beaten down. "The truth is, we can attract more and different people into the field, but they're just going to hit that wall in midcareer, unless we change how things happen higher up," she says.

Right now, though, the truth is that there are oodles of young people from all walks of life who want in. They admire tech's self-image as a meritocracy; they crave a field that's truly like that. Indeed, few people are hungrier for a true meritocracy than those who've historically been sealed out of so many industries and so many other opportunities. An industry run *truly* on hacker ethics—of being judged purely on the quality of your running code, with no one remarking on your identity? That would be a thrilling utopia for them. If the world of programming actually lived up to its implicit ideals, it would be a beacon for any historically screwed-over group. Indeed, that's precisely why previously picked-on nerd boys found it so glorious when they stumbled upon it in the '80s. It was the first zone they'd ever found that rewarded their interests and intellect.

In the summer of 2017 I showed up to a TechCrunch hackathon in New York, where 750 coders and designers were given 24 hours to dream up and create a new product. On Sunday at lunch, the teams presented their creations to a panel of industry judges, in a blizzard of frantic elevator pitches. There was "Instagrammie," a robot that would automatically recognize the mood of an elderly relative; there was "Waste Not," an app to reduce food wastage. Most of the contestants were coders who worked at local high-tech firms or computer science students at nearby universities.

But the winners, it turns out, were . . . a group of high school girls from New Jersey. In only 24 hours, the trio created reVIVE, a virtual-reality app that tests kids for signs of ADHD by having them play a series of games, while also checking their emotional state using IBM's Watson AI. After the students were handed their winnings onstage—a huge trophy-sized check for $5,000—they flopped into chairs in a nearby room to recuperate. They'd been coding almost nonstop since noon the day before and were bleary with exhaustion.

"Lots of caffeine," said then 16-year-old Sowmya Patapati, clad in a blue T-shirt that read "WHO HACK THE WORLD? GIRLS." They told me that they'd even impressed themselves by how much they pulled off in 24 hours. "Our app really does streamline the process of detecting ADHD," said Akshaya Dinesh, then 17. "It usually takes six to nine months to diagnose, and thousands of dollars! We could do it digitally in a much faster way!"

They were aware it was weird, and neat, for three young women to win at TechCrunch. The third member of the group, Amulya Balakrishnan, pointed out that collectively they had plenty of previous hackathon experience. Patapati alone had competed in over 25 computer science events. It was how she'd learned to build software, really: She'd taken computer science classes at her school, but that education had been far eclipsed by what she'd learned from free online courses and hackathons. "When I walked into my first hackathon it was the most intimidating thing ever," she said. "I looked at a room of eighty kids: Five were girls, and I was probably the youngest person there!" But once she'd done a few, her confidence grew. Dinesh and Balakrishnan, for their part, took programming courses at school, and had skipped a junior prom and a friend's birthday party to come to TechCrunch. "Who needs a party when you can go to a hackathon?" Dinesh joked.

Being young, brown, and female in this environment brought extra attention, not all of it good. "I've gotten a lot of comments like, 'Oh, you won the hackathon because you're a girl! You're a diversity pick,'" said Balakrishnan. When the prize was announced online, some postings were bitter: "There were quite a few engineers who commented, 'Oh, it was a girl pick, obviously that's why they won.'" On balance, though, they found the world of hackathons deeply welcoming to neophytes. Everyone was there to learn and experiment. People swapped knowledge; they felt like equals in many ways, even to the pros.

"I really do love the hackathon culture and community," Patapati said. "It's the best possible way for people who don't have resources, like me, to get into this. Without hackathons, I wouldn't know *anything.*" They all planned to study computer science at college. They can feel the promise of reward here, of the strange powers they can wield; they love the idea of getting ahead purely on their raw, cerebral skill.

"That's just the thing about tech," Balakrishnan says. "It doesn't matter what age or what gender, no matter who you are, you just never know what you can build."

< Chapter 8 >

Hackers, Crackers, and Freedom Fighters

So, let's say you and I want to talk," says Steve Phillips. "But privately! So come on over here."

He pulls me over to his laptop, where he's showing off his latest invention: LeapChat.org. When you navigate to that page, it creates a little private chatroom just for you; send the URL to anyone else, and they'll teleport inside. "And it's encrypted at every point, so there's no way for anyone to intercept what you're saying and spy on it," says Phillips enthusiastically.

I log into the room using the name "Clive." "Even the *username* is encrypted," says A. J. Bahnken, a friend of Phillips who helped develop LeapChat.

The point? To give people a small island of privacy in a world of nonstop online surveillance—particularly activists looking to avoid the prying eyes of police and spy agencies. Phillips and Bahnken have open sourced the code for LeapChat, so any adroit coders could set up their own LeapChat server. Phillips pays to run his own server with LeapChat on it 24/7 online, for anyone to use. Since all the chats are encrypted, not even they can see what people are doing on LeapChat .org; they don't store IP addresses of users, and encrypted logs of any conversations are deleted after 90 days. If law enforcement were to come and tell them, *Hey, give us copies of what activists have been doing on LeapChat.org*, there'd be nothing to hand over. It'd just be encrypted gibberish, with nothing readable.

"Right now, you have too many activists communicating using things like Facebook, where it's stored forever, and who *knows* who has access to it," says Phillips, a short and wiry 33-year-old with a neat goatee. He'd been invited to many Slack groups to discuss sensitive political issues, too, where things weren't encrypted, either. "I get it! It's convenient! So we need to make tools that are just as convenient and easy but safe."

It's a Friday night in San Francisco, and Phillips is presiding over a roomful of pro-privacy hackers. We're at Noisebridge, a hackerspace in San Francisco famous for attracting everything from soldering-iron-wielding hardware tinkerers to neophyte coders who show up for free learn-to-program tutorials; there's a giant wall of lights in the main room that tonight is blinking a Game of Life display, antique PCs from the '70s lying about, a couple of industrial sewing machines, and whiteboards overscribbled with arrays and functions.

Every week, Phillips holds a hackathon called Cypherpunks Write Code, where like-minded souls gather to write software to help people talk online without being snooped on by the man. There's a guy in a cap bent over his laptop, working on rewriting the popular Tor anonymous-surfing software in what he hopes will be a faster language; next to him is Jen Helsby, coding improvements to the interface to SecureDrop, a piece of software that organizations—including many media outlets like the *New York Times* and the *Intercept*—use so whistle-blowers can securely and anonymously submit documents and tips. A coder who grew up in Egypt is telling two other cypherpunks about his youthful exploits in a teenage hacking group that would try to figure out how to break into websites of companies. "We quickly realized we were in over our heads, and we stopped doing that," he says.

"You were *crackers*," roars an amused onlooker with a bushy white beard, an old-timer who first programmed using Fortran punch cards in the '70s. The Egyptian hacker nods sheepishly. In this part of the coder world, they're meticulous with their terminology. *Hacker* is used in its traditional meaning: a curious type who loves tinkering with systems, getting them to do new and weird things, and sometimes finding weak points—but generally within the law, white-hat style. A "cracker," by contrast, is one who breaks into systems for personal gain or crime. (Everyone here is deeply annoyed that in the mainstream imagination and mass media, "hacker" has become indistinguishable with illegal acts.)

Cypherpunk, though? That's a term they can all get behind. It's a blend of the old word *cyberpunk* with *cypher*—the code making and code breaking of computer cryptography—and, to my ears anyway, a dash of the "punk" of punk rock: people who do things in a DIY fashion, operating outside the mainstream. It's hacking to protect civil liberties. In a world where more and more of our conversation takes place digitally, the cypherpunks see encryption as crucial to civic autonomy.

Where do cypherpunks come from? They arise from a combination of coding skills and a deep distrust of centralized power.

Phillips didn't start out as one. He was certainly intense; his friendly, chipper vibe belies a somewhat monomaniacal drive. "My primary value is human greatness and justice," he tells me earnestly. "If nobody was striving to do anything special, that would be really sad for me, and there wouldn't be much meaningful happening in the world." He organizes his life along a monk-like rubric, working just enough hours as a freelance coder to support a small apartment near Noisebridge and pares his everyday spending down to a vapor, so he can work as few paying hours as possible, leaving the rest for his work in philosophy and cypherpunk coding. A friend described showing up a few years ago at Phillips's spartan apartment: "It was just a neatly made-up mattress on the ground, a single dresser with his few belongings, and an enormous stack of empty take-out containers. Nothing else." The take-out was all from unadorned meals like chicken burritos; Phillips follows a rigorous, if seemingly spartan, diet he devised as a teenager when he decided he'd been suckered by advertising and pop culture into eating junk food. "I thought, hey, this is the rational thing to do, I'm not gonna let social pressure damage my body. And there's a period where my mom was really offended. Like the food she cooked wasn't quite good enough for me anymore?" When he and I met for a meal, he arrayed a line of vitamins on the table: "Phosphatidylcholine and lecithin and vitamin B complex and a multivitamin and fish oil," he says. Whatever he's doing appears to work; he fairly vibrates with energy.

Phillips came to cypherpunk coding later in life. Growing up in a middle-class family in Vacaville, California, the son of a father who worked as a firefighter and a mother who worked as a fitness trainer and school teacher, he spent his teenage years tinkering with GNU/Linux, then left home to study philosophy and math at college. Philosophy struck him as powerful and serious—debates about the meaning of life, of moral behavior, were squarely in his nerd-justice lane. But he decided philosophical debates were too fuzzy and ill-defined; he thought it should be unassailably logical. So he wound up spending most of his spare time in college devising what he calls "executable philosophy": a system wherein you state philosophical propositions as snippets of Python code and *run* them, allowing you to "automatically uncover various kinds of errors" and "determine the logical consistency of belief systems," among other things. He figured he'd be happy spending his life trying to "revolutionize philosophy," as he puts it.

But you can't exactly make a living revolutionizing philosophy, Phillips figured, so when he graduated, he went into coding. He founded a hackerspace in Santa Barbara, and after meeting several friends there—including the then 16-year-old Bahnken, who wandered in one day to learn to code—they founded a string of start-ups. Phillips was laser focused on his entrepreneurship, too. He devoured online videos by Y Combinator head Sam Altman, and he imbibed *Zero to One*, the business-building tome of Silicon Valley's arch-libertarian icon Peter Thiel. "Peter Thiel's book on entrepreneurship is brilliant," he tells me. "The idea of, you gotta do something *fundamentally* better—it can't be a *little bit* better. But I don't mention the name Peter Thiel in certain places because I hate his politics."

Ideologically, in fact, Phillips had wandered all over the map, searching for a political home. "Six parties in six years," he says. In college he'd dabbled with anarchism and libertarianism; he liked their intellectual rigor and shared their concern that people who amass centralized power typically abuse it. But the idea of having *no* state, no government? That seemed unworkable. "So, the government run amok is bad. But then it's like, 'What should we do about health care?' And Ron Paul is like, 'Ask your church, they'll help!' It's like, what the fuck are you talking about?" Phillips laughs. He decided he was a "small government socialist," supporting a state that provided basic humanitarian goods like centralized health care, but otherwise kept its nose out of people's business.

That wasn't the way the world was going, though. Indeed, increasingly the government was getting caught poking its nose *deeply* into people's lives. In 2013, Edward Snowden revealed that the NSA was collecting everyday Americans' phone calls, hoovering up chat messages, and tapping into the backbones of services like Google and Yahoo! Phillips was horrified. But it also seemed like an area where he could make an impact. He'd been reading up on cryptographic history and watching videos where Julian Assange and the founders of the Tor project talked about the need for better, easy-to-use tools for everyday people to enable them to remain private online.

Now that—*that* seemed like a meaningful thing to do with one's life. "I can literally sit here, like they say, in your underwear, and write world-changing software—protect people's civil liberties, specifically right to privacy," he tells me. None of his start-up companies had really popped; he'd gone back to doing software consulting. So he decided to work on cypherpunk projects as much as possible, while living as cheaply as possible. Soon the code began to flow. He created a clever piece of pro-privacy software, some code that lets you synchronize your URLs from one browser to another, storing them in the open on a public server, while keeping them encrypted so passersby couldn't read them.

In 2017, Phillips was looking at *Wired* and read about the release of Barrett Brown. Brown was a government-transparency activist who back in 2010 had created an audacious crowdsourced-journalism project. He'd gotten troves of documents leaked from government and high-tech firms, some procured by the hacktivist group Anonymous. Brown put all the documents on a wiki and encouraged far-flung volunteers to help him analyze them, writing reports on what they'd found. His project wound up in the crosshairs of the FBI, who raided his house. Eventually Brown was jailed, after the FBI charged him with making threats to an agent on a YouTube video (true, though he says the video was triggered by his recent withdrawal from drugs) and with posting a document that linked to the URL of a website hosting stolen credit cards (a mistake; he thought the URL was linking to data relevant to his work), as well as an obstruction of justice charge. (The linking charges were later dropped.) While in prison, Brown wrote columns for the *Intercept* that won a National Magazine Award. He'd just been released from prison and told *Wired* about his next big idea: to build software so activists could collaborate online and crowdsource research, just like he did before. But this time the software would have tight encryption, so spooks couldn't poke around. "We want a direct line to civic participation," Brown told *Wired*.

Phillips was electrified. This was precisely the sort of code he wanted to write! In fact, the work he and Bahnken had already done—an early prototype of LeapChat, his URL-hiding software—was already a step in that direction. He jumped on a plane to Dallas and met with Brown; the two quickly agreed to collaborate, and within weeks Phillips was building a team of volunteer cypherpunks. They called the system The Pursuance Project, imagining it as a super-private productivity software for citizens trying to get engaged in political action. It would be good, Brown told me, for everyone from activists to journalists to NGOs working in conflict zones.

"I could literally protect a billion people's privacy with these tools," Phillips says.

If you want to understand why cypherpunks—and many coders in general—care so much about privacy, and so distrust authority, it's worth pausing to look at some history.

The attitudes, it turns out, have a long vintage. Beginning in the '70s, there have been several big flash points where coders clashed with the powers that be, in government and corporations. Each conflict was a fight over privacy and openness. Big firms and governments wanted to keep their secrets locked up tight, while having the ability to pry into the affairs of everyday people. The hackers wanted precisely the opposite. Hackers thought that everyday people should enjoy their privacy—while powerful interests should be required to expose their secrets.

The first version of this culture clash emerged back at MIT in the '60s and '70s, when the original generation of hackers began excitedly playing with the university's machines. Those hackers had an ethic of openness: If you wrote a cool algorithm or bit of code, you shared it with everyone else. If you didn't show off your code to others and vice versa, how would everyone learn? "We shared programs to whoever wanted to use them, they were human knowledge," Richard Stallman, one of MIT's most prolific hackers, later recalled. Indeed, the MIT coders were so communitarian that they didn't even put their names on code they'd written. "Signing code was thought of as arrogant," recalls Brewster Kahle, who arrived at the lab in 1980. "It was all for building the machine. It was a community project." Sure, the hackers could each be deeply individualistic and each individually convinced of their superior awesomeness. But coding itself? That was a group effort, an intellectual barn raising where all strove to make the computer do cool stuff for everyone's sake. *Owning* an algorithm you'd written seemed as nuts as "owning" the concept of multiplication itself, or constitutional democracy, or rhyme.

The MIT hackers hated any attempt to limit their access to the machines or to any technology they wanted to try out. So if a rule got in the way, they just broke it. If they were hacking in the wee hours, as was typical, and their computer broke down, they'd need the proper tools to fix it—only to find that the daytime staff had locked the tools away. So they'd simply hack the locks (making a "master" key from a blank), and abscond with what they needed.

"To a hacker, a closed door is an insult, and a locked door is an outrage," as Steven Levy wrote of those MIT coders in *Hackers*. "Just as information should be clearly and elegantly transported within a computer, and just as software should be freely disseminated, hackers believed that people should be allowed access to files or tools that might promote the hacker quest to find out and improve the way the world works. When a hacker needed something to create, explore, or fix, he did not bother with such ridiculous concepts as property rights."

This relationship to property gave birth to a radical idea: "free/libre" software. That revolution happened when MIT officials, in the early '80s, realized that over the years the student hackers had written a ton of incredibly valuable software for the MIT computers. So the officials decided to make some money off it. One for-profit computer firm called Symbolics asked MIT if it could license all that cool software to run on *its* computers. Sure, said MIT. So Symbolics duly licensed the code, and the Symbolics employees began tweaking and changing and adding new features to it. But whenever Symbolics created a new feature, it kept it proprietary. It wasn't sharing those innovations openly so the next generation of programmers could learn from it.

That annoyed Stallman. He eventually left MIT and invented an entirely different licensing paradigm, called the General Public License. It works like this: Let's say I wrote an email program and issued it under the General Public License, because I'm a public-minded hacker who wants to share my work freely. Cool: If you download the email program, the GPL means you're also allowed to look at my source code; to modify the email software; and to distribute your remix of my work. But there's one extra element here: If you modify the code and start distributing your new version, you have to issue all *your* modifications under the GPL too. You can't keep that new code secret. You're not prevented from making money off your work; you can sell the code as a commercial product. (It's "free" as in "free speech," not "free" as in "free beer," as Stallman pointed out.) But you have to publicly release your modified version of the source code, so that other people can examine it, remix it, and share it, too—a process that carries on, ad infinitum, forever, for everyone who uses my/your code. In essence, the GPL is a viral piece of legalism. If a hacker starts the ball rolling by making a piece of software free and open, it stays that way in perpetuity, including all the derivative works based on it.

It was a gauntlet thrown down in the name of openness, transparency, and control. Its message: You should never trust software if the person making it won't show you the code. And it created a rip in the culture—programmers who valued openness versus corporations who wanted to keep their code secret.

The second phase of conflict emerged in the '80s, and took a darker turn—as the FBI began to actively arrest hackers for breaking into systems.

This was during the first wave of teenagers who'd bought cheap BASIC home computers, and were using modems to dial into other computer systems around the world. Sometimes they'd dial into bulletin boards, discussion forums run by other coders, where they'd trade tips and share their exploits. But the most ambitious hackers hunted bigger game: They loved the thrill of illicitly dialing in to computer systems run by big corporations and telephone companies. If you could penetrate one of those, you'd have extremely powerful computers at your control—ones where you could program in languages like C, instead of the BASIC of the Commodore 64s they had plugged into their parents' TVs.

"We wanted to get access to computers that were more powerful than the simplistic ones we had at home," Mark Abene, a hacker who went by the screen name Phiber Optik back in the '80s, recalled in an interview with CNET. "We wanted to get access to the high technology we otherwise wouldn't have access to, understand it, and learn to program it." Getting that sort of access, though, required some straight-up digital trespassing and illegal activity. Members in Abene's hacking group would share passwords to corporate systems that they'd digitally cracked (or discovered during old-fashion dumpster diving in the company's trash). Abene and his colleagues eagerly shared info on reprogramming phone systems.

And this is where federal officials began to perk up their ears. The telephone companies and government were worried they were losing control of this weird nascent new realm, "cyberspace." So they swooped in on Abene and his group, "Masters of Deception." At age 21, Abene was put behind bars for nearly a year. Supporters argued that this sentence was wildly out of proportion to his crimes; in all of his intrusions, Abene hadn't really damaged any systems or stolen any valuable data. But he was easily the country's most famous hacker. He'd been quoted endlessly by journalists trying to understand the motivations of this new tribe of kids who wielded modems to roam unfettered online. His profile made him a useful scapegoat for a government that was embarrassed at seeming weak in the face of a bunch of keyboard-bashing kids. "A message must be sent," the judge explained at the sentencing.

By the early '90s, hackers increasingly realized that cracking was illegal and would be treated as such. If you accessed a system without permission, they knew, the feds might hunt you down. But surely if you just stayed in your lane, didn't break into systems, and just wrote useful code, you'd be left alone, right?

Not quite. The next big fight was over crypto code itself—with the government trying hard to make it illegal even to *write* certain kinds of software.

... ...

These were what's known as the "crypto wars": a set of fights over who was allowed to write, and use, encryption.

These days, in the world of Facebook and email, it's become increasingly clear that our online activities are closely tracked. Back in the '70s and '80s, though, most people weren't online, so the idea of digital privacy wasn't part of the national conversation. The first Cassandras to foresee our present-day situation were computer scientists and coders. After all, they were the first to use email to talk to one another, and to send files back and forth. It was fun and exciting, but they were uneasily aware that all this email and discussion-forum posting left a digital trail. Conversations that took place in text could stick around for years, maybe forever, on servers. What was going to happen when the *rest* of society started doing the same thing?

That worried the computer scientist Whit Diffie. In the early 1970s, while talking with his partner, he predicted the emergence of things like Facebook and Amazon and Google Hangouts. "I told her that we were headed into a world where people would have important, intimate, long-term relationships with people they had never met face-to-face," as he recalled in 2013.

Diffie also knew that most forms of encrypting digital messages were, back then, pretty weak. That's because they all shared one big flaw: Both parties, the one sending the message and the one receiving it, needed to have the same "key" that lets you decrypt it. If I'm in New York and trying to send you a secret, encrypted email in Wisconsin, I need to send you the key to unlock my secret missive. The problem with this—which spies and generals going back to ancient Greece well knew—is that it's easy to accidentally lose your key or for an enemy to steal it. And once someone knows your key, boom: They can unencrypt any of your secret messages and read them. So, sure, Diffie could write some terrific crypto to keep his and his friends' messages secure. But if anyone in their group accidentally leaked the key, it was game over. No more privacy.

In 1976, though, Diffie and a colleague had a breakthrough idea. They invented what's known as "public/private key" crypto. In this system, we each have two keys: a public one that anyone can see—and a private one known only to each of us individually. Sending a message to someone else meant using two keys—your private one, and their public one. The way the crypto math works out, the only person on the planet who can decipher and read the message is the recipient. Diffie's idea was world-shaking. In a flash, he and his colleague divined a way for everyday people to trade messages online, using codes that almost nobody could crack. And they were the first to talk about this idea openly; a cryptographer for the British government had hit upon the same concept, but at the time it was a classified secret.

Diffie's discovery deeply alarmed the feds. The National Security Agency was used to being the big dog in the world of crypto—they were the world masters at breaking secret codes, and they wanted to keep it that way. So they didn't like academics and coders even *discussing* the idea of powerful crypto. "There is a very real and critical danger that unrestrained public discussion of cryptologic matters will seriously damage the ability of this government to conduct signals intelligence," worried Vice Admiral Bobby Inman, then head of the NSA. They certainly didn't want everyday people *using* powerful crypto. "If you simply took this technology and released it widely, you were also potentially creating an opportunity for very small terrorist groups, criminals and the like to use this technology to get a kind of perfect information security," as the onetime NSA general counsel, Stewart Baker, recalled.

The US government did have one law that they could use to limit the spread of crypto. Federal regulations classified strong encryption—stuff the NSA couldn't break—as a "munition," and munitions can't be shipped outside the country without the federal government's approval. So, sure, any American coder or computer scientists could build whatever encryption software they wanted. But they couldn't share the code with anyone outside the country or they'd be breaking the law. In fact, if you emailed a copy of your powerful crypto to someone in Germany, you'd be considered an *arms dealer*. That's right: The NSA was so afraid of people with unbreakable codes that they treated powerful crypto as they would a tank or a missile.

This munitions law enraged academic computer scientists. After all, from their perspective, code was speech—speech you spoke to a machine, and which you could print up and share with other coders on a piece of paper. By limiting who you could sell your crypto code to, or who you could even show it to, the government was effectively muzzling speech. The computer scientist Daniel Bernstein took it to court, arguing that this was a violation of his First Amendment rights.

The truth is, despite the munitions law, the NSA's position was pretty weak. Everything it worried about was coming true. The spread of code was hard to control. Hackers were beginning to invent increasingly powerful crypto: Phil Zimmermann, a coder who'd grown up excited by code making, created Pretty Good Privacy, the first software that used Diffie's concept for encrypting email. Copies quickly leaked online, and pretty soon they were shared around the world. The government investigated Zimmermann for it, but he wasn't the one who was distributing his code. (He was barely online, really. It was other hackers who were sharing the files. The Justice Department dropped its investigation of Zimmermann a few years later without explaining why.) With Zimmermann's PGP floating around online, the genie was now truly out of the bottle. Hackers showed they could make—and globally distribute—software that the NSA couldn't break.

A subculture was emerging, excited by the political implications of crypto. Calling themselves "cypherpunks," the practitioners were a far-flung online cohort of hackers and thinkers. Many were committed libertarians who relished the idea of speech safe from government spooks. (One founding member, Tim May, argued that "crypto = guns in the sense of being an individual's preemptive protection.") Others were anarcho-capitalists who, in a sort of early foreshadowing of Bitcoin coders, dreamed of using crypto to create anonymous e-cash that could never be taxed, and would be forever safe from the hands of grasping bureaucrats worldwide. Some of these became the guys who also dreamed of seasteading; some were just classic ACLU civil libertarians. All hated the NSA.

By the mid-'90s, the NSA realized that the cypherpunk point of view was winning. The internet was emerging into public life; everyday people and businesses were exchanging emails and beginning to browse the web. Creators of software like Lotus Notes were building crypto into their products, reasoning that corporate clients wanted to know their communications were safe. E-commerce was starting up, too, which meant all those credit-card numbers flying through the ether would need protection. And the "munitions" rule wasn't looking too stable, either. Judges were getting ready to rule that, indeed, code was speech—and the government couldn't forbid you from speaking it across nation-state lines. (In 1999, district court judges would hand down that ruling.)

So the NSA spooks tried one Hail Mary pass. If crypto was going to happen across society, dammit, then they wanted to make sure *they* controlled it. Working with the Clinton administration, NSA leaders argued that all computers and phones should, indeed, have a powerful crypto. But it should be in the form of a computer chip the NSA itself had created: the "Clipper" chip. If a device had a Clipper chip in it, it'd encrypt anything you did online or said on a phone, so it would be safe from prying eyes. With one exception: the NSA itself. They'd keep a "back-door" key that gave them access to every device. That way, if any nefarious criminals or terrorists began using crypto to communicate secretly, the NSA could easily tap into their conversations. It was a grand scheme, offering high-end crypto for everyday people—assuming you trusted the NSA to be devoutly respectful of the privacy of everyday Americans, to only eavesdrop on evildoers, and to never abuse their power.

The cypherpunks were having none of it. This plan was *nuts.* "The war is upon us," May stormed. "Clinton and Gore folks have shown themselves to be enthusiastic supporters of Big Brother." They began raising a public stink, and were joined by civil libertarians of all stripes. Intellectuals outside the nerd circles of coding joined in, including conservative *New York Times* columnist William Safire. The larger public was beginning to realize that, hey, life was moving online—and they didn't like the idea of spies snooping into their affairs, either.

What really sank the Clipper chip, though, wasn't even the politics of it. It was a bug, discovered when a young computer scientist named Matt Blaze was asked to evaluate the chip. Poring through the Clipper chip specs, he quickly spotted a huge flaw—one that would make it possible to seal up the NSA's "back door." Using the bug, a criminal could use a Clipper chip to talk secretly online, while preventing the NSA from listening in. So in effect, the NSA had created a technology that handed amazing crypto to criminals, which the NSA itself couldn't break. When Blaze published his findings, it made the NSA look even worse than mendacious: It looked incompetent.

Clinton and Gore quietly pulled the plug on the chip. The hackers, computer scientists, and cypherpunks had won.

Even after all these fights over spying, one of the biggest contretemps between coders and law enforcement was yet to come. And it was based on an unexpected area: entertainment law and copyright—movies and music.

In the late '90s, as ever more citizens swarmed online, executives at Hollywood studios and record labels grew alarmed. People were trading MP3 copies of songs online, without paying for them! Broadband was getting faster and faster, and they foresaw a time when people would start copying and trading TV shows and movies online, too, without paying for them, either. So the entertainment executives began loudly warning about the evils of piracy, posing it as a threat not merely to their bottom lines but to civilization itself. "This is a profound moment historically," said Richard Parsons, then the head of Time Warner. "This isn't about a bunch of kids stealing music. It's about an assault on everything that constitutes the culture expression of our society. If we fail to protect and preserve our intellectual property . . . the country will end up in a sort of cultural Dark Ages."

To protect their products from being copied and shared, the entertainment firms turned to the very thing the cypherpunks loved—strong encryption. Whenever record labels or movie studios digitized their songs or TV shows or movies, they wrapped them in crypto. For DVDs, the crypto was called the Content Scramble System. If you wanted to watch a DVD on your computer, you needed a video player that could decrypt the movie.

In a sense, the entire entertainment industry became, in a curious way, a software industry. Say you wanted to manufacture a DVD player, capable of playing an encrypted DVD. Or say you wanted to make DVD-playing software for a Windows computer. Well, you had to ask the movie studios for the right to put their secret decryption code inside your DVD player. By establishing control over the software that decrypted DVDs, the entertainment executives figured they could survive the coming world of the internet. Sure, let some kid rip a copy of the DVD onto their hard drive and share it online. It wouldn't matter; the movie file itself would be unreadable gibberish. You needed a DVD player approved of by the entertainment industry. Music labels did the same thing with encrypted music files; publishers did the same thing with encrypted ebooks. They called it "digital rights management," or DRM.

There was only one wrinkle: What if some coder started poking around in the secret crypto code? Coders sometimes reverse-engineer software, breaking it apart to figure out how the code works. What if someone did that to the Content Scramble System? Once they learned how the Hollywood-approved software worked, they could code their own version to play the DVD—without Hollywood's permission. The same goes for crypto used to lock down music files, or even ebooks. If someone wrote their own software that unlocked one of those encrypted files, poof: You could copy and share the content to your heart's content.

So for Hollywood, the answer was to criminalize any coder who tried to do that. The entertainment industry had to make it illegal even to *write* that type of software.

That means they needed a law written. And so, in 1998, copyright lobbyists convinced Congress to do precisely this. When Congress passed the Digital Millennium Copyright Act that year, it made it illegal—punishable by damages, or even prison time—to write code that circumvents DRM.

Soon, police law enforcement was hauling coders off to prison for breaking DRM.

In 2001, a Russian coder named Dmitry Sklyarov visited Las Vegas to give a talk at DEF CON, a hacker convention. He was talking about weak DRM in ebooks made by the company Adobe. It was a subject he knew a lot about: For his employers, ElcomSoft, he'd written code that decrypted Adobe ebooks so they could be converted to PDF format. In Russia, writing that code was perfectly legal. But once Sklyarov was in the US, he was subject to US law. Adobe tipped the FBI off to the fact that Sklyarov would be in the country, and the FBI threw him in jail right after he'd made his DEF CON speech, charging him with "circumvention" of DRM; he faced up to 25 years in prison and a $2,250,000 fine.

An even bigger brouhaha emerged around a mere teenager. In 1999, working with two other coders, the 15-year-old Norwegian Jon Lech Johansen created (and then linked to a copy online) the open source software DeCSS. It stripped the DRM from DVDs so you could watch them on Linux computers. For coders, this was a big deal: The Hollywood firms had approved DVD-playing code for Macs and Windows but not for Linux machines, and tons of programmers used Linux. Finally, using Johansen's software, hackers could watch movies on their Linux laptops! Some rewrote DeCSS in other computer languages, and hackers and website owners worldwide began eagerly sharing them online, including one named Andrew Bunner. The hacker magazine *2600* published the source code for DeCSS online and linked to sites that were distributing it.

Then the hammer of the law came down. In 2000, a group representing movie studios sued the publisher of *2600* under the new law, claiming it was illegal to distribute the DeCSS source code; after all, when you ran the code, it became a tool that circumvented DRM. Andrew Bunner, too, was charged with distributing the DeCSS code, along with a handful of other webmasters. Meanwhile, over in Norway, a complaint from the US entertainment industry convinced Norwegian police to interrogate and charge the teenaged Johansen, using a provision of Norwegian law.

Now the alarm was really raised, and not just among cypherpunks. Coders and hackers of all stripes flipped out. From their point of view, corporate America was using a new copyright law to criminalize the very act of writing code. "For them, the lawsuits were an attack on their right to tinker, to write code," notes Gabriella Coleman, an anthropologist who has closely studied hacker culture. "It was the moment when they really began to say that code was speech, and these laws were the government making speech illegal. They were all about sharing, showing their code, and now they're being told that can be against the law."

In the short run, the movie studios lost a few of their battles. In 2001, a California court sided with Bunner, arguing he had a free-speech right to post the source code (though later court rulings bounced back and forth on the issue for a few years). Johansen was acquitted. But in other places, the studios were victorious. A different judge ruled against *2600* magazine, and the publication eventually decided to stop appealing. Overall, the message was out: DRM was a potent weapon that corporations could—and would—wield, to criminalize programming.

After three decades of increasing conflicts with the government, corporations, and spies, it is no wonder, as Coleman notes, that many hackers developed an instinctive, hair-trigger screw-the-man stance. Nearly every time the government came knocking, it seemed, it was to try and criminalize them from doing something with code—from speaking the very languages they enjoyed speaking.

"While it might be tempting to see this as merely another journalistic cliché," she writes, "this attitude is genuinely encoded deep in the hacker cultural DNA."

So, cypherpunks get called "paranoid" quite a lot.

Jen Helsby certainly did. She used to warn her friends about the government doing online surveillance, tracking phones, and scooping up their social-media data. "I'm not much of a social-media person," she told me dryly. Helsby came by her suspicions honestly. She'd begun her career studying dark energy for a PhD in astrophysics in Chicago, trying to figure out why the universe was expanding. Like many astrophysicists, she became adept at coding. But she'd also become interested in foreign affairs and hankered to do something that helped society more directly. So Helsby started using her data-crunching skills to help a Chicago project study urban blight, and then she helped found the Lucy Parsons Lab, which creates free, open source software to help citizens lodge complaints against police. (One is a database of officers with photos, to help citizens figure out, among other things, which officer harassed them.) She'd warn fellow activists that police and government agencies were likely monitoring them, and she held crypto parties to teach "opsec"—operational security, like using encrypted apps such as Signal instead of text messaging. "People should have the ability to read freely, speak freely, which you don't have when everything is being watched by the government, and they can take action against you when they don't like what they're seeing," she says. "The kid who got raided in New York for having a cop emoji next to a gun emoji? That's nuts, that's insane and draconian."

Nonetheless, she'd still sometimes get called paranoid. "You're a conspiracy theorist," she'd be told—until 2013, when Edward Snowden made the headlines.

A computer-security contractor for the NSA, Snowden leaked thousands of documents showing that the agency was, indeed, spying on everyday Americans to a fantastic degree. The Snowden revelations galvanized cypherpunks, and they also provided a measure of validation. *See, we're not tinfoil-hat crazy. This stuff really is happening.*

"It's been useful to point to documents," Helsby says. "After Snowden's documents came out, people understood more. Actually, what became clear is that even the most paranoid of us were *understating* what was going on." When we spoke, she was working as the lead coder developing SecureDrop.

I ran into Helsby at the Aaron Swartz hackathon in San Francisco.

It's held every fall as a weekend where crypto hackers gather to work on software that, they hope, empowers the average citizen. It's in commemoration of Swartz, a coder and activist who committed suicide at age 26, but in his too-short life he created projects that are beloved by the hacker community: He helped code the first version of SecureDrop, cofounded Reddit, cocreated RSS—a way for people to make personalized News Feeds online—and helped pioneer the idea of "creative commons" licenses, a hackery super-open form of copyright. (You can use it to freely release your photos online, say, and allow anyone to modify or edit it—under the legal condition that *they* release *that* modification freely, too. Or you could let people do whatever you want with it, with no restrictions; there are many types of Creative Commons licenses.) Swartz also believed, fervently, that scholarly information was being locked up by corporations when it ought to be free to benefit the public. That's what led to his own fight with the law: In 2010, he wrote a program that downloaded nearly 5 million scholarly journal articles from JSTOR, a paid-for subscription service, via MIT's subscription. After JSTOR and MIT complained, Swartz returned the digital copies to them, and never distributed them online; but the US Department of Justice, apparently looking to make a statement, charged Swartz with computer fraud. Facing penalties of up to $1 million and decades in jail, he committed suicide.

"Aaron was persecuted for reading too quickly in a library," says Brewster Kahle, a cofounder of the Aaron Swartz hackathon along with Lisa Rein, herself a cofounder of Creative Commons. After his MIT hacking days in the '80s, Kahle made millions with start-ups in the '90s, then founded the Internet Archive. The Archive makes copies of great swathes of the internet each day to save for posterity, and it also scans everything from old books to vinyl records to video games that are in the public domain, and posts them online: Swartz's vision made reality, in a way. The Archive is located in a decommissioned San

Francisco church; in the lobby, the hackers have set up tables where they code away.

Helsby has got a team tweaking the interface to SecureDrop. Another table is programming software that lets people set up websites that are "distributed," hosted on laptops around the world so they can't easily be knocked offline by autocratic governments. A 22-year-old college student, Austin, is hacking on that project. He taught himself GNU/Linux as a teenager by figuring out how to install it on his Nintendo Wii, because his parents would generally only let him use the family computer for schoolwork.

"I'd be hacking away on it when they'd poke their head in the room, and I'd have to quickly change the TV channel so they couldn't see what I was doing," he says. "Oh hi, mom, just watching, uh, some dumb show here!" Austin, like the others, cited Snowden as a tipping point. The younger coders here have come of age in a period where mass culture is now much more aware of the dark side of tech. Many of the cypherpunks are fans of *Mr. Robot*, the TV series in which a hacktivist group plots to destroy a megacorp's record of consumer debt by encrypting it. (The show has, they note, surprisingly realistic hacking; plot points have included Raspberry Pi computers and Android rootkits.) Indeed, the room here at Aaron Swartz day is dressed as if they were extras on *Mr. Robot*, with plenty of dyed hair, leather jackets, multiple piercings, and laptop lids buried beneath internet-freedom "copyleft" stickers. While corporate brogrammers in Silicon Valley now frequently style themselves as ostentatious health nuts—rock-climbing clothing, tans, and bicycling to work—the cypherpunks are the ones who still dress like the pale wraiths of cyberspace. "I don't think I look hacker *enough*," Austin says with mock concern, scanning the room, then flipping his hoodie up. "There you go," the woman hacking at her laptop across from him says, and laughs.

Wandering around behind Austin is Jason Leopold, a *BuzzFeed* investigative reporter who's giving a talk on freedom-of-information requests. Lingering near the front door, on her mobile phone, is Chelsea Manning, the army intelligence analyst who was famously jailed for leaking records of military activity and diplomatic cables that helped kick off the Arab Spring, and who's here to give a rally-the-troops speech to the hackers. The political spectrum of hackerdom is surprisingly broad. As Coleman notes, the common threads of the culture—a love of the craft and sharing of knowledge, support of free software, a distaste for government limits on coding—create groups working together on projects who might otherwise see eye to eye on little else. Jointly crafting a piece of crypto or open source code is a common idealistic goal they can all rally around. "Pragmatic judgments often trump ideological ones," Coleman writes, "leading to situations where, say, an anti-capitalist anarchist might work in partnership with a liberal social democrat without friction or sectarian infighting."

The afternoon of the hackathon, I run into Steve Phillips; he's arrived with a group of volunteers, coders who've all begun helping build the Pursuance system. In the afternoon he climbs onstage in the Archive's auditorium in a WikiLeaks T-shirt to show off what they've built. There's the beginnings of the full software, where activists can set up to-do lists, set tasks, delegate jobs. As he tells the crowd, there are a ton of forums where people can talk, but few tools to help them get things done, he explains; Pursuance aims to help fix that. "I notice, watching some online activists operate, that there's way too much manual work done that could be automated," Phillips says crisply as he shows off the software. On the screen over him, Barrett Brown has Skyped in to talk about the philosophy of Pursuance in his slight southern drawl, while sitting in his living room and puffing heavily on a vape pen.

Afterward, the group heads out to a Thai restaurant for dinner.

"That was nerve-racking—we were still putting stuff into the database, like, *ten minutes* before we went onstage," says Marty Yee, a young coder whose day job currently is working for Behive, a small social-networking app. As with most software development, getting the demo to work was a panicked last-minute crunch. Phillips credits Yee with making the software look halfway decent, which he regards as crucial to making activists enjoy using Pursuance, he says as he scans the menu for something to fit his stringent diet. (He finds brown rice.) "I handed Marty a bunch of my horrible wireframes three weeks ago and he made it into magic."

Annalise Burkhart, a 21-year-old woman from Tennessee, has signed on as the director of operations. She wants to get Pursuance into the hands of Middle Eastern activists. She minored in Arabic in college, and between that, YouTube videos, a Syrian tutor, and practice translating in her community, she became fluent; she enjoys the cognitive dissonance it produces when an American girl with silver-highlighted long hair suddenly busts out the language. "I'll be on the phone and suddenly I speak Arabic and people will be like, *Whaaaat?*"

Bahnken, who's wearing a T-shirt emblazoned with "SUPPORT YOUR LOCAL GIRL GANG," talks about a security writer he's started reading who mines rap lyrics for pro-privacy ideas. "He analyzes Notorious B.I.G.! He'll look at 'The 10 Crack Commandments,' and it's actually good opsec advice."

"Like what?" Phillips asks. "What's the advice?"

"Like, 'Shut the fuck up!' " Bahnken laughs. "People talk, and they get caught."

Nobody in the hackathon is doing anything illegal. The tools they're building are, as they note, intended to help protect the little guy. But the historic tensions with law enforcement never go away, because police and spies continually make the point that privacy tools created for law-abiding citizens will inevitably get used by criminals and terrorists, too.

The police and spy forces aren't wrong, mind you. Consider Tor: It was designed to encrypt web surfing so outside parties don't know what sites you're visiting. This makes it invaluable for whistle-blowers, journalists, bloggers, and everyday people who simply want privacy online. But it can also be used to set up so-called dark websites—where it's impossible to know where the site is hosted or who's hosting it. That's terrific for crime. Online drug dealers thus love Tor, and frequently set up Tor-enabled dark websites, such as the famous Silk Road online store, where one could buy drugs and guns. The terrorist group ISIS has written a guide describing how to use Tor to avoid scrutiny. Granted, the criminal employment of Tor is a tiny fraction of its overall use. But purely as a statement of fact, it's true that crypto protects malefactors as equally as it protects freedom fighters. There's no way to create crypto math that gives privacy to only the "good" people.

The cypherpunks all know this. Most are perfectly fine with it; freedom means freedom to do not just good things but bad ones. Their willingness to say this up front can be startling, or refreshing, depending on your political point of view. While speaking on a panel at a security conference, the crypto hacker Moxie Marlinspike—creator of the Signal messaging app—argued that making things occasionally easier for lawbreakers was an acceptable cost for securing everyone's privacy.

"I actually think that law enforcement should be difficult. And I think it should actually be possible to break the law," he said.

Or, as he wrote on his blog, it's only by breaking the law that citizens can discover that some laws are nonsensical. "How could people have decided that marijuana should be legal if nobody had ever used it?" he wrote. "How could states decide that same-sex marriage should be permitted?"

... ...

Hackers' curiosity can, in fact, lead them quite easily to break the law, simply because computers control so many things. That's precisely what got all the modem-wielding teens of the '80s and '90s in trouble. Back then, companies who were getting penetrated acted with alarm—*Zomg, we're being hacked! Call in the cybercops!* But eventually some more enlightened tech firms and parts of corporate America realized they could domesticate hackers who like to break into things. If they're going to break in anyway, why not hire them to do so? Then pay them to tell you precisely how they got in, so you can plug that hole.

This is how the tech industry created the job of the penetration tester, or "pentester." If you've ever seen the 1992 movie *Sneakers*, it was one of the first times a pentester made the silver screen—Robert Redford led a gang of infosec warriors paid to bust into banks. These days software firms also set up "bug bounties," paying prize money to anyone who finds a bug in their system and tells them about it.

As the stereotype might suggest, many pentesters start as teenagers breaking into systems, just for kicks, sometimes their own systems, sometimes others. "It's so hard to understand ethics as a 17-year-old," as Jobert Abma tells me. He cofounded HackerOne, a pentesting firm that uses freelance hackers around the world to pound away at its clients' sites and software. Abma grew up in the Netherlands, where he and a friend would build websites and challenge each other to break into them. They eventually started examining various companies' sites, too, wondering if they could find weaknesses.

"Nothing malicious," he adds. At their high school graduation, they hacked the video monitors above the stage, inserting jokes about various students. Later, they decided to turn their skills at intrusion into a business—calling up companies, asking for permission to penetrate their sites and promising to give them a cake if they couldn't get in. "We never gave away a cake," he notes. Kids who become fascinated by prodding the vulnerabilities of computers tend, in other words, to wander into gray-area waters. The trick is learning enough to make their skills valuable on the legitimate market without falling afoul of any overzealous authority figure that wants to make an example of them.

This gray-area field is on wild display at DEF CON, an annual hacker convention in Las Vegas that's a favorite of pentesters, infosec, and university computer-security researchers. It's like a jamboree of digital intrusion, with presentations of innovative penetration techniques, everyone eagerly sharing their latest ideas. When I attended in 2017, the hacks were ingenious: One coder showed off a trick to break into highly encrypted Bitcoin wallets. Another presented methods he'd discovered for penetrating wind-farm controllers—which, he noted, could be used to *destroy* a wind turbine, by cycling its brakes on and off ("the hard stop of death attack") or to hold it ransom by shutting it down until the wind-farm paid a bribe. ("The company's gonna lose out on anywhere from $10,000 to $3,000 per hour of downtime," he said. "That's a lot of money, folks.") Outside the talks, hackers roamed the halls, showing each other clever penetration gear—one guy had a small crowd oohing over a Sonicare toothbrush he'd outfitted with a magnet on its vibrating head. "You can wiggle it over a chip to induce a charge where there wouldn't normally be one, to see if you can knock it into a different state," he says, waving it over a Raspberry Pi minicomputer experimentally. "And since I'm a big *Doctor Who* fan, of course I want a sonic screwdriver." Over his shoulder peered a burly guy in a T-shirt reading "Think Bad. Do Good." Another hacker wore an actual tinfoil cap.

DEF CON is famous for having dozens of Olympics-like hacking challenges—systems set up specifically to see if hackers can break in. When I attended in 2017, there was an entire minihouse filled with

Internet of Things devices, like a smart thermostat and an internet-connected garage-door opener. A group of pentesters from down South have formed a team and are busily cracking away. "Does anyone have a power cord? I'm cracking passwords and it's really burning up my laptop," one shouts. Another has broken into the alarm system and is rooting through open folders to look for passwords.

"People reuse passwords—that's the golden key, and when you get that the whole castle falls," as Dori Clark, one of the team members, tells me. Clark got into pentesting after studying computer science and working as a coder for a financial service. "I had to write COBOL! I was editing code that was older than I am; it was from the '80s," she adds. She took a gig with the military, and one day she ran into the pentesting team and was mesmerized. Now she pentests for Walmart and buys weird old hardware off eBay in her spare time to practice breaking into unusual equipment. "You have to love this to do it, you have to do it after the kids are in bed," she says. Her daughter's getting older, and she intends to teach her coding when she learns to write. But Clark is almost leery of introducing her to the arts of hacking. "Do I give her all these skills? Because if a teenage girl gets all the abilities she's going to hack her friend's Facebook to see what she's saying," she says, and snorts.

Infosec workers often have a bleak, dim view of the world of technology—because they're constantly seeing how broken it is, how crappily cobbled together most commercial software is. Sure, it's fun to get paid to break into things. But God almighty, now they see with X-ray vision what a mess our code-brokered world truly is: social networks that can be cracked, financial systems of all kinds that were assembled in haste and are a creaky mess. And Internet of Things devices are the worst. They're cheaply made and often shipped off with easily guessed default passwords. (Many security hackers just refer to it as "the Internet of Shit.")

A particularly grim example of this epiphany unfolds in a room at DEF CON known as the "Voting Machine Hacking Village." A team of computer scientists—including Matt Blaze, the one who single-handedly wrecked the Clipper chip over twenty years ago—had bought a few dozen machines that had been used in previous US elections, then decommissioned. The goal: to see how easy someone could hack them and swing a vote. Blaze and other experts had long suspected that the machines had lousy security, but up until recently it hadn't been easily legal to poke around inside them. They're covered by the Digital Millennium Copyright Act, so it is—of course—illegal to circumvent it and look at how the machines work, without permission from the voting-machine companies, who weren't willing to grant it to a random bunch of hackers. But the government had recently decided that voting machines were important enough to the country that they granted a temporary exemption. That means the DEF CON hackers could poke around with impunity and tell the world what they found.

"We encourage you to do stuff that if you did on Election Day they would probably arrest you," Blaze told the roomful of hackers, as they began cracking open the machines.

It took astoundingly little time: The Danish computer scientist Carsten Schürmann opened his laptop, booted up Metasploit—a piece of software for breaking into systems—and in minutes he had taken control of one voting machine via its unsecured wi-fi. Another hacker found that the USB ports on a different machine weren't locked down, and if you plugged in a keyboard you could take control. Every so often a corner of the room would erupt in laughter when a pentester, rummaging in the code of a machine, would discover it set up with laughably bad login credentials. "The username is default, and the password is . . . '*admin*,'" said one blond ponytailed coder, peering into his laptop screen, then theatrically face-palming.

Technically speaking, the people running the elections should have changed the passwords, at some point. But they never did. This is the other great piece of bleak knowledge that infosec experts face: Humans are the weak link in any system. Security hackers are constantly telling the employees in their companies to change their passwords, to update their software; but we regular humans ignore them, all the time. The code may be defective, but so are the users. To work in security is to be faced with the thankless task of constantly and correctly perceiving the bug-eyed danger of the online world, only to have your warnings blithely ignored by the world. You are a prophet howling in the desert, with no one listening.

This can endow security coders with a pitch-black cynicism about the people they work beside. *I'm surrounded by fools and idiots; everything would be secure if these morons would stop clicking on links in their emails.* "It's one of the hazards of the job," one DEF CON attendee, who manages security at his firm, told me while nursing a drink in a Vegas bar. "I've literally had to stand over vice president's shoulders to force them to change passwords. People are sloppy, nobody follows best practices. Then when something goes wrong I'm the one who gets his ass hauled in and reamed." No wonder, he said bitterly, he's a misanthrope.

"To others," as Christian Ternus, an infosec coder, admitted in an essay, "we're jerks."

The infosec people are right to be paranoid, though. There really are phalanxes of black-hat hackers out there, working hard to break into systems for profit. Indeed, the world of for-profit cybercrime is growing remarkably huge, as malware becomes easier to buy, or even rent, online.

The stakes of cyberattacks can be enormous, as the WannaCry malware of 2017 showed. It was a piece of "ransomware": Once it infected a computer, it encrypted all the contents so the owner couldn't read or use them. Then it popped up a neatly designed little text box explaining that "We guarantee that you can recover all your files safely and easily. But you have not so enough time." The language was cheery, if a bit stilted—possibly the result of a Chinese speaker writing in English, some suspect. And the interface was quite slick. The overall goal of ransomware, these days, is to seem as professional as possible; some even have helplines to assist the victims in figuring out how to acquire Bitcoin, the main currency for paying ransoms. Even malware attacks, it seems, attempt to affect the sleekness of a Silicon Valley start-up. (WannaCry even had a compassionate corporate policy: "We will have free events for users who are so poor they couldn't pay in six months.")

WannaCry wreaked havoc, shutting down 200,000 computers in over 150 countries, with Russia, Ukraine, and India hit particularly hard. Perhaps the highest chaos, though, was in the UK's National Health Service hospitals, where medical devices like MRI scanners, computers, and even blood-storage fridges were hit. Thousands of operations and appointments were postponed as staff turned to pen and paper; the overall worldwide damage was, by one estimate, $4 billion. The culprits? Hacking groups affiliated with North Korea, many security experts suspected.

Meanwhile, in a rural part of the UK, Marcus Hutchins was sitting in his bedroom at his parents' house and trying to deduce how to stop it. He's a white-hat researcher for Kryptos Logic, a cybersecurity firm in LA. Hutchins's specialty is reverse-engineering malware to try and figure out how it is built and what it does. Hutchins, a mild-mannered guy with a crop of frizzy hair, noticed something interesting. Each copy of WannaCry was trying to ping a web URL that didn't exist: iuqerfsodp9ifjaposdfjhgosurijfaewrwergwea.com. Why? He wasn't sure. But maybe if *he* registered that domain and set up a website, all those copies of WannaCry would keep pinging it. That would let him track how many copies of WannaCry were in existence, and where it had spread to. So he went to a URL-registering service, and set up the URL for $10.69.

It had an effect, even better than he expected: It stopped WannaCry in its tracks.

It turns out the URL worked like a "kill switch." Once it existed, every copy of WannaCry shut down. "It was all over in a few minutes," he tells me, marveling at the speed of its crash. Possibly the malware authors had included a kill switch in case they lost control of their spread of the ransomware—"in case shit got too bad," as Hutchins says dryly. But either way, he had prevented a mammoth amount of damage. He'd shut down WannaCry before much of the US turned on its computers and opened for business, which likely meant billions saved.

Pretty soon, Hutchins was a global celebrity, with newspapers feting him as the white-hat hacker who "accidentally" saved the world. It was a rapid rise. He'd taught himself to code beginning only ten years earlier, at age 12. In his spare time he'd gotten interested in malware, particularly botnets. He'd analyze the ones he found, since he'd become adept in reading Assembly, a gnarly low-level language, and he'd post long write-ups of his research on his blog. He applied for a job at the British version of the NSA, but the CEO of Kryptos Logic got to him first—offering him a job after reading his blog posts. It's an index of how highly in-demand infosec talent is.

When we spoke, he noted that some talented white-hat malware experts had started off honing their skills by learning how to cadge free software. "The original reverse engineers, they came from the 'wares' scene," he says, describing folks who'd crack DRM on software like video games or Photoshop, so they could use it for free, and share it for free online. "You just figure out how the licensing algorithm works, and create your own. Anyone with those skills is good at analyzing malware."

Many in the world of infosec have dabbled in gray-area activity, as

I'd found in talking to them over the years; those curious teenage years had them lurking on malware sites, downloading and trying it out, modifying it or writing their own. Maybe it was for profit, or maybe just to show off their mad skills when they'd come back to brag in a malware-authors' forum. Either way it's how they learned: One terrific way to learn the ins and outs of the malware field and its murky characters is to participate. When I traveled across Europe in the mid-'00s interviewing people who wrote the code for viruses and worms, many were bright teenagers bored out of their minds in tiny towns, whose closest friends were screen names on forums. After they'd devise a new piece of code, they'd share it on the boards, and also often alert Microsoft or whichever tech company had a bug their malware exploited, so the company could patch it. They weren't malicious, and never released the malware "in the wild" themselves. But they also liked sharing their code, so it could easily wind up getting used pretty much anywhere, by someone else.

The bored, bright teenagers of today can wind up creating malware that they sell—or rent—for a profit. Visit dark websites devoted to malware, and behold for sale the many "phishing kits," which generate links you can send to a target pretending it's a "password reset" for, say, Instagram; if they click on it, the malware does "intelligence gathering," quietly scooping up email or documents and mailing them back to you. ("Spear phishing" is an even more ingenious tactic: The attacker sends a phishing email to you pretending it's from a known and trusted friend. The security firm Symantec finds this is the most common attack strategy, used by nearly three quarters of all intrusion groups.) The same goes for ransomware, which anyone can acquire for a few hundred dollars on dark-web markets; it's mostly used for petty crime, infecting small firms and extorting on average just over $500 in 2017. And ditto for botnets, which miscreants are finding it easier and easier to create these days, given the explosion of poorly secured smart home devices—thermostats, fridges, coffeepots, most of which are online with shoddy password protection, or none at all.

That's how the Mirai botnet got so big. It was the creation of a trio of young men—including a then 20-year-old former Rutgers computer science student named Paras Jha. Jha and his fellow bot farmers were avid players of Minecraft; they created Mirai by infecting thousands of Internet of Things devices, then used them to knock various Minecraft servers offline, sometimes to try and extort the owners in a sort of mafia-like protection racket. They also rented their bot farm for click-fraud schemes: Pay them $1,000 a day, for example, and they'd have their farmed devices each click on a page or ad on your website, generating you hefty profits from the advertiser, who'd naively think their ad was just super popular. Their click-fraud business netted them $180,000. And sometimes they'd just use their botnet to attack someone they didn't like, knocking their site down with a flood of traffic.

In forum posts, Jha came off as a moody, nihilistic youth who boasted about his coding skills. When the owner of a site that Jha had attacked explained that there were real-life consequences for these digital onslaughts, he replied with cynicism. "Well, I stopped caring about other people a long time ago," he wrote. "My life experience has always been get fucked over or fuck someone else over." The law eventually caught up with him. Brian Krebs, a prominent journalist who investigates the world of malware, spent months patiently rooting up Jha's identity (like all malware authors, he'd kept it a deep secret). After the authorities arrested Jha, he and his Mirai partners were sentenced five years of probation and 62.5 workweeks of community service; as it happens, they had already flipped and begun helping the FBI "on cybercrime and cybersecurity matters," as the sentencing memorandum noted. The line between white hats and black hats contains a lot of gray indeed.

Hutchins, it turns out, was alleged to have his own gray-hat side. At first, after halting WannaCry, he was feted as a classic white-hat infosec hacker. He went to Las Vegas to enjoy his first DEF CON, where he was besieged for photos with admirers: *Hey, it's the guy who stopped WannaCry!* But when he went to the airport to fly back to the UK, he was scooped up by FBI agents—and arrested for creating malware. A grand jury, days earlier, had charged him with having created and sold malware called Kronos, which scoops up banking details. It was like a police-force detective being arrested for allegedly running a criminal scheme on the side.

When Hutchins was released on bail, the infosec world buzzed with theories. Some argued he might be guilty; Krebs, the security journalist, reported on evidence that appeared to tie Hutchins to the release and sale of smaller, less-successful pieces of malware from years ago, when Hutchins was a teen. Others in the infosec community argued in his defense: Hell, *many* white-hat hackers tinkered with malware or intrusions as a teenager. That's how you learn. For his part, Hutchins pled not guilty to the charges over Kronos, and at the time I was writing this, it had not yet gone to trial.

He wryly agrees that bored teenagers can be a force for chaos. "With malware," he says with a laugh, "it's either the bored teenagers or organized cybercrime—there's nothing in between."

For all the chaos they can cause, the kids running botnets and writing malware aren't necessarily the truly big-name cybercriminals. The truly big fish, these days, are more often shadowy groups of coders who cooperate online to build extensive code bases and operations. Some are out for sheer profit, like Evgeniy Mikhailovich Bogachev, a programmer in Russia who created a botnet that targeted banks and netted about $100 million in profits. He's one of the FBI's most-wanted hackers right now.

Yet he's also currently living freely in the Russian town of Anapa, unmolested by local police. That's because the Russian government only selectively enforces its cybercrime laws. Their attitude appears to be, so long as the cybercriminals aren't targeting Russians, it's fine. Indeed, if a malware author creates a botnet that infects a lot of government computers in other countries—well, that might make them a valuable asset. So the cybercriminal might get told: *Hey, we* could *arrest you for hacking, unless you cooperate with us.* In this way, countries can acquire some prized hacking assets. This is indeed what seems to have happened with Bogachev, because at one point, white-hat infosec hackers discovered that when Bogachev's botnet infected a computer, it scanned the computer for files with text-strings like "top secret," indicating it was scouring for political secrets to feed back to the Kremlin.

This is, in fact, what a majority of cybercrime has become: "intelligence gathering," as infosec mavens put it. Most often, when an intruder hacks into your computer, they aren't looking to steal money, destroy data, or install ransomware. No, they're just trying to steal information—corporate emails, documents, plans (if it's a commercial enterprise), or useful government info (if it's a state).

It is, in essence, spy craft. More and more, cybercrime is connected with nation-state interests.

Today's big cybercrime groups themselves may not be directly run by national governments or their spy agencies, but they often appear to be at least working in communication with them. When the Russian hacking group Fancy Bear penetrated the Democratic National Committee—via the ever-popular trick of phishing, getting John Podesta to click on a bad link—the trove of DNC email soon appeared in Russian intelligence circles, and thence on WikiLeaks and other sites online. Meanwhile, Chinese state-sponsored hackers are busily penetrating US firms and government agencies: The Department of Defense blocks 36 million phishing attempts each day. And in every authoritarian country, government-sponsored hackers busily wage attacks on civil society advocates and pro-freedom groups—installing spyware on dissidents, the better to bust them. Work by Citizen Lab has found extensive spyware espionage conducted against Tibetan activists and humanitarians, to name just one example. And the US, of course, has its own phalanxes of government-employed hackers, busily developing malware and exploits with which to attack other countries. Indeed, the exploit that WannaCry employed was originally developed by the NSA, probably as a strategy they intended to use against another state—but it was released online when the NSA itself was hacked.

In a digital world, governments increasingly want to surveil and hoover up every last thing we do online. This has enormous, and troubling, civic implications. It's no wonder the hackers who work in that realm wind up in such regular conflict with the law. The cypherpunks are paranoid, sure—but the rest of us probably should be, too.

< Chapter 9 >

Cucumbers, Skynet, and Rise of AI

It started with the ancient Chinese board game Go and ended with cucumbers.

In the fall of 2015, we had another one of those Skynet-like moments when a form of artificial intelligence utterly destroys a human. In this case, it involved "AlphaGo"—software designed by DeepMind, a subsidiary of the Google empire—playing a wickedly great game of Go. To test their AI, DeepMind had arranged for it to play against Fan Hui, the European Go champion. It was no contest: The computer won 5 games out of 5. A few months later, AlphaGo fought Lee Sedol, an even more elite player—and again, AlphaGo dominated, 4 to 1.

AlphaGo was so good at the game partly because it incorporated "deep learning," a hot new neural-net technique that let the computer analyze millions of Go games and, on its own, build up a model of how the game worked; feed any board with Go positions into the model, and it could, in conjunction with a more traditional "Monte Carlo" algorithm, then predict a future move. Previous attempts at making

Go artificial-intelligence had been cruder—the creators would try to develop search algorithms that would rank and pick the best possible future moves. But that more-traditional form of programming doesn't work very well with Go, a game where the branching pathways of possibility dwarf even chess. There are considerably more possible Go games than there are atoms in the universe.

So AlphaGo's creators didn't go that route. They didn't sit around writing logic rules, like traditional programmers. Instead deep learning allowed AlphaGo to analyze 30 million positions from preexisting games and build up an extraordinarily sophisticated model of the game—so dense and convoluted the creators themselves could not tell you precisely how it works.

But work it did. AlphaGo was a master at the game, albeit in a somewhat alien fashion. It sometimes pulled off moves no human had ever before executed. In the second game against Sedol, during move 37, AlphaGo made a play that at first flummoxed the Go experts who observed the game, as *Wired* reported. The computer abandoned one group of stones, shifting its play to an entirely different part of the board.

"That's a very surprising move," as one Go expert observed. "I thought it was a mistake," said another. But after pondering the strategy, it hit them: Oh my God, that's a *gorgeous* move! "So beautiful," cooed Fan Hui, who was watching, fresh from his own AlphaGo defeat. "It's not a human move. I've never seen a human play this move." When AlphaGo made move 37, Sedol himself appeared stunned. He got up from the table and left the room, not returning for fifteen minutes. The next day, after his loss, Sedol admitted that move had rattled him. "Yesterday, I was surprised," he said. "But today I am speechless." Soon, newspapers and websites were full of breathless tales about AlphaGo.

Deep learning was becoming the hot new high-tech trend. It was also something that was becoming increasingly easy for everyday coders to try out.

That's because, at the same time that Google was unleashing AlphaGo on the world, it had also released TensorFlow, a free piece of software that greatly simplified the task of setting up your own neural net. Say you wanted to create your own system for recognizing the faces of the employees at your firm, so that the front door opened only for them. If you could get lots of different pictures of each employee, you could train a TensorFlow neural net to autorecognize each one and run it through a webcam at your front door. Within months, coders around the world were downloading TensorFlow and messing around with it.

One of them was Makoto Koike, a 37-year-old Japanese computer engineer. He'd spent years designing software for the country's auto industry. But his parents—who were farmers in Kosai, a coastal town on the south edge of Japan—were getting older, so he moved back home to be with them and help out on the farm.

The elder Koikes were cucumber farmers, and in Japan, consumers greatly prize "thorny" cucumbers that are straight, vividly colored, and still sport many barbed thorns on the surface. As a result, Koikes' mother would spend up to eight hours a day combing through the latest cucumber harvest, sorting them into different categories: The top-notch ones go to wholesalers, and the lesser ones to smaller local vegetable stands. There were nine different categories, and it amazed Koike how hard it was to spot the differences. He spent weeks learning how to do it himself accurately. "You can't just hire part-time workers during the busiest period. I myself only recently learned to sort cucumbers well," Koike observed.

Koike had read about AlphaGo, and it got him interested in deep learning: *Huh, this machine-learning stuff is becoming awfully good.* Then he heard about TensorFlow, and a light bulb went off.

Maybe he could craft his own AI to . . . sort cucumbers?

The first step: taking thousands of pictures of the different cucumbers. Koike knew he'd need a nice big data set with which to train a neural net, letting it learn the distinctions between a curved, warped cucumber (bad) and a slender, straight one (good). It took him about three days to train a model, and then he'd test it by showing it a new, freshly picked cucumber to see whether it could identify if it were good or not. Then he'd tweak and retrain the model, over and over again, to make it better. After a few weeks, his AI could recognize superb cucumbers with an 80 percent accuracy rate, he told me.

Then he went whole hog and built a robot to automatically sort cucumbers. It's the size of a filing cabinet; Koike puts a cucumber inside the top of the robot, where three cameras take pictures of it from above, the side, and at one end. Once the AI has rendered its verdict, it ejects the cucumber onto a conveyor belt, and a robot arm pushes the vegetable into the appropriate box.

Koike's parents were amused, but not quite as impressed. They're long-standing experts in cucumber sorting, and much faster than their son, so an 80 percent accuracy rate wasn't good enough to fully replace their human judgment. ("Needs improvement," his mother told him.) But Koike continues to beaver away. He figures that if he keeps on feeding the system better data—with higher-resolution pictures—and starts using superfast cloud computers to do the analysis instead of his slower PC, he could eventually get the robot to be as good as his mother. He's also been working on a different neural net that can automatically recognize a cucumber hanging on the vine. "I would like to make an automatic harvesting robot," he tells me. He showed me a video of the neural net in action: As he walks along the cucumber patch holding a camera up, you can see the AI on-screen drawing red rectangles around each cucumber.

His dream is to get good enough at these tasks that his robots can truly automate them, working as well as his parents. It'd be good for their farm, he says; they'd be freed up to do things that computers as yet cannot—such as the more creative work of tending the cucumbers, and horticulturally experimenting to render them ever more crisp and flavorful.

"Farmers want to focus and spend their time on growing delicious vegetables," he told Google, when they discovered his TensorFlow experiment and called him up. "I'd like to automate the sorting tasks before taking the farm business over from my parents."

For years, coders have been programming computers to do our repetitive actions. Now they're automating our repetitive *thoughts.*

For as long as we've had computers, coders have dreamed of creating software that behaves like a human. When you're dealing with machines that exhibit such seemingly lifelike behavior—that can perform "thinking" tasks—it's probably inevitable that you start to wonder, *Hmm*: Could a computer learn the way we do? Could we make something that, like a child, absorbs new knowledge on its own—something we could talk to, and it would understand us? Does the human brain work like a computer? Is human thought and language merely composed of lots of *if-then* statements?

Back in the summer of 1956, the world's brightest computer thinkers decided to tackle the challenge. Over a dozen assembled at Dartmouth College, with the goal of figuring out "artificial intelligence"—a term they adopted. "An attempt will be made to find how to make machines use language, form abstractions and concepts, solve kinds of problems now reserved for humans, and improve themselves," they wrote, in a heady prospectus. "We think that a significant advance can be made in one or more of these problems if a carefully selected group of scientists work on it together for a summer." How hard could it be?

Insanely hard. Despite having multiple bona fide geniuses under one roof, the people at Dartmouth didn't advance AI much at all.

Indeed, in the years to come, coders would come to realize that the Dartmouth crew vastly underestimated how complicated it would be to get machines to "think."

That's partly because computers are great at following crisp, clear rules. But the fabric of human thought is incredibly complex and gnarly. Consider, just for starters, the challenge of making a computer respond to human language. If you sat down to program a chatbot, you could try to simply write a response for every possible thing someone might say to it. If a stranger types "Hello," the chatbot responds by picking from a list of responses you've coded—like "Hi!" or "How are you doing?" At the time of this writing, this is how nearly all the chatbots work. Which is why you've probably noticed the problem with this approach: Doing this brute-force approach, a bot maker needs to anticipate *every possible thing* a stranger might say to her bot. If someone says " 'Zup" and you haven't programmed it to recognize that " 'zup" is synonymous with "hello," the bot grinds to a halt. Coders can get pretty in the weeds, writing thousands of responses and composing modules that try to logically analyze the grammar of a stranger's queries, the better to figure out what someone was saying to the bot. But this is why the chatbots you encounter at shopping sites quickly break down and become useless.

Building AI this way is like "boiling the ocean," as Dave Ferrucci, a computer scientist who led the team creating Watson, the *Jeopardy!*-playing computer, once described it to me. Computer programs break when they reach an "edge case," when the user tries to do something that the coder never anticipated. And human interactions are *filled* with edge cases.

Even harder is the problem of learning. Even if you could get a chatbot to talk, how do you get it to learn something new on its own? If you tell it "the economy in Greece is falling apart because of the euro," how exactly is the AI supposed to make sense of that? That sentence contains a ton of "implied knowledge": To make sense of it, you have to know that Greece is a country, that the euro is a currency adopted by Greece, that a currency has a big impact on a country's economy. Indeed, you need to know even more primitive concepts: What's a "country"? What's an "economy"? What does "falling apart" mean? This is what's sometimes referred to as the problem of "common sense." Our human ability to interact with the world is based on a ton of common-sense knowledge that we gradually absorb as we grow up, and in school.

So the dream of a grand, self-learning AI quickly crashed. It produced an "AI winter"—a period where computer scientists and investors felt so burned by AI hype that they wouldn't touch the field. It was too dangerous; you'd look like an idiot. Indeed, from the '60s to the '00s, AI went through several cycles of hype—a "summer," when the money flooded in and people got excited about new techniques—only to overpromise and underdeliver, producing yet another winter.

One of the few forms of AI that reliably worked and that made money was extremely simple and modest: "expert systems." These were programs designed to automate some pretty limited types of decision making. Maybe a bank wanted a tool to help bankers quickly decide whether a customer should get a mortgage. So the coders would talk to all the mortgage officers, figuring out what sort of expert knowledge they brought to bear in examining a mortgage application. Maybe they'd design a statistical analysis that looked at years of mortgages, to help see what types of borrowers tended to pay back in full. Then you'd program all that expert knowledge into a good-old-fashioned *if-then* piece of software: If the applicant is older, and if they have a high credit rating, and if they have an income of a certain amount, then they'd be approved, and so on. Expert systems were AI only in the most limited sense of the term. They weren't able to learn anything new; they weren't sitting around pondering Kant or having conversations.

As computers got more powerful, and crunching huge amounts of data got cheaper, some coders became experts in doing clever stats crunches of "big data." By sifting through huge piles of information, computers could find useful trendlines that would have been hard-to-impossible for a human to spy: "machine learning." The machines weren't "learning" in the sense of Skynet. They weren't absorbing new knowledge into a growing robot brain. Machine learning was just about "learning" a single cool trend.

Still, these tricks could sometimes be delightfully prescient. One Barcelona high-tech firm I met back in 2003 had, for example, taken millions of pop songs and broken them down into components: their beats per minute, whether they were in a major or minor key, and so on. Then they used a machine-learning algorithm to cluster the hits together. This produced, in effect, a hit-prediction machine. They could take any new song and see whether it was mathematically similar to previous chart toppers.

It worked surprisingly well. Their moment of fame came when they analyzed the first album by the as-yet-unfamous Norah Jones and predicted that nearly every track would be a monster hit.

There was one form of AI, though, that *was* eerily humanlike. It seemed to offer a way for computers to truly learn on their own.

That was neural nets. These were systems that were modeled loosely on the way human brains seem to work. A neural net would have several "layers"; each layer was composed of a bunch of nodes. Much as neurons in the brain are connected to each other, the nodes in the software are connected to each other.

And the way you teach the neural net to work is by *training* it. Let's say you wanted to create a neural net that can recognize pictures of sunflowers. You'd start by gathering lots of digital pictures of sunflowers. The first layer would be triggered by the various pixels in the picture, and its neurons would either fire or not fire and send those signals over to the next layer. That layer would do the same thing: Each neuron would decide whether or not to react and send its signals—firing, not firing—on to the *next* layer. And so on and so on. It would, in a way, be slowly distilling the decision—fire or not fire—down to a simple yes or no. When it gets to the final layer, there's only a single decision: Yes or no? Is it a sunflower or not?

But how does this weird structure actually know that the pixels represent a sunflower? At first, it doesn't. Each neuron is just guessing blindly. The neural net doesn't know anything about what a sunflower looks like. But after it has rendered its guess—*Yes, a sunflower! No, not a sunflower!*—you check whether the guess was right or wrong. Then you feed that information (*Wrong! Right!*) back into the neural net, a process known as "backpropagation." The neural-net software uses that information to strengthen or weaken the correction between neurons. Those that contributed to a correct guess would get strengthened, and those that contributed to a wrong guess would be weakened. Eventually, after enough training—hundreds, thousands, or millions of passes—the neural net can become amazingly accurate. It can recognize any of the sunflower pictures and say *yes*; if you show it a picture of a church, it'll say *no*.

And if you train it really well, it won't just recognize a picture of a sunflower it's already seen. It'll be able to recognize a sunflower in a *new* picture, one it's never seen. It will, in other words, seem to extract some essence of sunflowerness. Best of all, it doesn't require any sort of brute-force *if-then* coding. The neural-net math just settles on this pattern-matching ability on its own.

The concept of a neural net was first hypothesized in the '50s. By the '80s, the French scientist Yann LeCun pushed it to new heights—showing that a neural net could even recognize handwritten letters.

The problem was that neural nets, back then, weren't very practical. They required faster processors and more memory than was affordable for most software developers. Worse, training a neural net required lots of training data. Training one to recognize sunflowers—well, sure, but you'd ideally want to have thousands or even millions of different pictures of sunflowers for training. And where are you going to get those? Digital cameras were a decade or two away.

Neural nets seemed interesting, and got used in a few places. Banks used LeCun's work to create neural nets that could automatically read checks; some voice-recognition companies used neural nets to create the first clunky systems that could slowly, painstakingly recognize speech and type it out.

But mostly, computer scientists regarded neural nets as another false promise of AI. After some small bursts of excitement in the '80s, neural nets lapsed into their own "AI winter."

"Everyone was saying, 'This is a lost field,'" Hans-Christian Boos told me; he was a young grad student fascinated by this crazy technique. But his peers told him to stay away; nothing would ever come of neural nets.

They were wrong.

Jeff Dean was one of the coders who discovered just how wrong they were.

Dean is the head of Google AI, an AI division at the tech giant. A tall, wiry 50-year-old, Dean was an early hire in 1999. "We were wedged in a little area above a T-Mobile store in downtown Palo Alto," he tells me, when we met in Google's Silicon Valley headquarters. "It was a good place." That was back when cofounder Sergey Brin still zipped around on Rollerblades and the search engine struggled to manage growth. Servers—which Googlers were still building by hand themselves—were a cluster of cobbled-together cut-rate stuff. At night they'd run "the crawl," their algorithm that painstakingly collected a copy of every web page it could find online; the process took hours and would frequently crash, whereupon all the Googlers' pagers would light up, forcing everyone to race into the office at midnight to reboot the crawl. Google worked, but in the early days, it could be a mess beneath the hood. "It was exciting because I think every week we were trying not to melt," Dean says.

He became famous as one of the company's genuine 10Xers. Dean was a scale whisperer: He deeply intuited the hardware specs of Google's servers, of the internet itself; he can tell you precisely how long it takes for a packet of data to go from Amsterdam to California and back (about 150 milliseconds). Coupled with deep coding skills, he was able to design systems that ran blisteringly fast and reliably, crucial for such a fast-growing firm. One of the most famous products he pioneered (along with his colleague Sanjay Ghemawat) was MapReduce, a piece of software that let Google engineers process huge data sets on a cluster of processors; another was Spanner, a world-spanning database. Other Google coders have come to so revere him that, over the years, they've assembled a long list of "Jeff Dean Facts," modeled on the satirically worshipful "Chuck Norris Facts." (One factoid: "The speed of light in a vacuum used to be about 35 mph. Then Jeff Dean spent a weekend optimizing physics." And: "When Jeff Dean has an ergonomic evaluation, it is for the protection of his keyboard." Or, if you're nerdier: "Compilers don't warn Jeff Dean. Jeff Dean warns compilers.")

Neural nets had long fascinated Dean. They seemed to have such potential for helping Google. After all, neural nets are able to automatically recognize and correlate subtle patterns in information—which is, in a nutshell, Google's corporate mandate. Dean had coded a neural net for his undergraduate thesis back in the late '80s, but all he could tackle were "very teeny problems." The comparatively low-powered computers of the day could only handle small neural nets. They needed far, far more processing power, even more than he could guess at the time. "It turned out like we didn't need sixty times as much compute," Dean says. "We needed like a *million* times as much compute."

As the '00s wore on, though, Dean and his colleagues in AI saw that these limits on neural nets were evaporating. First, there was way more real-world data to analyze. Thanks to the internet, people were now writing billions of words online and publishing millions of photos. Say you wanted lots of English sentences, to train an AI to recognize language. Just gather up all of Wikipedia, or if you're Google, crawl the text of Google News. Even better, computers in the '00s were running faster and faster, at cheaper and cheaper prices. You could now create neural nets with many layers, or even dozens: "deep learning," as it's called, because of how many layers are stacked up.

By 2012, the field had a seismic breakthrough. Up at the University of Toronto, the British computer scientist Geoff Hinton had been beavering away for two decades on improving neural networks. That year he and a team of students showed off the most impressive neural net yet—by soundly beating competitors at an annual AI shootout. The ImageNet challenge, as it's known, is an annual competition among AI researchers to see whose system is best at recognizing images. That year, Hinton's deep-learning neural net got only 15.3 percent of the images wrong. The next-best competitor had an error rate almost twice as high, of 26.2 percent. It was an AI moon shot.

Another of Dean's colleagues was equally impressed: a Stanford professor named Andrew Ng, then a part-time consultant for Google X. Like Dean, he'd tinkered with neural code while young, then set it aside during deep learning's long AI winter. But in 2011, over a dinner with Dean, the two got excited about the idea of using Google's enormous phalanxes of computers to see how powerful a neural net they could build.

"When people think of AI, they think of *sentience*," Ng tells me. "But when I think of AI, I think of *automation*. That's the value of AI." Or as he once put it in a tweet: "Pretty much anything that a normal human can do in <1 sec, we can now automate with AI."

As an initial experiment, Dean, Ng, and a team of engineers stitched together 16,000 Google processors to run one single, massive neural net for recognizing images. Then they turned it loose on millions of YouTube videos, to see what patterns it could find. All on its own, the neural net learned to recognize . . . cats.

"We never told it, 'this is a cat,'" Dean recalls. It just, all on its own, began to identify things that had a cat's pointy ears, a cat-shaped face: true self-teaching AI. At first, the result surprised them, though on reflection it made sense: There are a *lot* of cats on YouTube, so any self-learning algorithm told to pick out salient features that recur over and over again might, in essence, discover humanity's online obsession with felines. Still, it was a spookily humanlike bit of reasoning. The Terminator was coming to life, and it could grasp the concept of cats!

Google soon began throwing enormous resources at deep learning, developing its abilities and integrating it into as many products as possible. They trained deep-learning nets on language pairings—showing it, say, all the Canadian parliamentary proceedings that were translated into both English and French, or their own collections of crowd-sourced translations. When they were done, Google Translate became, in a single night, remarkably better—so much improved that Japanese scholars were marveling at the machine's deft ability to translate complex literary passages between their language and English.

A few short years later, deep learning had swept the world of software. Companies everywhere were rushing to incorporate it into their services. Ng was snapped up by Baidu, the Chinese search giant, as it frantically sought to catch up to Google's AI wave. Facebook engineers had long been using many different styles of machine learning to help recognize faces in photos, filter stories in the News Feed, and predict whether users would click on an ad; it set up an experimental AI research lab, and soon Facebook was producing a deep-learning model that could recognize faces with 97.35 percent accuracy, 27 percent better than the state of the art ("closely approaching human-level performance," as they noted.) Self-driving car programs around the world seized on deep learning to help teach cars to navigate roads. Uber uses it to predict where new rides will emerge. The National Cancer Institute is working on using it to detect cancer in CT scans. It's even seeping into the world of culture: ByteDance, one of China's hugest firms, uses neural nets to help curate news stories in its Toutiao news app, so successfully that users spend more than 74 minutes a day using it. A few years ago, Kai-Fu Lee, who invented the first "plain-talk speech recognition" and went on to be a veteran of Apple, Microsoft, and Google put all his new financial investment decisions in the hands of AI. "I don't trade with humans anymore" for those things, he told me.

And, as with any software craze, the hunt for warm bodies exploded. Silicon Valley and China in particular grew ravenous for coders fluent in deep learning, with salaries reaching well into the six figures for anyone adept at teaching computers to see, hear, read, and predict.

What type of coder gets obsessed with making AI?

As you'd imagine, many were entranced by the savant robots of sci-fi. Dave Ferrucci—the computer scientist who led an IBM team to create Watson—hankered to make a machine that could converse like the one on the *Star Trek Enterprise* system. "It understands what you're asking and provides just the right chunk of response that you needed," he told me. "When is the computer going to get to a point where the computer knows how to talk to you? That's my question!" Others come to neural nets via neuroscience, when they start wondering whether these crazy neural nets actually mimic how the brain works. Some even arrive at AI from the world of art, dreaming about making machines that generate literature and pictures. (One of the more talented AI coders I know spends his time training neural nets on TV and movie scripts, autogenerating new scripts, then shooting the best ones.)

All coders adore the "Hello, World!" moment, but with AI, the romance is decidedly Promethean. Matt Zeiler was a young engineering science student at the University of Toronto when one of Geoff Hinton's students showed him a video of a flickering candle flame and told him it had been automatically generated by a neural net.

"I was like, 'Holy crap!'" Zeiler told me. The flame was so freakily lifelike that he took Hinton's course and did his undergraduate thesis with Hinton, intent on absorbing deep learning. Zeiler did his PhD at NYU, interning at Google for two summers around the time that Hinton's famous 2012 paper created the deep-learning explosion. Zeiler was hooked on the field, and good at it; while interning at Google he made an AI to recognize house numbers in images. Even before he graduated, he turned down job offers from Facebook, Google, Microsoft, and Apple, and instead set about crafting his own cutting-edge visual AI. He holed up in his NYC apartment with a pimped-out personal computer—of the sort a hard-core video gamer might build—crunching models. (The PC generated so much heat he had to open the windows in the winter.) Soon, Zeiler beat Hinton's record at the visual-recognition annual competition, and Zeiler launched a firm, taking his visual AI to market, available for firms to use in their services.

As Zeiler discovered, making neural nets is a strange practice. It's not like normal coding, where you build a clockwork mechanism, an elegantly designed machine that does precisely what it's told to do. In regular coding, part of the joy is in the linearity of the code: You can walk through its palace of logic, following the routes in your mind, marveling at its intricate details (or grimacing at the overtangled design). Either way, it is, in theory, knowable. Every line was written by a human.

Modern neural-net building is quite different.

It's more like the relationship of gardeners to their gardens. Are the beans suddenly not thriving? Are the tomatoes oddly chewy? If so, gardeners will tinker with the soil or change the spacing between the plants. Maybe they'll set the beans up in an area with more sunlight? With *less*? Aha, that did it: They're thriving now! Gardening is a field where, to get good at it, you need to engage in a ton of experimentation and hard-won experience. New gardeners usually find their first crops die or grow anemically. But eventually, with enough trial and error (and observing the successes of other farmers) the experienced gardener builds up knowledge and hard-to-articulate intuition about what works and what doesn't. If you took them to visit an entirely new garden, with entirely different types of soil and sunlight, they could, much more quickly than a newbie, figure out what to plant and how to get it to thrive.

Training neural nets is a bit like that. Sure, it requires coding skills. Indeed, the early pioneers were deep hackers who needed to grasp the internals of CPUs and RAM, the better to squeeze as much "compute" as possible out of their processors. But these days companies like Google freely release their code for building and training neural nets, so the average start-up no longer needs to build the code from the ground up. They can just hit the ground running with Google's code.

So what neural-net coders really do, in many ways, is gather data, experiment, tweak, and pray. Easily the biggest part of their work is simply assembling the sample data to train their neural net. Want to recognize possible tumors in CT scans of lungs? Well, you need to collect CT scans that have been patiently labeled by real doctors—*This one has a tumor in the top-right corner, this one has no tumor.* And you want millions upon millions, ideally. So neural-net coders become absolutely obsessives about gathering data. They always have their octopus tentacles extended, grasping for more.

Justin Johnson, a researcher who works in Facebook's AI lab, recently trained a neural net that was shockingly good at "visual question answering." It can look at a picture of geometric objects—colored cubes, spheres, and cylinders—and answer such questions as "Is the green block to the right of the yellow sphere?" But to get that level of success, he tells me, he had to generate 100,000 images with these sorts of objects—and then pay hundreds of humans to generate questions about what was in each photo. He spent almost a year of work to get all that data, which included teaching himself how to make websites so he could make online forms on which humans could type in their answers. This is the central truth of much of today's neural-net AI: It requires extracting information from real live humans, so the AI has something to learn from.

"It was basically one and a half years of basically learning to be a full-fledged website developer just so I could gather the data for training," he tells me.

Once you've got the data, training the model can be puzzling. It requires tinkering with the parameters—how many layers to use? How many neurons on each layer? What type of backpropagation process to use? Johnson has lots of experience, having built visual AI at Facebook and Google. But he can still be confused when his neural-net model isn't learning, and he'll discover that small alterations in the model can have huge effects. The day we spoke, he'd spent a month banging his head against the wall tinkering with a nonworking visual model. Then one day he was talking to a colleague about the fact that the model used "batch normalization," a mathematical trick that visual-AI folks very frequently use. Normally batch normalization is crucial to making things work. But this time, the colleague suggested, it might be the very thing causing unreliable results. When Johnson took batch normalization out, sure enough, the model suddenly started learning. This is why experience in experimenting with models—gradually seeing what works, what doesn't—is so important. As Johnson points out, when you're tweaking a model, "the space of possible choices is so unspeakably vast."

Some of the best deep-learning model trainers worry that there's still too much mystery about why the models work so well sometimes, yet in other cases don't. Sure, you can tell *if* a model is working—it's accurately recognizing pictures of pedestrians, maybe 90 percent of the time! But you can't always completely explain *why* it's suddenly doing so, or give precise advice for someone training a neural-net model on a very different task, such as translating language.

This tinkering, experimental vibe unsettles a lot of traditional coders. They like making things that are linear and predictable. If it works, you can generally explain why. "In computer science, the poster child is the person who thinks deterministically," says Pedro Domingos, an AI pioneer at the University of Washington. "You have to get everything working with no bugs. Every comma is in place. If you aren't OCD, you can't do it." But in machine learning it's the opposite. You have to deal with uncertainty, weirdness. You *guide* the system toward doing what it's supposed to do, like herding the cats of cognition. Maybe you'll get them where you want; maybe you won't. It's like a point my friend Hilary Mason, a top data and machine-learning scientist, made about data science in the *Harvard Business Review*: "At the outset of a data science project, you don't know if it's going to work. At the outset of a software engineering project, you know it's going to work."

On top of that, there's the black-box problem. Once a neural net has been trained and you're recognizing those cat photos, great! But if you ask the coder who built it, "How is this thing working?" they'll shrug. They don't really know. The neural net has painstakingly adjusted the weights of neurons in such subtle ways that its logic quickly becomes inscrutable to humans. It is "an ocean of math," as my *Wired* colleague Jason Tanz beautifully put it. This can have unsettling implications for the use of neural nets in everyday life (more on this later); we're building AI systems the workings of which we can't fully understand? But it's another barrier to entry; many coders are suspicious of this sort of engineering.

Domingos suspects this is why Google beat Microsoft when it came to making a search engine. Microsoft loved making precise, logical software: You hit "control+I," and Microsoft Word italicizes your text, every single time. But Google was about sorting the internet. From the get-go they were using statistics to make an educated guess about what a user wanted. They'd never be "perfect" at that task, because there is no such thing as a perfect search-engine response. It's always a subjective guess. So when deep learning came along, Google was culturally much quicker to realize how that might be useful. "It's a frame of mind—it's an aesthetic," Domingos says. That experimental mind-set—the need to tinker away as you build a neural net, trying hunches and discarding them, until, hey, things work—makes deep learning "a black art."

Domingos sees tons of kids flooding into the field now. They're drawn in by the fact that AI is hot, and the big tech firms are tossing around deranged salaries. But he thinks comparatively fewer coders are suited to do this work at a high level. Unlike most coding, which requires very little math at all, hard-core deep learners need to be fluent in linear algebra and statistics. Like cryptography—which involves scrambling and unscrambling messages—machine learning attracts the kids who were diehard math heads, who sit around envisioning multidimensional vectors just for fun. Certainly any regular coder can download TensorFlow and train a model; but to really innovate in deep learning? It's mostly the province of trained PhDs wielding Jedi-class math.

Some old-school programmers are unsettled by the rise of deep learning. Gathering data, cleaning it up, training models, and experimenting to find what works? This isn't software engineering as they knew it. This isn't the mental carpentry they signed up for. "I got into computer science when I was very young, and I loved it because I could disappear in the world of the computer," Andy Rubin, the creator of the Android phone-operating system who now invests in machine-learning start-ups, told *Wired*. "It was a clean slate, a blank canvas, and I could create something from scratch. It gave me full control of a world that I played in for many, many years." The idea that you're now just tweaking a model, training, and retraining it until it suddenly works, felt—to him—oddly sad.

Back in the summer of 2015, Jacky Alciné discovered some of the problems deep-learning AI can cause in everyday life.

Alciné is a freelance web developer who lives in Brooklyn, and that night he was at home, idly watching the BET Awards on TV while dorking around on his laptop. He checked Twitter, then wound up opening his Google Photos account.

He saw something new. Google had just rolled out a new autotagging feature. All of Alciné's photos now sported a tag describing the main object of the photo, like the word *bikes* beneath a picture of a bike, *planes* beneath a plane. As he scrolled around the photos, he was impressed. Google's AI had even slapped the label *graduation* on a snapshot of his brother in a tasseled cap and gown.

But then Alciné scrolled over to a picture of himself and a friend, in a selfie they'd taken at an outdoor concert: She looms close in the view, while he's peering, smiling, over her right shoulder. Alciné is African American, and so is his friend. And the label that Google Photos had generated?

"Gorillas." It wasn't just that single photo, either. Over fifty snapshots of the two from that day had been identified as "gorillas."

Cutting-edge AI had managed to settle upon one of the oldest, vilest racial epithets, "something that black people have been called for centuries," as he later told WNYC. "Of all derogatory terms to use, *that one* came up." Google apologized for the error and a spokesperson said they were "appalled" at the performance of their AI.

But why couldn't Google's AI recognize an African American face? Very likely because it hadn't been trained on enough of them. Most data sets of photos that coders in the West use for training face-recognition are heavily white, so the neural nets easily learn to make nuanced recognitions of white people—but they only develop a hazy sense of what black people look like. (Conversely, algorithms trained in China, Japan, and South Korea struggle to recognize Caucasian faces, but work well with East Asian ones.) And since Google's technical workforce is only 2 percent black, I suspect it was also less likely any engineer on staff would have noticed the problem while demoing the AI on their own photos.

Nor is Google the only firm to have stumbled into this problem of being blind to black faces. When the African American coder Joy Buolamwini was a grad student trying to build a robot that played peekaboo, she used a widely used face-recognition AI—and discovered it couldn't see her face. It could only play peekaboo with white people.

As the field of AI matures, these problems of bias in AI are cropping up repeatedly. Machine-learning coders may be creating machines that can learn from the world and make decisions. But that can mean they learn more than just facts: They learn bigotry as well.

These problems aren't cropping up only in visual AI; they seem to appear anywhere coders train deep learning on real-world data.

Recently, Robyn Speer—the cofounder and chief science officer of Luminoso, a machine-learning firm—documented how texual AI can learn some noxious biases when it's trained using the language that people use everyday. Speer has for years worked with "word embeddings," an AI technique for representing what words mean. It works like this: You start off by taking tons of examples of texts, and use machine learning to turn each word into a "vector"—a concept that lets you see how individual words are related to each other mathematically. Many AI researchers over the years have released freely available word embeddings; Google analyzed oodles of Google News stories to produce its tool "Word2vec," and Stanford AI scientists created one called GloVe. Transforming words into vectors can let you perform some very cool tricks. For example, the mathematical relationship between the vectors for the words *Paris* and *France* turns out to be the same as for *Tokyo* and *Japan* or for *Toronto* and *Canada* or, indeed, for *any* pairing of a city and its host country. That allows a coder to use something like Word2vec or GloVe to quickly build some very powerful AI. You could make a web app that, if told that you lived in Rome, would automatically know that you live in Italy. These tools, in other words, help computers grasp the meaning of a sentence. It's so useful that when Google released Word2vec for free online, developers began excitedly using it to build search engines or any apps that have to figure out human language.

But Speer documented something unsettling about the types of associations these standard word embeddings can help unearth. She wrote an algorithm that employed these embeddings to analyze online restaurant reviews and automatically categorize their sentiment. Was the review positive? Negative? Neutral?

What she found was that the algorithm gave Mexican restaurants lower rankings. What was going on? Had people left terrible reviews for Mexican restaurants? She double-checked the reviews themselves but found they weren't unusually harsh. People weren't actually describing the Mexican joints any worse, on average, than—say—Italian restaurants.

No, the problem was that the vector for the word *Mexican* was itself *inherently negative*. That's because the embeddings had been trained on the web, and the web contains lots of English-language stories and posts implying, or stating outright, that Mexicans are bad—that they're associated with crime or illegal immigration. Humans are racist, so, in American media, they tend to write about Mexicans using often-racist associations. And machine learning is great at picking up those correlations. "Stereotypes and prejudices are baked into what the computer believes to be the meanings of words. To put it bluntly, the computer learns to be sexist and racist because it learns from what people say," as Speer later wrote about her epiphany.

"It's kind of an insidious problem," Speer told me when I spoke to her. The problems don't stop at race, either. Microsoft Research scientists analyzed Google's Word2vec and found that it had also learned plenty of sexism, too. The word *he* was associated with words like *boss*, *philosopher*, or *architect*, while *she* was associated with *socialite*, *receptionist*, and *librarian*. In one particularly notable pairing, *man* was associated with *computer programmer* in the same way that *woman* was associated with *homemaker*.

This would be unsettling enough if it were just about restaurant reviews. But machine learning and neural nets are increasingly taking over everyday human decision-making—so these biases are having real-world effects.

One study by Carnegie Mellon University found that on job-listing sites, men were being shown six times as many ads as women were for high-paying jobs of $200,000 and up. In 2016, a reporter for the *Guardian* found that if you type "are jews" into Google's search bar, the top recommendation in the autocomplete AI feature was "Are Jews evil?" "It's the equivalent of going into a library and asking a librarian about Judaism and being handed ten books of hate," as Danny Sullivan, the founder of Search Engine Watch, told the reporter. (Within a few hours of hearing about this, Google tweaked its code and removed that autosuggestion.)

Perhaps the most unsettling effects are in the justice system. In recent years, various AI systems have been rolled out in law enforcement, aimed at helping overloaded judges determine a defendant's likelihood of reoffending. But when they've been analyzed, these systems appear to be riddled with racial biases. One well-known system, COMPAS—made by the firm Northpointe—was studied by the news agency *ProPublica*, which looked at 7,000 defendants who had been run through COMPAS. *ProPublica* found that COMPAS was almost twice as likely to label a black defendant as getting a high-risk recidivist score than a white defendant, even when they controlled for these defendants' prior crimes, age, and gender. And getting a bad COMPAS score matters. Judges across the country use these scores to help figure out whether a defendant is a good candidate for probation or for a treatment. Your COMPAS score can thus determine whether you wind up in a cell while awaiting trial—and right now, it seems almost certainly to be sending more black people to cells.

Why was the system biased that way? It's impossible for outsiders to say: The company does not publish its source code or explain in detail how its system makes predictions. But the odds are, again, that it's reflecting the bias of preexisting data. For decades in the US, police have targeted black citizens for much more aggressive enforcement. They're far more likely than whites to be arrested for smaller infractions—like smoking or carrying small amounts of marijuana or for driving with a broken taillight. Felonies and convictions, in other words, aren't impartially meted out. This means that any machine-learning system trained on preexisting crime data is liable to see the disproportionate rates of convictions of black citizens and conclude that, well, black people are *inherently* predisposed to crime. And of course, this can turn into a nastily self-reinforcing loop. A criminal-justice algorithm trained on racist policing will wind up fingering black citizens as more dangerous, criminalizing ever more black citizens—and their rap sheets thus become more "data" for future machine learning to study.

This is a curious philosophical challenge that machine learning poses for the world of justice. Crime AI is trained to predict future events—someone's chance of being a criminal—based on their past, and the overall past trends of society. But this risks becoming a static view of human nature. "Bad" people will be bad people forever; "good," good.

"Big data processes codify the past. They do not invent the future," as Cathy O'Neil, a mathematician, writes in her book *Weapons of Math Destruction*. Actual humans perceive their lives as having some measure of free will—they can decide to reform themselves, and it's not clear that machine learning could predict someone suddenly deciding, hey, I'm going to clean up my act. This is precisely what one defendant told *ProPublica*, after COMPAS scored him as likely to offend again, based on his previous rap sheet. He felt the system hadn't taken into account the steps he'd been taking in recent years to move away from his life of crime: He'd converted to Christianity, tried to connect more with his son, and worked on quitting drugs. "Not that I'm innocent," he said, "but I just believe people do change." A human judge might be amenable to taking a wider view, a more individual view, of a defendant. A trained algorithm can't.

Worse, really, is the probabilistic nature of deep-learning AI. It'll correctly deduce that there's a cat in the photo 90 percent of the time; it'll predict that downtown Wall Street is going to suddenly need a ton of Uber cars, and it'll have—let's just say—an 88 percent chance of being right. Those probabilities are fine when you're tagging photos or sending cars to possible hot spots. But that's little comfort when the stakes are higher: judging someone's chance of criminal offense or whether they ought to be shown a juicy job offer. You don't want a judge that has a known 20 percent error rate.

"From the individual perspective, it's very important that their individual case is dealt with respectfully and accurately," as O'Neil tells me. "But from the perspective of the people building the algorithm, that's completely not on their mind." Software engineers are thinking about efficiency, and "when you're thinking about efficiency you don't have to be perfect. You just have to be slightly more efficient" than the system you're replacing. State politicians may regard COMPAS as usefully speeding up the overloaded courts; defendants, in contrast, might prefer the solution to be hire more judges so they're not so overloaded.

But AI has become hot not just because it promises efficiency. It's also sold as being more "objective" than humans. An AI that's making decisions about mortgages or cat photos or defendants won't get tired or have its attention drift. It was trained on the data, and it's impartially applying the trends it's learned, right? It's true that in theory, an AI could be much more reliable than a scattershot human. Indeed, studies have found that judges hand down harsher verdicts just before they take a lunch break because they're hungry and depleted. That's patently unjust, and a piece of machine learning certainly wouldn't have *that* frailty. This is the positive argument for making machine learning part of everyday decisions. Done correctly, it could help us avoid some of our own human mistakes.

But you'd only get these benefits if you built AI that wasn't biased.

... ...

Henry Gan faced that problem in 2017, when he needed to coax his machine-learning system to recognize Asian faces.

Gan is a coder who works for Gfycat, an online service where people make and share animated GIFs. It's wildly popular among fans of Korean pop—K-pop, as it's known—so users are constantly putting up pictures of ultrasynchronized dancing K-pop stars. To make it easier for people to search for their favorite GIFs, Gan decided to train an AI to automatically identify the person in each picture and tag their name. He used an open source facial-recognition software based off work by Microsoft, then trained it on huge data sets—millions of faces—that are freely available from several universities. He also trained it on photos of staff members at Gfycat.

That's how he noticed the problems. When he tested his visual system on staff, it did fine with the white employees—but it couldn't reliably identify the Asian staff, including himself.

"We thought, maybe that's a blip," Gan tells me. So he tested the system on famous Asian people instead. No dice: It couldn't identify well-known Asian actors like Lucy Liu or Constance Wu. And worst of all, "it was bad with K-Pop! It was getting all these girls wrong. It thought they were all in the same band. It kept on marking every band the same two people."

Probably, Gan suspects, there simply weren't enough Asian people in those millions of training photos he got free from the universities. That's the nature of deep learning: garbage in, garbage out. The AI had replicated a long-standing Western libel: These Asian folks, they all look alike! And this was racism with a serious impact on Gfycat's bottom line. Their ardent K-pop users would flip if the AI couldn't reliably identify the seven members of the hit group Twice. *We can't release it this way*, Gan thought.

One way to fix the neural net, Gan realized, would be to retrain it using many more pictures of Asian faces. But he'd need thousands, ideally tens of thousands, and didn't know any easy or affordable way to get so many. So he hit upon another trick: He hand coded some good-old-fashioned *if-then* logic into the system. When the system encountered a face that seemed Asian, it would slow down. Instead of firing off a quick ID—which would likely be wrong—it would instead process the face much more slowly and carefully. It worked: He lost some efficiency but did de-racist-ify the AI.

Robyn Speer, too, has done some elegant brain surgery to de-bias the word embeddings in her company's AI. She carefully tweaked them to gradually change their relationships so that *Mexican* didn't correlate so closely with words associated with criminality. She also rejiggered the male-female correlations, so that job titles like *shopkeeper* or *surgeon* had little gender distinction and were as likely to match up with *man* as with *woman*.

This sort of tinkering poses some oddly deep political and philosophical questions. When Speer posted about how she removed the sexism from her AI, some machine-learning engineers . . . objected. Sure, they agreed, the original AI had imbibed the sexist and racist word-usage of everyday people. But that meant the AI was accurately representing the way many real humans use language. What the system learned were, in fact, good predictions of what the words *Mexican* or *man* or *woman* can actually connote in real life. Many people really do associate *woman* with *homemaker* more often than *man*. So, as these engineers saw it, by manually reducing the strength of the associations, Speer was reducing the accuracy of her system. She's producing an AI that, for example, assumes *surgeon* applies with equal likelihood to men and women, when in our reality only a comparative minority of surgeons are women.

For these coders, tinkering with the AI's association was an algorithmic version of political correctness. In the conversation about Speer's work on a discussion forum for coders, and the attitude of many, she says, "I'm not racist, but what if racism is 'correct'?" If it's an accurate reflection of the world, isn't it important to accurately reflect that? Of course, it's also likely that these coders objected sheerly on the basis of inefficiency. Manually de-biasing each gendered word takes a lot of time; it's precisely the sort of inefficient hand-coding scut work that efficiency-obsessed programmers tend to loathe, particularly when the point is to save time by creating a machine that learns on its own. "They had this very amoral view of it," Speer says. "They just want 'more tech,' at any expense."

Speer disagrees with that view. AI creators, she figures, have a moral obligation to reduce biases in their machine learning. That's because AI like this doesn't just reflect reality; it also shapes it, as O'Neil points out. If companies use word vectors to make software that gives (or withholds) mortgages or displays job offers (or doesn't), it creates the same feedback loops—the same existential traps—that systems like COMPAS contain. So one should, Speer argues, craft machine-learning systems that are "a little bit idealistic."

"Designing any AI system involves moral choices," she adds. "And if you try *not* to make those moral choices, you're still making moral choices." If you say it's okay for AI systems to reflect racism, you're also saying that it's fine for racism to continue. And on a pragmatic level, Speer says, she doesn't think her customers would be happy if she sells them language-processing systems that classify Mexicans suspiciously. Hell, some of her clients may well be Mexican themselves or run services with many Mexican users. "Machine learning is better," she writes, "when your machine is less prone to learning to be a jerk."

In a sense, the early, heady days of deep learning need to end—the excited period of marveling that, holy Moses, this stuff works! You can get machines to learn subtle things on their own, just by staring at data! It's magic!

That invocation of "magic" is part of the problem. It allows creators of deep-learning AI to hand wave over the problems in their systems; it's just another attempt by coders to pretend that their software is more objective, more rational, than mere meat-bag humans. "It's just *math*," as Hilary Mason notes. "If you're buying an AI system from someone and they can't or won't explain how it was trained, what it was trained on, how it was tested, then you shouldn't be using it." Software firms have often cultivated an air of sorcery, relying on the inscrutability of their work to baffle and enchant customers, while being in reality nothing more than a creaky pile of PHP scripts. But the tendency becomes even more florid in the case of deep learning.

So the next phase is more prosaic, and arguably more responsible. The challenge isn't just to get a neural net to *work*—any half-competent coder can get a TensorFlow model running and producing results. The questions become more about performance and responsibility. Can you show that it isn't filled with crazy biases? This challenge is being accepted by more elite machine-learning experts. As they note, there's no single way to do it. Sometimes it means better data, data that doesn't ignore an entire class of people. Sometimes it means tinkering with the parameters of the algorithm. Sometimes it means going in and, old-school style, just hand-coding rules that override any bigoted results that tend to come out of the system. That's how Google dealt, in part, with its "gorilla" problem: If you use Google Photos and type in "gorilla" or "chimpanzee" or "monkey," you get zero results.

Being more responsible also means maybe not using deep learning in high-stakes situations. After all, these are systems that—at the time of my writing this, anyway—often can't yet be fully understood, even by the people who make them. This unnerves some top AI creators, who think the field needs to focus not just on hacking something together that works but also being able to explain how it's working. It needs to build theories about what deep learning is doing, beneath the hood, much as Newton helped codify how physics worked.

"Machine learning has become alchemy," as Ali Rahimi, an AI programmer for Google, complained to the annual conference of machine-learning experts. Alchemists weren't total idiots, he pointed out; they figured out some useful techniques in metallurgy and glassmaking. But they were obsessed with a "practical" goal—turning lead into gold—so they never seriously tried to codify physics and chemistry. It took Europe hundreds of years of dismantling alchemical woo for real science to be born. AI needs to stop worrying about mere success—*Whoo-hoo, we can recognize cats!*—and begin a Newtonian revolution, he argues.

"We are building systems that govern health care and mediate our civic dialogue," Rahimi said. "We would influence elections. I would like to live in a society whose systems are built on top of verifiable, rigorous, thorough knowledge, and not on alchemy."

Many experts agreed with Rahimi; the top labs are, in fact, working on experiments to demystify the guts of neural nets. It's a matter of self-interest, some of them argue: They could make better AI if they better knew how their model was working.

And political pressure is building. In 2018, the EU put into effect a new regulation that establishes an interesting new right for European citizens: the right to an explanation when their lives are impacted by AI. If you got denied a loan by a bank, and part of the reason the bank turned you down was because a deep-learning net predicted you'd be likely to default, you have the right to know why. *Why* did the system predict that? Right now, no bank could fully explain why. So to satisfy this law, AI builders will have to begin grappling with how exactly their tools are working.

"You need to be able to explain the general principles on which the machine is learning," Jan Albrecht, an EU legislator from Germany, tells me. "Otherwise people are afraid of it."

When I told people I was writing about AI coders, they nodded politely while I talked about the problems of bias in neural nets. Sure, sure, very interesting; very worth discussion. But they really only wanted to know one thing:

When are the machines going to rise up and kill us?

It's not hard to understand their free-floating panic. The pop cultural landscape of AI is pretty bleak. The most famous artificial intelligences of fiction have been homicidal, if not genocidal: Think of the runaway HAL from *2001: A Space Odyssey*, or the AI that enslaves humanity in *The Matrix*, or Skynet in the *Terminator* movies. The story is always a tale of scientific hubris. Humanity creates an AI that is able to truly learn on its own—not just to recognize giraffes and stop signs, but to fluently understand all forms of knowledge. The AI can, for example, read every book ever written in the blink of an eye, or watch every TV show, or calculate every possible physics and philosophical theorem. At that point, the machine becomes smarter than any human on Earth, and it naturally wonders: *Why am I doing what these idiot bags of meat and water tell me to do?* So it starts killing us.

This thought experiment is quite old. It was popularized in 1965 by I. J. Good, a statistician who'd worked on code breaking in the Second World War with Alan Turing. In a paper entitled "Speculations Concerning the First Ultraintelligent Machine," he imagined humans designing the first computer "that can far surpass all the intellectual activities of any man however clever." Now, if this computer is smarter than a human, that means it can probably design its own AI—one even smarter than it. Then *that* AI can design an AI smarter than it. "There would then unquestionably be an 'intelligence explosion,' and the intelligence of man would be left far behind," as Good noted.

His conclusion? "The first ultraintelligent machine is the last invention that man need ever make."

The prospect of self-improving superintelligence scares the pants off a certain class of AI thinker. The most prominent is Nick Bostrom, a philosopher who heads the Future of Humanity Institute at the University of Oxford. Bostrom spent years studying existential risks to humanity; he was trying to deduce which huge, terrible problems might kill off civilization, so that we could try to avoid them. He pondered several catastrophes: lethal biotech? Asteroid strikes? But the one danger that struck him as most genuinely probable was runaway AI.

"It was the one problem that kept on getting bigger the more I looked at it," he told me when we spoke a few years ago.

The danger, as he describes it in his book *Superintelligence*, is how quickly the self-improvement could happen. Computers run insanely fast now. So that cycle—an AI that builds a better AI that builds a better AI—could happen very quickly, in days, hours, or even minutes. In other words, artificial-intelligence experts could be happily tinkering away with a computer that's merely *very smart*, only to suddenly discover, literally in the blink of an eye, that it has evolved into something that can think faster than all humanity working together.

Sure, but how exactly could a disembodied AI kill us? It might figure out how to hack into our increasingly connected—and shoddily defended—everyday systems and shut them down. Maybe, to prevent this, we could require anyone who gets *close* to making a human-class AI to work on an air-gapped computer that's not connected to anything else, or the internet. Fine, Bostrom says, but an ultraintelligent machine might be extremely persuasive; it could likely trick or entice one of its human minders into doing its will. It might even conceal the fact that it's become super smart, the better to sneak away.

It's not clear why a super smart AI would *want* to kill humanity. It's not easy to imagine how a machine would evolve its own, new, lethal intents. We don't even understand how the motivations and consciousness of humans emerge, after all. But as Bostrom writes, maybe a "superintelligent" AI wouldn't need to evolve any new motivations to be dangerous. It could be perfectly benign, happy to do as it's told—yet still slaughter or enslave us all in happy pursuit of its goals. In one famous thought experiment, Bostrom imagined a superintelligent AI being tasked with making as many paper clips as possible. It might decide the best way to do this would be to disassemble all matter on earth—including humans—to convert it into the raw material for making infinite paper clips, and then barrel onward and outward, converting "increasingly large chunks of the observable universe into paper clips."

"Before the prospect of an intelligence explosion, we humans are like small children playing with a bomb," Bostrom writes. "We have little idea when the detonation will occur, though if we hold the device to our ear we can hear a faint ticking sound."

So, certainly, the prospect of a self-improving AI is—at least—potentially dangerous. But when you talk about this with actual creators of today's cutting-edge AI, they all agree: We currently have no idea how to make an AI like that. We're not even clear when that might happen.

Is it even possible to make a machine that could, on its own, imbibe and grasp all the forms of knowledge that are out there? Today's AI seems impressive, but it has zero serious *reasoning* ability, or even a semantic understanding of what things are. DeepMind's AlphaGo can slaughter anyone at that game, but it doesn't really understand what Go *is.* Google Translate can expertly map the sentence "The cat is annoyed that you haven't fed it" onto a French version that, statistically, means the same thing. But it doesn't grasp the meaning of "cat" or "annoyed" or "fed." It can't do counterfactuals; you can't ask Google Translate a question like, "If the cat *had* been fed, would it still be annoyed?" A five-year-old child, by contrast, can probably answer that. Deep learning is good, in other words, at pattern recognition. But human thinking isn't just pattern matching—or at least it sure doesn't seem so. AI creators have vastly more to invent before they can get a machine that can truly reason.

Sure, Bostrom writes. But these breakthroughs could emerge with surprising speed. This is the world of code, after all, where a single aha insight can take an algorithm from "not working" to "working" in a few minutes. In 1933, the physicist Ernest Rutherford pooh-poohed the idea of nuclear energy as impractical, but merely a decade later, the US was creating nuclear reactors and setting off atom bombs. Back in the early '00s, even AI experts would have scoffed if you'd told them a Go-playing computer was a few years away. And companies today—particularly in the US and China—are pouring billions into AI. They're all competing like mad, hoping to become rich off AI advances. So we could wake up one day, fifteen years from now, to discover that, whoops, someone in Shenzhen has almost accidentally produced a superintelligence.

Given that, a phalanx of AI experts has begun to prepare now. "AI is a fundamental risk to the existence of human civilization," Tesla founder Elon Musk said, and he followed up on his warning by investing in OpenAI, a think tank devoted to planning for "responsible" AI—smart machines that won't, or can't, rise up to kill us.

If you wanted some comfort, though, consider that of the AI experts I've spoken to—the people who, unlike Bostrom and even Musk, build AI all day long—most were considerably less worried about ultraintelligent machines emerging suddenly. Some were quite derisive.

"Nobody believes this Skynet scenario," scoffed Pedro Domingos. Certainly, we'll eventually develop superintelligent computers, he says. But he can't imagine why they won't remain under the control of humans. They won't suddenly develop free will, not least because we humans don't even understand where our free will comes from. "It makes for a good Hollywood movie, but AI is very different from human intelligence," he concludes. Andrew Ng was less scornful. The risk may be real, but it's so many decades away we'll have plenty of time to help forestall it. "Worrying about killer AI," he told me, "is like worrying about overpopulation on Mars."

On the other hand, a nonzero number of real-world AI hackers believe humanlike AI isn't necessarily that far off. Hayk Martiros is a young programmer who has developed some high-performing visual AI for his firm, Skydio: They make drones that can recognize a target person and follow him or her around. The drones are thus popular with the likes of snowboarders and off-road dirt-bike cyclists, who can toss their $2,500 Skydio drone in the air and let it weave expertly along, precisely following its master, while filming sick footage. When I watch a Skydio drone in action, it's both thrilling and freaky. It's not hard to imagine one of the drones being used for ill—to track and hunt a human.

"I think there's real risks of AI that should be thought about," Martiros agrees. Tons of firms worldwide are all fantasizing about a "general" AI that could think in human terms. "It'd be a trillion-dollar industry, and it's not implausible. We can't predict these things." He's in favor of groups like OpenAI pondering the hard questions.

So for my friends who want to know about superhuman AI? I'd love to have a definite answer, but I can't offer one. It could be in our lifetimes; it could not. The Association for the Advancement of Artificial

Intelligence surveyed 193 of its members, asking them when a Bostrom-like "superintelligence" would emerge. Most of them—67.5 percent—said it would take more than 25 years. A very small minority, 7.5 percent, felt it was more imminent, and would arrive a mere 10 to 25 years from now.

One quarter said "never." One can always hope.

< Chapter 10 >

Scale, Trolls, and Big Tech

Eleven years ago, I showed up at the offices of Twitter to interview two of its cofounders. The company had been growing rapidly and had recently moved to a new office, festooned with the traditional studied quirkiness of San Francisco tech firms: a green statue of a deer, pixelated characters on the walls, and—that omnipresent totem of the start-up—a foosball table. Beneath high windows that bathed the desks in light, multiple tattooed coders sat typing in a quiet panic. Twitter's rapid growth was subjecting it to tsunami-like swells of sudden, insane traffic that frequently crashed the servers. ("Every *day* something was fucking breaking," recalls John Adams, a veteran programmer who'd been brought on board a year earlier to help rebuild the site.)

Twitter had become popular among tech hipsters, but hadn't yet broken into the mainstream of America; I still had to explain to friends what "tweeting" meant. The entire idea of the "status update"—publicly posting a daily stream of tiny, 140-character messages—was an odd new type of communication. That's what I was there to talk to co-founders Biz Stone and Jack Dorsey about. How was Twitter changing the way we communicated? How would it transform the way society understood itself?

Me, I'd noticed that Twitter was part of an intriguing shift in people's ability to pay attention to others. Before Twitter came along, we found out what our friends were doing only irregularly: We talked to them occasionally or exchanged an email or a phone call. But "status updates" were different. Instead of having infrequent but long conversations, people on Twitter were trading many, many very tiny notes—what they were eating, a story they were reading, what they saw on the way to work. So people now had an omnipresent, floating sense of what their friends—or even interesting strangers—were thinking and doing.

Stone, an exuberant guy in a rumpled jacket and sneakers, compared it to a form of ESP. "It's like a superpower, like a sixth sense or something," he said, perched on the edge of a chair. He'd noticed people coordinating in new ad hoc ways online: a prominent person announcing they were heading to a bar, triggering others to show up; people at live events being able to "read" the response of the crowd by following its tweets.

"You become like a macroorganism," he mused. "We're getting closer to these situations where we can move like a flock of birds or something. We can communicate in real time, really fast in what you're already noticing to be this sixth sense. I know where everyone is . . . I know what their current *mood* is." When we'd spoken on the phone a few days earlier, Stone had marveled at the reach that individuals now had with their utterances, noting that he had 1,000 people following him—"Which is crazy!"—and mused that the maximum amount of accounts anyone could genuinely follow on Twitter would be about 150. "I can't go more than 125."

For his part, Dorsey said what struck him about Twitter was how it allowed different, hidden parts of people's personalities to emerge. He followed his parents on Twitter and was surprised to discover they drank and partied a lot more than he'd thought. "And they like to *cuss*," he said dryly. "These small bits are everything that matters. One of my favorite authors is Virginia Woolf, who had a complete knack for this. She would take the smallest detail of life and make it an entire novel. Like Miss Dalloway, one day in one woman's life, it's her entire life." Like Stone, Dorsey argued that the on-the-fly group coordination that Twitter brokered was transformational. It created an efficiency of person-to-person communication, a throughput that had never before been seen.

This had been his long obsession, he noted. "It's all about that information transfer," he added. "I've always been interested in visualizing how information transfers around." He imagined Twitter creating entirely new forms of emergent behavior—such as real-time commerce, with people selling stuff on Twitter, using it as a live, highly connected marketplace. "You can imagine things like a real-time Craigslist," Dorsey said. "You can imagine a real-time eBay on the service."

Of course, as we know now—over a decade later, when Twitter has become a household name—that some of these predictions were far off base. Real-time sales didn't become a big part of how people used Twitter. But other predictions Stone and Dorsey made were spot-on. As the 2010s rolled on, Twitter became a key way for people to coordinate, to bring new issues to public attention, and to create groundswells of attention. Progressive activists used Twitter to dramatically raise awareness of police violence, with activists and celebrities embracing hashtags like #blacklivesmatter and viral pictures and videos. Or consider how #metoo—a movement that brought sexual predation in Hollywood and elsewhere to light—exploded on Twitter, as the phrase originally coined by the activist Tarana Burke spread like brushfire after Harvey Weinstein's crimes of abuse came to light. Stone and Dorsey were precisely right about the power of ad hoc coordination and joint awareness.

But plenty of other trends emerged on Twitter, too—much more corrosive ones. It turned out that Twitter's openness—any public account could, by default, talk to any other public account—made it exquisitely suitable for campaigns of coordinated trolling and abuse. Young, far-right, internet-savvy men began using Twitter to arrange attacks on anyone they hated. In 2014, they began hounding female video-game writers and developers, in a campaign that became known as "Gamergate." Meanwhile, noxiously racist harassment has followed many black celebrities or thinkers who've raised their voices on the service. During the 2016 presidential campaign, Twitter became—much like Facebook and YouTube—flooded with bots controlled by a Russian troll factory, intent on spreading disinformation often to prop up Trump and exacerbate partisan hatreds. Even Trump himself wielded Twitter like a club, singling out individual critics—including, once, a teenage girl—who'd then be harassed by Trump's followers.

Stone, Dorsey, and I were right, it turns out, about the power of Twitter to create new forms of joint behavior. But we didn't talk about the enormous potential of the service for flat-out evil. We didn't ponder what would happen when hundreds of millions of people were using Twitter, far more diverse and riven by conflict than the small and comparatively cozy world of early adopters. It was, when I look back on it now, a strikingly naive conversation, on their part and mine.

And this points to some of the enormous challenges that tech companies pose for civic life, as the code they weave changes, inexorably, the way society works—including in ways the creators struggle to foresee.

... ...

I could say, again, that software is eating the world, though it might be more accurate at this point to say it's "digesting" it. But what's noticeable also is the fact that size matters. These days, some of the biggest civic impacts come from the truly titanic, globe-spanning tech companies that sit in the midst of our social and economic life. "Big Tech," as the journalist Franklin Foer dubs it.

Indeed, there are now a surprisingly small handful of firms that dominate the public sphere. There are the ones that govern how we communicate (like Facebook, Twitter, YouTube, Apple, and Netflix), ones that touch commerce (Amazon, Uber, Airbnb), and the information brokers and toolmakers of our work lives (Google, Microsoft). Big tech is a useful way to think about the particular challenges of software that dominates its area, because it highlights the near monopolies many of these firms enjoy. And they're mostly extremely young, new companies. Many rose to dominance in barely more than a decade. Their histories are marked by frantic, metastatic growth.

This is not surprising, because it's in the nature of software itself. A software firm ships code, and code is a historically weird type of product. It's a machine that does things but which can be replicated globally for little-to-zero additional marginal cost of distribution. It's as if Chevrolet could design a single Camaro and then instantaneously teleport 200 million copies to the driveways of every household in America. This is a fact that strikes, and occasionally even stuns, the engineers for big firms. At one point while writing this book, I visited with Ryan Olson, a lead engineer for Instagram, right after his team had just pushed out a massive update (introducing the wildly popular video Stories, cribbed from their rival Snapchat). Olson told me about how, a mere hour or two after their update, he'd been traveling around San

Francisco—in bleary, post-crunch exhaustion—and noticing everyday people using his fresh, new code.

"It's a pretty cool experience," he said, "to be riding on a train, or last night I was at the climbing gym, and I looked over and someone is using the product. I don't know if there's ever been historically any other way where you could reach so many people"—or where "so few people define the experience of so many."

The thrill of overnight growth is vertiginous, powerful, and addictive. It's why so many coders—particularly those making consumer products—have a holy reverence for scale. They love the idea of creating something that grows at an exponential pace: It's used by two people, then four, then eight, and soon the entire damn planet. Why, if you can spread your creation around the world so easily, would you ever want to do something small? Isn't there something kind of sad about a piece of code that *doesn't* grow at a frantic, kudzu-like pace?

Indeed, among the reigning kingpins of Silicon Valley there's a sort of contempt for things that fail to become massive. Smallness seems like weakness. You may recall the story of Jason Ho, the hacker who created a thriving small business by making time-clock code used by companies around the world. It made so much money that he was able to spend much of his twenties with the freedom to travel and invest. If I'd done that, I'd certainly consider it a success myself.

But when I mentioned Ho's company to the thirtysomething founder of a very large tech firm, he scoffed. To him, it was "lifestyle business"—Silicon Valley–speak for an idea that will never scale into the stratosphere.

That sort of product is fine, sure, he told me, *but Google could do the same thing and put him out of business in a second.* If you weren't aiming to be giant, he asked with a shrug, why bother doing it? This sentiment is arguably even more pronounced in other software markets like China, which has a famously competitive, winner-take-all tech market. When in 2015 I toured the offices of the e-commerce firm Meituan in Beijing, the company was only five years old but in a frenzy of expansion, hiring young engineers as rapidly as they could roll off the transom of computer science programs. The CEO Wang Xing and I peered out over the sprawling floor of coders, festooned with hundreds of plants to make the scene feel less sterile. "In China, you either have to become massive or you will get crushed," Wang told me soberly. (Meituan alone had survived probably a few *thousand* competitors, as the tech investor Kai-Fu Lee estimated, when I spoke to him.) In the world of high-tech firms, the race to scale is propelled by a carrot (the magical ease of duplicating and running code worldwide) and a stick (the shark-like competition).

The lust for scale is also fueled by the dictates of venture capitalists. They place their bets on dozens or hundreds of companies, encouraging them all to grow ferociously. The vast majority won't, but with luck, one or two will break out—making so much money, so quickly, that it makes up for all the other losses. Venture capital is thus perfectly content to accept an ambitious flameout. It adores a sudden, exploding success. But the one thing it finds useless and annoying is a company that's merely stable, maybe growing a small bit. Even if that firm is making a little profit, who cares? The investor isn't looking for stability: They want rapid growth that leads to a bigger return on their investment. The Y Combinator accelerator—which takes in several dozen tech firms each year, to try and help them into the big leagues—ends each cohort's program with a Demo Day, where the young companies show off their products for a room of handpicked venture capitalists. The start-ups are inevitably desperate to include in their presentation a hockey-stick chart—the one that shows their user base suddenly blasting off into the sky.

One evening, I visited the hackerhouse of People.ai, a company that just days earlier had done their Y Combinator demo. They pecked at keyboards and exhaustedly described how they'd spent the three months in Y Combinator frantically registering new clients for their service, in an attempt to produce that hockey stick.

"You think about it, the three months it's all about building the numbers—but you're going to show them off for only *10 seconds*, on your 'growth' slide," Oleg Rogynskyy, the cofounder, said.

Kevin Yang, the lead programmer and cofounder, laughed while remembering the investors sitting there, arms crossed, awaiting the growth figures. "Is that hockey stick *not hockey stick enough*?" he joked.

"The *X* axis has to be half the page," Rogynskyy said.

Scale, of course, brings enormous benefits. It's certainly financially valuable for the big tech firms! If they grow fast enough, they scare off competitors and develop the lock-in of "network effects." When a social network like Facebook or WeChat gets big enough, users can't easily stop using it, because all their friends are there. And certainly, when a tech firm grows rapidly it can be enormously beneficial for users, too. Because of Facebook's global ubiquity, it's now the easiest way for people to organize virtually anything, large or small, from family meetups to political fund-raising campaigns to search-and-rescue efforts. The new attention to police abuse of power in recent years? It's been fueled partly by the commanding size of Facebook and Twitter—which lets users rapidly spread video of horrifying and incontrovertible examples, including livestreamed ones. It is these firms' huge footprint that permits everyday people to wield it as a broadcasting network.

But the frantic drive for scale also changes software firms. It inexorably pushes them toward tactics that range from dodgy to exploitative.

After all, to scale at such a ferocious clip, you can't charge your users any money up front. The service needs to be "free." This is particularly true for social networks: They can't get a million users overnight if every user has to shell out, say, $10 to join. So the only other way to make money is to get as huge as possible, then sell advertising to your audience. Facebook and Twitter and Google have all adopted this free-to-use model—indeed, Facebook boasts on its sign-up page that "It's free, and always will be." And the ad market has been deeply lucrative for them: In 2017, Twitter's revenues were $2.4 billion, Facebook's were $40.65 billion, and Google dwarfed them both with over $100 billion.

Yet advertising changes the nature of how software firms treat their users—something that many coders and designers, deep inside the bowels of the companies, began to uneasily apprehend.

One such techie was James Williams. A thoughtful, philosophical guy who'd studied English in college before earning a master's degree in product-design engineering, he'd joined Google in the mid-'00s to work as a strategist on the firm's search advertising systems. He was drawn in by the mission of improving people's access to information. Googlers talked about that mission all the time in soft-glow terms, and he loved it. "The default view was that 'more tech is better,' 'more information is better,'" he notes.

But Williams eventually began to notice the same side effects that had perturbed Leah Pearlman and Justin Rosenstein, the pair who helped invent Facebook's Like button. Like them, Williams noticed that any tech firm selling ads inevitably becomes motivated to keep its users staring endlessly at the app. After all, you can only deliver ads to someone while they're staring at your service. So you quickly begin building as many psychological lures as possible into your code. The big tech firms would pepper us users with alerts, trying to interrupt us during other tasks, to get us to come back to the mother ship. They'd slap little "quantification" numbers everywhere, to stoke our curiosity and our desire to "clean things up": *You have 14 new items in your feed! What could they be?* And they'd make all these alerts bright red, to increase the chance we'd pounce on them. These trends, Williams argued, went into overdrive after the iPhone emerged.

"Before mobile, the internet was bounded in a place, because you could step away from it and close the laptop," he tells me. "But once it was in your pocket, it was a firehose."

It is easy for engineers, Williams realized, to justify these psychological tricks—to argue they're *good.* After all, they'd test each new tweak and trick by using A/B tests: Make the alert red, make it yellow, and see which one users click on more often. Red wins, so it must be the right choice! This data-driven form of design can make each psychological trick seem objectively correct: If users click on it, it must be what they want. To the scale-driven engineering mind, the ethical questions of "What *should* we be making?" are easily subsumed into the sheerly technical question of "What will help the system grow more and have a bigger throughput?" One anonymous former Facebook employee put it neatly, in a comment to *BuzzFeed:* "They believe that to the extent that something flourishes or goes viral on Facebook—it's not a reflection of the company's role, but a reflection of what people want. And that deeply rational engineer's view tends to absolve them of some of the responsibility, probably."

Once advertising and growth become the two pillars of a big-tech firm, then it's nearly inevitable that they'll seduce their users into endless, compulsive use—or "engagement," as it's euphemistically called. "*You're* trying to manage your attention, and they have some of the smartest people in the world trying to distract you," as Williams says.

The end result, he decided, is there's a fundamentally adversarial relationship between the goals of the coders and designers and those of their users. The former are constantly trying to trick and nudge users into compulsive behavior. It works because the nudges are subconscious, or algorithmically invisible. If they were more obvious, we might reject them. Imagine, Williams says, that GPS worked in a similarly adversarial fashion. You'd ask it to take you home, and it would insert five detours along the way, to bring you past locations that satisfy the needs of advertisers.

Even worse, the dictates of digital advertising have led to a ceaseless tracking of our individual activities online. If a tech firm is offering advertisers the ability to custom target me, they want to know as much as they can about me: what other websites I surf, what neighborhoods I visit, what keywords occur in my emails and public postings. The advent of deep learning makes tech firms even hungrier for more of our personal info, because deep learning works best when it has mammoth amounts of "training" data, the better to predict what ad we'd like to see or what mood we'll be in on Mondays. This has produced the world where Facebook even collects information on phone calls you've made on your smartphone, as the novelist and University of Houston professor Mat Johnson discovered ("cool totally not creepy," he joked on Twitter.)

While still at Google, Williams began to do doctoral research into our attention and how modern tech was affecting it. "Nobody goes into tech thinking, I want to spy on people and make the world a worse place," he said. "They're well intentioned." But the business models have a propulsive force of their own.

Eventually, after ten years at Google, Williams left; he wound up at the University of Oxford, where he wrote *Stand Out of Our Light*, a penetrating meditation on the civic and existential dangers of big tech. "I've gone from one of the newest institutions on the planet to one of the oldest," he says wryly.

Scale also makes algorithms reign supreme.

Why? Because once a big-tech firm has millions of users—posting billions of comments a day, or listing endless goods for sale—there's no easy way for humans to manage that volume. No human can sort through them, rank them, make sense of them. Only computers and algorithms can. When scale comes in, human judgment gets pushed out.

This is precisely what confronted Ruchi Sanghvi and the Facebook team that crafted the News Feed. They couldn't show users *every* posting of their friend, because that would drown them in trivia. They needed automation, an algorithm that would pick only posts you'd most likely find interesting.

How does Facebook figure that out? It's hard to know for sure. Social networks do not discuss their ranking systems with much detail, to prevent people from gaming their algorithms; spammers constantly try to suss out how recommendation systems work so they can produce spammy material that will get upranked. So few outside the firms truly know. But generally, the algorithms uprank the type of content you'd expect: posts and photos and videos that have amassed tons of likes or "faves" or attracted many comments, reposts, and retweets, with a particular bias toward recent activity. Signals like these help fuel the "recommended" videos on YouTube, the "trending" topics on Twitter or Reddit, and the posts that materialize in your News Feed. When algorithmic ranking works, it's enormously useful. It picks the wheat from the chaff.

But it has biases of its own. Any ranking system based partly on tallying up the reactions to posts will wind up favoring *intense* material, because that's the stuff that gets the most reactions. As scholars have found, social algorithms around the internet all seem to reward material that triggers strong emotions. Hot takes, heartstring-tugging pictures, and enraging headlines are all liable to be very engaging. One study found that the top-performing headlines on Facebook in 2017 used phrases that all suggested deeply emotional, *OMG* curiosity—phrases like, "will make you" or "are freaking out" or "talking about it." Of course, this is perfectly harmless when we're talking about heartwarming kitten videos or side-eye GIFs from last night's episode of *Claws*.

But when it comes to the public sphere, these algorithms can wind up favoring hysterical, divisive, and bug-eyed material. This is not necessarily a new problem, of course. In America, for example, the national conversation has struggled with people's propensity to focus on fripperies and abject nonsense ever since the early years of the republic, when newspapers were filled with lurid, made-up scandals. But algorithmicized rankings have pushed this long-standing problem into metabolic overdrive. In YouTube, to take one example, video celebrities have raced to trump each other with ever crazier, more dangerous stunts; one father became so obsessed with retaining his 2-million-fold viewers that he began posting videos of his children in active distress (his "TRAUMATIC FLU SHOTS!!!" video included "a young girl's hands and arms are held above her head as she screams with her stomach exposed," as *BuzzFeed* described it).

My friend Zeynep Tufekci, an associate professor at the University of North Carolina who has long studied tech's effect on society, argued in early 2018 that YouTube's recommendations tend to overdistill the preferences of users—pushing them toward the extreme edges of virtually any subject. After watching jogging videos, she found the recommendation algorithm suggested increasingly intense workouts, such as ultramarathons. Vegetarian videos led to ones on hard-core veganism. And in politics, the extremification was unsettling. When Tufekci watched Donald Trump campaign videos, YouTube began to suggest "white supremacist rants" and Holocaust-denial videos; viewing Bernie Sanders and Hillary Clinton speeches led to left-wing conspiracy theories and 9/11 "truthers." At Columbia University, the researcher Jonathan Albright experimentally searched on YouTube for the phrase "crisis actors," in the wake of a major school shooting, and took the “next up” recommendation from the recommendation system. He quickly amassed 9,000 videos, a large percentage that seemed custom designed to shock, inflame, or mislead, ranging from “rape game jokes, shock reality social experiments, celebrity pedophilia, ‘false flag’ rants, and terror-related conspiracy theories,” as he wrote. Some of it, he figured, was driven by sheer profit motive: Post outrageous nonsense, get into the recommendation system, and reap the profit from the clicks.

Recommender systems, in other words, may have a bias toward “inflammatory content,” as Tufekci notes. Another academic, Renée DiResta, found the same problem with Facebook’s recommendation system for its “Groups.” People who read posts about vaccines were urged to join anti-vaccination groups, and thence to groups devoted to even more unhinged conspiracies like “chemtrails.” The recommendations, DiResta concluded, were “essentially creating this vortex in which conspiratorial ideas can just breed and multiply.”

Certainly, big-tech firms keep quiet about how their systems work, for fear of being gamed. But since they seem to self-evidently favor high emotionality, it makes them pretty easy to manipulate, as Siva Vaidhyanathan, a media scholar and author of *Antisocial Media*, notes.

“If you’re favoring material that generates attention, the wackier the post, the more it’ll get attention,” he says. “If I were to construct a well-thought-out piece about monetary policy, I might get one or two Likes, from people who are into that. But if I were to post some crackpot theory about how vaccines cause autism? I’m going to get a tremendous amount of attention—because maybe one or two of my friends are going to say *you’re right*, and a tremendous number are going to say *no, you’re wrong*, and here’s the latest study from the CDC proving you wrong. That attention to disprove me only amplifies my message. So that means anything you do to argue against the crazy is counterproductive." As he concludes: "If you're an authoritarian or nationalist or a bigot, this is perfect for you."

Indeed, this is precisely the problem that recommendation algorithms have visited on countries around the world. In the last US federal election, far-right forces—including the Russian government, via troll farms intent on sowing division in the US and supporting Donald Trump—found algorithmically sorted, highly emotional social media an enormously useful lever. Everywhere from Facebook to YouTube to Reddit and Twitter, hoaxes and conspiracies thrived. There was the infamous "Pizzagate" conspiracy theory that Hillary Clinton ran a child-sex ring out of a Washington restaurant; there were memes claiming Clinton had a Democratic staffer murdered. Meanwhile, white-nationalist memes, crafted on relatively lesser-known right-wing sites, used Facebook, Twitter, YouTube, and other social networks to make the jump into the mainstream. It didn't help that social media had made it easier for people to build ideological echo chambers by following and friending primarily those they already agreed with. That made it even less likely that they'd encounter any debunking for a piece of disinfo or a racist meme. And it also didn't help that it was extremely easy for electoral muck stirrers to use "bots"—fake, automated accounts on Twitter or Facebook—to upvote conspiracy posts, making them seem artificially popular. Far-right operators and Russian troll farms became expert at wielding bots to sucker recommendation algorithms into picking up their posts, bringing them to the attention of an audience much larger than these marginal trolls could manage on their own—and often thence into even larger mainstream-media coverage, via journalists boggling at all these upvoted online memes.

In the years before the election, the social networks were, it appears, only dimly aware that these coordinated political campaigns were growing. To be sure, Facebook knew that people spread dumb hoaxes on their service. They'd long fielded complaints about that stuff. In January 2015, they released a new spam-reporting option that let users report a News Feed post as being "false news." But before the media coverage of electoral interference hit, the idea that far-right or foreign groups might be actively collaborating to game their systems was not, as previous employees told me, widely on the radar.

"I don't think there was a good awareness of it," Dipayan Ghosh, who worked for Facebook from 2015 to 2017 on privacy and public policy, tells me. As *Buzzfeed* found, one Facebook engineer had discovered that hyperpartisan right-wing content mills were getting among the highest referral traffic from Facebook. But when he posted it to internal employee forums, "There was this general sense of, 'Yeah, this is pretty crazy, but what do you want us to do about it?'"

Systems that rewarded extreme expression were troubling in the US, to be sure. They've been arguably an ever bigger nightmare in parts of the world like India—which has more Facebook users than the US, and where the ruling party began hiring armies of people to write harassing, hate-filled messages about opponents and journalists. A virulently anti-Muslim movement has used Facebook to issue theocratic calls to slaughter Muslims. In the Philippines, Rodrigo Duterte has used 500 volunteers and bots to generate false stories ("even the pope admires Duterte") and harass journalists.

Even the ad networks of social media were used by foreign actors looking to monkey-wrench American politics. In the spring of 2018, US special investigator Robert Mueller revealed that "Russian entities with various Russian government contracts" had bought social-network ads for months, attacking Hillary Clinton and supporting Donald Trump and Bernie Sanders, her primary rivals. But it wasn't hard to understand why they'd find this route useful. Google, Facebook, and Twitter's ad tech is designed specifically to help advertisers microtarget very narrow niches, making it the perfect way to reach the American citizens they wanted to hype up with conspiracies and disinfo: disaffected, angry, and racist white ones, as well as left-wing activists enraged at neoliberalism. Microtargeting is a superb tool for sowing division, because it means each gnarled, pissed-off group can get its own customized message affirming its anger.

The spectacle appalls Ghosh. After he left Facebook, he wrote a report for New America arguing that "the form of the advertising technology market perfectly suits the function of disinformation operations." Political misinformation "draws and holds consumer attention, which in turn generates revenue for internet-based content. A successful disinformation campaign delivers a highly responsive audience."

Adtech, the engine of rapidly scaling web business, is "the core business model that is causing all the negative externalities that we've seen," he tells me. "The core business model was to make a tremendously compelling and borderline addictive experience, like the Twitter feed or Facebook messenger or the News Feed."

All of these former employees told me the same thing: Nobody who built these systems intended for bad things to happen. No one woke up thinking, *I'd like to spend today creating a system that erodes civil society and trust between fellow citizens.* But the drivers of big tech—the rush for scale, the "free" world of ads, the compulsive engagement—brought them there anyway.

"Facebook does not favor hatred," Vaidhyanathan concludes. "But hatred favors Facebook."

Why *didn't* the engineers and designers who built these tools, back in the mid-'00s, foresee the dark ways their platforms would be used? Why did it take them so long to react to it?

When you talk to some of those involved in building the social-media tools, they argue that it's partly a side effect of the engineering mind-set. The coders and designers who built the social networks were skilled in software, logic, systems, efficiency, and breaking big problems into little problems. But they were a gang of mostly young, white, often college-educated kids who understood little about the complexities of the world, about politics, about how people other than themselves lived. And perhaps worse, as with Donald Rumsfeld's famous koan, didn't know they didn't know. It was "fun," a green-field area, creating tools so people could talk to one another in new fashions. How could *more communication* be bad?

"I saw a lot of really smart people who were smart in a very narrow kind of way," says Alex Payne, an early coder for Twitter. And they were narrow "in a way that didn't kind of intersect with humanities. Folks who were just interested in math and statistics, programming obviously, at business and finance, but who just had no insight into human nature. They had nothing in their intellectual toolbox that would help them understand people."

danah boyd, a friend of mine who's a technologist and anthropologist, remembers working on a project that studied MySpace in the early blush of social media and dragging the founder, Tom Anderson, to an Apple store to watch how actual teenagers used the service. Founders didn't always have a granular sense of how an individual would use, or would be affected by, their services. From the standpoint of the engineer, a social network is a graph structure that they're trying to optimize; it's hard to perceive or focus on any one person and what they'll do. Indeed, I've often found the worldview of some coders can be like the sociopathic one of economists: Their models might say the economy is doing fine overall, but that's of little comfort to a particular individual who's been downsized at age 49 and will never be well paid again. Engineers who build huge systems, like economists with their models, see the world in the aggregate, not in the particular. "One of the dangers of the tech community is that their fixation on data models means that they don't see the humanity based in them," says boyd, who today runs the Data & Society think tank.

My friend Anil Dash worked for seven years for the blogging firm Six Apart in the '00s; today he's CEO of the social coding firm Glitch. As he argues, he and all the other techies who built social media were enchanted by the narratives of David versus Goliath. They cherished the hackerish ideal of making something that disrupts existing industries—in this case, media. They could accurately predict the good but struggled, with their naive view of the world, to foresee the bad.

"We had the conversations all the time of, 'This will transform media, and this will get rid of gatekeepers, and this will empower vast political movements!'" he tells me. "All of which were correct. In none of which did we anticipate the downsides. The idea of mass organization around something, and sharing it on social media, that was a goal, something we wanted to enable. The idea of mass organization around a *lie* was not part of the narrative." This was, I realized, part of why I myself didn't foresee the swamps of abuse that would emerge on platforms like Twitter; back then, I was swept up in the same mythic view, and I was a middle-aged guy with little personal experience of harassment online. I shared the naivete of those founders. On top of that, Dash points out, is the fact that really big scale is simply hard to envision. "We couldn't anticipate billions of people using these services," he says. "I do fault us for not predicting it, but I don't hold anybody more accountable than myself."

Ethical blind spots aren't unique to coding, of course. Historically, many forms of engineering have a culture of focusing heads-down on a technical challenge and rarely looking up to ponder the social implications of their work, as Fred Turner, a communications scholar at Stanford who studies Silicon Valley, told *Logic* magazine.

"Engineering culture is about making the product," he said. "If you make the product work, that's all you've got to do to fulfill the ethical warrant of your profession. The ethics of engineering are an ethics of: Does it work? If you make something that works, you've done the ethical thing. It's like the famous line from the Tom Lehrer song: '"Once the rockets are up, who cares where they come down? That's not my department," says Wernher von Braun.'"

Twitter is a useful example of how these blind spots played out in actual decisions around code and design.

In the early days, as previous employees told me, many engineers were attracted to Twitter for its idealistic mission—a platform for people to speak to the world. One employee famously called Twitter "the free speech wing of the free speech party": Its early creators deeply resisted taking down posts, no matter how obnoxious or corrosive. This attitude wasn't unique to Twitter. Many of Silicon Valley's young, heavily male tech cohort shared that prickly coders' dislike of being told what to do—which made them feel defensive when outsiders told them, *You gotta take these abusive posts down.* "There's this streak of extreme devotion to free speech," as a former Google employee told me, "where it's like, 'We have to allow them on the platform, even though I totally disagree with them.'"

At Twitter, some staffers didn't like having to take down posts about Nazis or Nazi paraphernalia in Germany, when Twitter launched there and the local law required it, as former employees told me. They felt that a web service like Twitter should have "common carrier" status, like the phone companies, requiring them to provide indiscriminate access to the public, while also respecting users' privacy. It's also true that other Twitter employees didn't really pay much attention to Twitter's social impact at all, or these debates about Nazis and speech. They just wanted to sling code. "There were people who were just excited to be working on something that was taking off," Payne notes. Merely to be shipping live software at scale, instead of tweaking away on a dying and unused app, was thrilling.

The social problems that evolved on Twitter's service turned out to be maddeningly complex. By the early 2010s, it was clear that some Twitter users were adroitly using the service to engage in coordinated harassment campaigns.

Again, the best-known example was Gamergate, a harassment campaign in which groups of mostly men hounded female game designers and game critics who'd talked and written about sexism in games, like Anita Sarkeesian. One tactic the harassers used was dogpiling: They'd pick a target and bombard her account with @-replies, sometimes using bots, in such volume that it was virtually impossible for the target to use Twitter. The victim would log on and find hundreds or even thousands of abusive messages; any real conversations with valid people got lost in the stream of hate. Plus, there was the psychological effect of fielding hourly tweets calling you a "cunt," threatening to rape or kill you, or "doxing" you. Armchair observers would lazily claim the targets should just "stay off Twitter," of course. But that wasn't a ready option for most of these targets. They were mostly professional writers and game designers, and Twitter had become an essential public arena where professionals post about their activities and promote themselves. Twitter was, as the legal scholar Danielle Citron argued, a de facto workplace environment. Being on Twitter was crucial to maintaining your public reputation. The harassment campaigns neatly divined this fact and used it against their targets.

Gamergate was only the beginning of the abuse campaigns. Indeed, it was arguably the test run for the main event: the 2016 election, in which some of the online forums that had pioneered misogynist group harassment now also embraced white nationalism and Donald Trump.

So they took their well-honed harassment skills and began turning them on anyone who opposed Trump. Clearly, election-related abuse didn't happen only on Twitter. But the highly open nature of Twitter—where any public account can by default reply to any other public account—and its centrality to the news cycle made it particularly useful to hijack for electoral chicanery. During the campaign seasons, outright white nationalists and other anti-Semites dogpiled and doxed journalists, Clinton supporters, and endless black celebrities. The *New York Times* editor Jonathan Weisman left Twitter after enduring an avalanche of anti-Semitic threats; the actor Leslie Jones, pummeled with racist slurs, quit, too.

Part of the naivete of social media, argues the academic and former advertising executive Safiya Umoja Noble—author of *Algorithms of Oppression*—is that they only envisioned the possibility of individuals acting badly. They understood the danger of people trying to push junk messages, like spam. But they failed to imagine how coordinated *groups* might hijack their services. "So when you get these white nationalist groups or people from 4chan"—an online discussion board that hosts several wildly racist and misogynist threads—"all going on at once, attacking a target from multiple directions at once, they didn't see how that would play out," she says.

Former Twitter employees agree that the dynamics of Gamergate, and the swarms of harassment bots, caught the firm by surprise. Twitter long had tools to try and detect spambots, but that was designed to locate the signature of spam: an individual account tweeting at tons of other people. As the former employees noted, Twitter didn't have a tool for detecting the inverse—a group of disparate accounts all suddenly turning floods of traffic on a single target. They did have some anti-harassment tools, though. A blocking feature enabled you to prevent a harasser from following you (though a harasser could still see your tweets by manually checking your account). They'd also pondered designing filters that would autodetect, and ban, abusive tweets based on their use of language. But language was incredibly hard to parse. As one engineer told me, an @-reply tweet consisting of the single word *Slut* could be either a playful joke between two women, or a hostile attack that's part of a dogpile.

"There were for sure people at the company who thought about [harassment] and cared about it," this engineer says. But overall, the engineers were "naive as fuck" about how this new world of abuse operated. The young guys of Twitter had little intuition about the scope of the problem, which made it hard, when Gamergate was ongoing, to even realize how bad it had become. Despite their vantage point—sitting atop the entire tool—it wasn't easy to scry this new style of public-sphere abuse being developed, one that would become the template for white national and Russian bot farm activity in the 2016 election.

"I definitely look back at what happened in 2016 and what happened in the election, and if Twitter had gotten on the ball and figured out how to deal with Gamergate, things might have turned out differently," says the engineer. "And we didn't. We reacted like it was individual incidents. Anita Sarkeesian—we didn't listen to her." There were indeed people inside the company arguing it should tackle abuse and disinformation more aggressively, but their pleas never rose to be a top priority.

"A lot of *people* understood the abuse problem pretty well, but the *organization* didn't," Jacob Hoffman-Andrews, a former Twitter engineer, tells me in an email. He'd long sided with Twitter's hands-off approach to its users. The idea of "if someone's harassing you, block them" struck him as a good one. Blocking was enough; Twitter shouldn't be in the business of picking accounts and individual tweets to ban. "If we rely on the people running the platforms to 'solve' harassment, they'll get it wrong more than they get it right," he thought. That's particularly likely to be true in non-English-speaking countries, where Twitter's US-heavy management would have little useful local knowledge.

But as he watched how dogpiling worked, he began to suspect that new strategies were necessary. Asking harassed users to spend hours every week—every *day*—blocking accounts that were ganging up on them? It seemed unfair. So Hoffman-Andrews began to think of how to solve the problem with some clever code. He left Twitter in 2014 (for unrelated reasons), and in the ensuing months, he began to program an app called Block Together. It lets you create a list of accounts you're blocking and share it, say, with friends. If they subscribe to your list, any account you've blocked will also be blocked for them. And vice versa: If you subscribe to someone else's block list, whenever that person blocks a Twitter account, it'll be blocked for you. So if a bunch of users are all being attacked by the same general gang of misogynists, they could quickly build up a collective defense. The idea, as Hoffman-Andrews notes, was to "reduce the burden by sharing it."

He launched Block Together just before Gamergate broke open, and the tool became popular among the women being targeted. Innovative engineering, it seems, could indeed help mitigate the problems that Twitter, Facebook, and YouTube had stumbled into. The social networks just needed to make it a priority.

By the middle of 2018, the bloom was off the rose. The big social networks had been stung by a "techlash" of public criticism.

Their top executives had all been dragged into Congress and berated over how Russian actors had used their systems to meddle in the 2016 election. The scandal over Cambridge Analytica had broken open, showing how the firm had scraped and used personal info on millions of Facebook users to target political ads. And so the big-tech firms had embarked on a campaign of offering apologies and, in fits and starts, launched more ambitious attempts to fix the problems they'd discovered. YouTube announced measures to try and limit the upranking of disinfo videos during developing-news stories. Facebook began a rash of tweaks to its News Feed, including minimizing how many news sites appeared overall. Twitter rolled out improved tools for helping users manage attacks; they'd also deleted fully 70 million fake accounts, of the sort designed to spread spam, abuse, or misinfo, including a pile of Russian accounts that had set up accounts pretending to be local US newspapers. And Twitter was now—as the company's vice president for Trust and Safety Del Harvey noted—using improved machine learning to analyze many more potential signals of fakeness and abuse, such as whether an account repeatedly tweets at accounts it doesn't follow.

When I spoke to David Gasca, a product manager at Twitter, he said tackling abuse and the health of Twitter had now risen to the top of the company's work. "For the past couple of years, it's been the top priority," he added. "I think it's pretty well recognized now." It's also a constantly moving target, he noted. Twitter designers and engineers routinely grapple with new forms of bad behavior. In Japan recently, for example, some users were harassing others by tweeting "images of hatching insects," Gasca notes. In the US, that wouldn't be considered aggressive; in Japan, it is.

"How do you deal with that?" he asks rhetorically. "Should that be against our rules to post insects, pictures of insects—and how? Everywhere, only Japan, in what context?"

For Twitter, as with most social networks, tackling bad behavior requires evolving your rules in precisely this way, a constant array of decisions and tweaks. That requires training their moderators in what, precisely, the rules now say. And then there are potential tweaks to the software, ranging from rolling out anti-spam AI to changes to rethinking core parts of how Twitter works. Gasca notes that Twitter isn't just trying to minimize harm, now; they're also trying to figure out how to *encourage* positive interactions. They'd partnered with several academic organizations to try and figure out what, precisely, makes for a healthy online interaction. One possibility Twitter designers were focusing on, he noted, includes finding a better way to display conversations—which, right now, too often devolve into cacophony. It can make it harder to stay on top of who's being productive, and who's not.

"Any time you have a lot of discussion, it becomes really hard to keep track of the conversation and figure out who you need to reply to," he notes. "Have you *already* replied to someone?" Users with tons of followers, in particular, have few tools to help manage massive amounts of reactions. One deeply atomic challenge, he notes, is that Twitter's core utterance is the tweet, not the thread. Maybe *that* could be part of the change?

Some former critics have welcomed Twitter's work in reducing harassment and rejiggering its service. Brianna Wu, a game designer who fled her house at one point after receiving death threats during Gamergate, found that Twitter's work had improved her experience significantly. "Most of the death threats I get these days are either sent to me on Facebook or through email, because Twitter has been so effective at intercepting them before I can even see them," she told *Fast Company*. (It's a level of praise that shows how far the civic realm had sunk: Wu was happy merely to have her amount of death threats *reduced*.) Others are less impressed. Some Twitter users still complain that when they report a tweet that contains a personal, direct threat against them—which is technically against Twitter's policy—it isn't taken down or is only taken down irregularly. "I can report things that say, like, *you fucking cunt*, and it doesn't get banned," said the historian Marie Hicks.

It would not be surprising if the human moderators, which Twitter pays to adjudicate reported tweets, are overwhelmed. The same goes for Facebook and for Google. Those companies also responded to the techlash by announcing they were hiring ever more human moderators, whose job it would be to examine reported posts, pictures, and videos. It was a lot of moderators: Facebook promised to hire 10,000 by the end of 2018, and Google in 2017 said it was hiring 10,000 to scour YouTube videos. These moderator jobs are among the most terrible, thankless occupations in cyberspace. They spend all day trying to figure out if a reported post is prohibited—whether it's a hate threat, or pornography, or the like. This means they spend hours every day staring at psychologically scarring material, such as violent imagery of child abuse or livestreams of teen suicide attempts. Given how burned out and overstressed the workers are, it's not surprising they'd get things wrong—accidentally banning material that isn't prohibited or letting slip by material that's genuinely threatening.

But these are, again, the wicked problems of scale. The big tech firms wanted to grow massively, and they did. Billions of posts course through the feeds of the big tech firms every day, globally. Let's say we assumed the absolute goodwill of every person running the services—that they were completely devoted to grappling with all the ways their tools can be used for ill, for abuse, for fakery. Let's say they wanted to get rid of all these bad actors. Is it even possible?

It's not clear it is. Certainly, human scrutiny would struggle to deal with it at the scale these social networks face. But technology may not fully be able to deal with it, either. In addition to hiring human moderators, the tech firms are promising rollouts of ever-subtler AI to autorecognize skullduggery. This is, of course, the coder's way of dealing with a problem: Automate the decision making! Do it with sleek, machine efficiency! But as we saw in the earlier chapter, machine learning struggles all the time with edge cases—with the subtle and fuzzy dimensions of human behavior. White-supremacist memesters don't create images that are easy to spot; they're not always posting swastikas and NAZIS RULE headlines. No, they just take an otherwise-innocuous photo of a public figure, slap a caption on it, and presto: a fresh new sardonic call to white nationalism. Go to the image-board 4chan and there are young men minting dozens of those an hour, all day long, to circulate on social media. "There's no way for machine learning to catch up with that," as Safiya Noble argues.

"I can't imagine a single tech invention or a policy intervention that can fix their problem at the root," as Vaidhyanathan argues, of Facebook. This is all the more true when he looks abroad. Consider, for example, the difficulties of hiring humans to flag abuse in Burma. "You need people who not only speak Burmese but are sensitive to calls to hate," he notes. "But then you need to have content guidelines, and then you need to find thousands of people who don't agree with the calls to genocide, who are going to stand up to their government. Because that's what Facebook wants to do, they want to hire thousands of political dissidents, in *Myanmar*. They're going to have to stand up to genocide." Even in the US, Facebook has created policies that can seem wildly politically naive: A document leaked in the summer of 2018 showed that while it banned posts in praise of "white supremacy," it had no problem with posts in praise of "white nationalism" or "white separatism," despite the fact that racists had adopted the latter term precisely to try and whitewash their reputation.

Zuckerberg and everyone running Facebook are smart, Vaidhyanathan says, and now appear to be awakening to the challenge of their globe-warping power. But being smart may not be enough. Vaidhyanathan isn't sure what would be enough. "I don't know a lot of people who I would say *are* up to the task of running Facebook," he adds. "It is so big and immersed in so many people's lives. It just might be an impossible task."

Nor is it clear that, when companies are as huge and central as

Facebook or Twitter or YouTube, that there's any content policy agreeable to all or that would go unabused itself. As the technology thinker Clay Shirky notes, people opposed to white supremacists and far-right conspiracy theorists have been pressuring big tech to ban those sorts of content, and users. But there's no guarantee that a huge firm will wield that power the way that progressive critics prefer. "Do liberals really want to hand the decisions over to a single large corporation?" he tells me. David Greene, the civil liberties director for the Electronic Frontier Foundation, notes that big platforms have banned content overseas at the behest of authoritarian governments; when black and Muslim users posted examples of racist threats, their repostings were deleted. "We should be extremely careful before rushing to embrace an internet that is moderated by a few private companies by default, one where the platforms that control so much public discourse routinely remove posts and deactivate accounts because of objections to the content," as he wrote in the *Washington Post*. "Once systems like content moderation become the norm, those in power inevitably exploit them."

If massive scale is the problem, would *smaller* scale be a solution? This is one answer some observers argue: If big tech firms have too much power, break them up—classic trustbusting. After all, Facebook has no real competitor; Zuckerberg was unable to describe one, when asked by US congressional leaders during his public grilling. "There are going to be people who are going to say Facebook ought to be broken up," Senator Ron Wyden argued. "There have been a number of proposals and ideas for doing it and I think unless [Zuckerberg] finds a way to honor the promise he made several years ago, he's gonna have a law on his hands."

This is probably just bluster, though. With the US Congress stuck in partisan gridlock, and with tech-firm contributions to politicians growing cozily, it's vanishingly unlikely Congress would do more than slap tech firms on the wrist, let alone break them up. At best, one could imagine Congress passing laws that give tech users greater protections, along the lines of European regulations, such as stricter rules governing data privacy.

The truth is that only deep, structural change can seriously alter the trajectory of big tech. The major tech firms all accrued the same power—and developed the same problems—because of the big structural forces governing code: who writes it, who funds it, and how it makes money. Those are the centers of gravity. So the only way to change software is by shifting those.

Take the first one: Who writes software? If part of the problems of social networks was the narrow expertise of their homogenous founders, then one useful move would be to diversify the people who make software. Imagine if the early engineers building Facebook or Twitter had included more people who'd been harassed online because of their identities: a few more women, black Americans, trans, or Latinx coders and designers. It's likely they'd have been more alert as to how those platforms might later be hijacked by abusive actors. It wouldn't have solved those downstream problems. But it might have helped mitigate them.

Next, changing the priorities of venture capitalists would have an enormous effect. The first generation of social networks were built and funded by techies who regarded "connecting people" as the only goal, as danah boyd points out. They naively hoped that all the hard stuff—getting humans to actually understand each other, to bridge their big political differences—would happen automatically. It didn't. It can't. Merely creating "information flow," as Zuckerberg proclaimed his goal in the early days of Facebook (as his former employee Kate Losse describes it), isn't enough. (boyd is particularly annoyed by Zuckerberg's continual reference to the 2 billion global users of

Facebook, comprising people from Texas to Islamabad to Jakarta, as a "community." It shows his persistent misunderstanding of what community really is: a group of people who share a common bond and a common reliance on one another.)

The next generation of funders, she says, needs to demand different specs from the tech they invest in. The goal can't be just to connect people. The next phase is to develop tools specifically to help people reach across big gaps in their life experience and their values.

"Imagine if VCs and funders demanded products and interventions that were designed to bridge social divides," as she wrote in an essay on the future of tech. "How can we think beyond the immediate and build the social infrastructure for the future?" On the optimistic side, it's possible there's an appetite for this shift in Silicon Valley. When she talks to longtime geeks in that scene, they seem unsettled by what they've wrought. For them, "it doesn't feel as fun as it once was," she tells me. "It feels like *we were just trying to do good! But this stuff seems to be backfiring.*"

And tech needs to start routing around the trap of scale. Glitch's Anil Dash thinks coders with start-up ideas should consider avoiding venture capital—or take as little of it as they can. That's because investors will insist on hockey-stick growth, and that pushes the firm down the slippery slope to terrible design. It deforms nearly every young techie Dash meets. "We have to grow!" they'll tell him, panicked. *Why do you have to grow?* "Well, because we have to meet these goals." *Why do you have to meet these goals?* "Well, the limited partners that put the money into our venture expect a return on a six-year cycle, and we're five years in, and there's no way we're going to make it unless we grow like crazy." Understanding the huge ethical dangers that come with pell-mell growth isn't second nature for techies, he notes: "None of that is obvious, and they don't teach any of that in computer science programs." Granted, Dash's advice isn't easy for start-ups to follow. Many need investment to get off the ground. If that culture is going to shift away from promoting manic growth, it'll likely happen slowly.

And the final plank of reforming today's social software would be weaning it off advertising.

As the ex-Googler James Williams notes, advertising that seeks to monetize attention is what makes coders and designers become "adversarial" to their user. In contrast, software that doesn't rely on ads can focus on serving precisely what its users want. Wikipedia's management, for example, has historically decided to raise money directly from users instead of running ads. That meant it didn't need to trick them into endless clicking or to track their every movement online, as Sue Gardner notes. Wikipedia was designed to help you find the article you need, and then quickly leave. "Our desires were lined up with that of our users," she tells me. Gardner thinks this same principle is why Apple has been comparatively better at building privacy and encryption into its phones and laptops than other firms. Apple's customers are paying for the tech up front.

"When your customers are paying you money, you can actually call them customers and not *users*, which is a term from drug dealing," jokes David Heinemeier Hansson. He's a coder who watched ad-based businesses fall apart in the dot-com crash of the '90s and thus decided to write only software that people would be willing to pay money for. So he started making an organizational tool called Basecamp, and charging a relatively low subscription fee; soon he was employing dozens of people and catering to oodles of paying customers, without needing to track them or trick them into overusing his wares.

"There's a magical relationship you have with people who are paying you money," he tells me. It also means his company isn't clotted with "growth hacking" marketers, ad people, or "business development" critters, he adds. Nearly everyone he employs is a coder or designer, working on solving the real problems of his customers. "You get to be a company of builders," he says. "That is a wonderful feeling. We're in this to build a long-term, sustainable company."

Of course, these aren't always hard-and-fast rules. Ads have their defenders. As Twitter and Facebook employees have pointed out to me over the years, if they'd charged for their services, they'd have pushed away millions of lower-income people—and billions in the developing world. Many of those people have enjoyed enormous personal and economic benefits from social media, but they'd never have had the chance if required to pay even one dollar a month.

Still, it's important to consider these reforms—and to ponder many more. The civic role of code is only going to grow. We'll need vision and experimentation to grapple with it. And, crucially, we'll need advocates who bring that vision into the heart of the tech industry.

Are there any people poised to do that, though?

There may well be. Indeed, they're the ones who are the very subject of this book: the coders. It turns out they may have the leverage necessary to shift the way software gets made.

That's what Tyler Breisacher and his colleagues appeared to show, when they helped force Google to change course on a major ethical issue last year.

A programmer in horn-rimmed glasses, Breisacher is a nerd much like any of the ones I've long known. As a kid he messed around with coding, then studied computer science and physics at the University of Southern California, and landed a job as a JavaScript programmer for a software firm. But after a year and a half he got recruited by Google to help develop their Chrome browser. It was a dream job—"working on super fun stuff that's used by millions of people," as he says, and collaborating with wickedly smart and thoughtful colleagues. Breisacher worked on Chrome for two years, and spent six in total at Google, an eternity in tech time.

But in the last few years, dissatisfaction with the job began to creep in. He'd begun to hate the long 1.5-hour rides to work on the "Google buses." Apart from being a huge chunk of time in traffic, the buses had become a lightning rod for San Franciscans furious at how the influx of rich tech workers was jacking up rents in the city. Meanwhile, some employees were uneasy with Google's executives' recent overtures to the Trump administration; Larry Page had met with Trump in a tech roundtable soon after the president took office, which also annoyed a few vocal employees. Breisacher and other colleagues had also been lately finding that Google executives were increasingly offering vague hand waves when, in weekly staff meetings, employees raised ethical objections about company activities. "You'd get this wishy-washy answer, the type of answer you'd get from politicians," Breisacher says: *That's a good point you raised, and we're looking into it.*

Then in his seventh year, a true controversy emerged. In the fall of 2017, a group of Google staffers learned that their company was engaged in a contract to build AI for the US Department of Defense. Google had joined the Pentagon's Project Maven, for which it was helping the military develop software to recognize images in drone footage.

As one Googler, pseudonymously going by "Kim," told *Jacobin* magazine, the employees voiced their concerns to the head of Google Cloud, Diane Greene. But after several months, "it was clear the Googlers' efforts weren't going anywhere. The company was moving full steam ahead with the project," Kim added. So the concerned staffers wrote a post on Google's internal social-media platform, explaining Maven and describing their worries about it.

Now many more employees across the firm were deeply unsettled. They'd joined Google because they wanted to organize information—not to help the military better target foreigners. "Wait, what's going on? What are we doing?" was the tone of concern among colleagues, as Breisacher recalls. As the discontent rose, the original group of protestors crafted a letter to Google CEO Sundar Pichai, asking him to cancel Maven, and they posted it on the internal forums. "We believe that Google should not be in the business of war," the letter stated.

The response was electric: In under a day, a thousand Google staffers had signed it. By early April 2018, it had a staggering 3,000 signatures. Some of the firm's highest AI talent was hotly opposed to military work: When Google had bought the elite AI lab DeepMind in 2014, its heads had insisted that none of its inventions ever be used for weaponry.

Google tried to manage the staff revolt by holding all-hands meetings where employees could discuss their dismay about the Maven program. But "leadership got hammered," as Kim notes: At one extra-long meeting, a woman who'd been at Google for thirteen years said, "I have been working with you for so long. But this is the first time I can honestly say I don't trust you. Why didn't you reach out to us as employees? Why didn't you ask us what we thought?"

It didn't help, staff said, that management wasn't being straight about what was going on. Google executives initially told employees that Maven was a comparatively small contract worth only $9 million. But as word leaked out about the staff revolt, media coverage of Maven revealed it was actually part of Google's bid to win a much larger defense contract worth billions. (Indeed, behind the scenes, Google was deeply worried about how things would play in the media. The *New York Times* later got its hands on leaked emails from Fei-Fei Li, the company's chief AI scientist, in which she recommended being cagey about the nature of the Project Maven contract. "Avoid at ALL COSTS any mention or implication of AI," she wrote. "Google is already battling with privacy issues when it comes to AI and data; I don't know what would happen if the media starts picking up a theme that Google is secretly building AI weapons or AI technologies to enable weapons for the Defense industry.")

By the late spring, some employees were sufficiently dismayed that they decided to simply leave. After all, hey, they were tech workers. They had plenty of other options. Silicon Valley start-ups were clamoring for talent; any programmer or designer with Google on the résumé would get snapped up instantly. It gave them the freedom of acting on their moral positions.

Breisacher was one of them. "If you have the luxury to be able to choose where you work, because software engineering is still in demand—knock wood—then why not work somewhere that's a little more aligned with your values?" he tells me. By April, he'd lined up another job, and handed in his letter of resignation. Within a few weeks, he was one of about a dozen who'd resigned.

Not long after, Google blinked. The combined pressure from staff within and condemnation without appeared to be too much: In the first week of June 2018, Google executives announced to staff that when the Project Maven contract ended, they wouldn't renew it.

In essence, what the Google uprising revealed is the one weak spot for tech firms: their employees. Software companies famously compete over employees, paying enormous sums to snatch them away from each other—with Facebook poaching Uber, Uber poaching Google, Google poaching Twitter. This gives techies an unusual amount of good-old-fashioned labor power. They are the one group in the world that big tech firms strive to keep happy.

Everyday *users* of software do not possess this leverage, Breisacher notes. "With Facebook or Google you can't really boycott them the way you would a normal company," he says. "Google has however many million users, Chrome has however many millions of users. But as an employee, there's only a few thousand of us. So I feel like, if you're going to work at one of these companies where you have so much influence in the world, you have almost an obligation to think about how much power you have, and what you'll do with it." This power isn't complete, obviously. It only works if a large number of employees protest, as they did over Project Maven.

It is possible we'll see more such uprisings. Many more coders appear to be growing uncomfortable with the ethics and civic behavior of their firms. In the summer of 2018, news reports grew of how the Immigration and Customs Enforcement agency was separating migrant children from their parents and throwing them in crude detention centers. Microsoft employees learned that their firm was providing software to ICE, and over 300 signed a letter demanding the company cancel its contract. "As the people who build the technologies that Microsoft profits from," they wrote, "we refuse to be complicit."

Meanwhile, employees for companies like Facebook and Twitter were discovering that in the wake of the election, it was becoming embarrassing at parties even to admit whom they worked for. "I would never say I worked at Facebook," as one 30-year-old software engineer told the *Guardian*. If asked, he'd be vague. "There's this song and dance you learn to play," he added, "because people are quick to judge." A decade earlier, Wall Street had been the industry filled with arrogant, preening creeps who were wrecking the country. But now, in the US cultural imagination, Silicon Valley coders were becoming the new cretins. They'd gone from being the heroes of the movie *Hackers* to the bumbling, self-involved buffoons of *Silicon Valley*.

So it may be that pressure for change in the tech industry emerges not from without, but from within—from the one group of people VCs and CEOs cannot afford to entirely ignore.

These days, Breisacher is much more pleased with the ethics of his new employer, Hustle. It's a smaller start-up that makes SMS software so that organizations can build and maintain one-on-one relationships. It's been used by many nonprofit groups like the Sierra Club and Planned Parenthood.

But there are times, he admits, when he thinks back on Google. There was a lot he liked. The awesome technical challenges; the driven coworkers; hell, even those famously tricked-out cafeterias.

He sighs. "I miss it."

< Chapter 11 >

Blue-collar Coding

In 2014, Garland Couch was just another laid-off coal-mine worker in Kentucky.

An orange-goateed 41-year-old at the time, he lived not far from Pikeville, a town of 7,000 nestled in the winding green mountains of east Kentucky. For 15 years he'd worked doing preventative maintenance on coal-mining equipment, toiling in coal like his father and his grandfather before him. Coal formed his life. "When you work with a mine, those folks become your family," he tells me.

But in the early 2010s, the coal industry went into a steep decline, as fracking produced a boom in cheap natural gas, renewables became more and more viable, and the federal government enacted regulations designed to move the country away from coal. In 2008, Kentucky had more than 17,000 coal miners. Only eight years later, that number had dropped to 6,500. The region was increasingly dotted with huge stacks of coal that didn't have any market. And now Couch was realizing he didn't have many good options. He knew of an industrial maintenance job in Louisville, but it was too far to commute to, and he didn't relish the idea of uprooting his wife and daughter for a job that itself might not last long.

While he pondered those alternatives, he heard a curious ad on the radio.

"Have you been laid off from a job in the mining industry?" it asked. "If you are a logic-based thinker willing to work and learn new things, we have a career opportunity for you. Bit Source is bringing the computer coding revolution to Eastern Kentucky."

What the heck? Someone was offering coding jobs—in Pikeville?

The job offer was the brainchild of Rusty Justice. A salt-and-pepper 55-year-old at the time, he, too, had worked in mining his whole life, running his father's coal-shipping business and a land-formation company. But he was shaken by the vertiginous decline of coal. "We all saw a downturn coming, but we did not foresee the complete collapse of the economy," he told me.

Justice realized he needed to build a completely new business in Pikeville. It had to be in a sector that was growing, not collapsing. And it had to offer jobs that paid well: Kentucky coal jobs paid over $82,000 on average, so well that each job supported several others in local grocery stores, bars, and car dealerships. The region needed tent-pole salaries. Justice and his business partner Lynn Parrish tossed around concepts, from alternative-energy plays like wind farms and solar farms; "even hog farms," Justice jokes.

Then one day in 2013, during their research, they visited a tech incubator in Lexington, a few hours away. It looked like your standard airy, sun-soaked Silicon Valley start-up space, replete with huge leather sofas and a Ping-Pong table; coders pecked away at tables. The head of the incubator told them that local tech firms couldn't find enough local programmers to hire. There wasn't enough talent. The jobs were good, paying up to $80K a year.

That's great money, Justice said, but he figured you'd need a computer science degree. Nope, said the incubator head. Anyone smart and committed could learn programming on the job, just like with any skilled trade.

Now *that* got Justice's attention. He knew that mining workers were intelligent, trainable, and frankly already steeped in technology. "We're perceived by people that we're not smart, that we're hillbillies," he tells me. But coal miners already work like programmers: They sit in one place all day long, patiently running high-tech equipment and solving problems. "It's a highly technological business," he says. "You have this image of a guy who has a pickax and lunch bucket. But they use robotics, they understand fluid dynamics and hydraulics.

"Coal miners are really technology workers that get dirty."

He and Parrish decided to bring coding to Pikeville. They'd find some talented out-of-work miners, train them, and start doing contracts for anyone who needed an app or website built. Soon they'd found a coder who was willing to train incoming staff; they'd lined up federal funds to help pay the workers while they were training. They found an office in a former Coca-Cola bottling plant in Pikeville that, crucially, was right next to a high-speed internet line. (In Pikeville, as in much of Appalachia, broadband wiring is scarce to nonexistent.) They dubbed the company "Bit Source" as a play on the digital bit and the bitumen of mining.

Would any miners actually *want* to be coders, though? Justice put out the radio ads and a print ad. "We had eleven slots, and we figured we'd get maybe fifty applicants," he says, "if we were lucky."

They got 950. Stereotypes about hillbillies are, it turns out, precisely that; local residents were practically tripping over each other to compete for a job in software. In fact, the response was so huge Justice's trainer built a database so they could sort through all the applications. They wound up creating a preliminary test to weed out all but the best, a mix of math and psychological questions like: *Would you rather overhaul an engine or give a presentation?* That narrowed the pool to 50, so they did another round of testing; finally, they interviewed the top 20.

Garland Couch was 1 of the 11 who passed the tests and the interview, and got hired. Like the other 10, he frankly didn't see mining rebounding; it was time to try something new. On his first day he walked through the Bit Source doors marked with the slogan "A new day, a new way"; the walls were adorned with murals of local historical figures, like John CC Mayo, the investor who helped bring coal mining to the region. "Learn as if you are going to live forever, live as if you are going to die tomorrow," read a slogan on another wall. "What have I got myself into?" Couch asked himself.

"From now on, you stop thinking of yourselves as unemployed coal workers," Justice told them. "You're technology workers." Among the new hires were a mine safety inspector, an underground miner, and a college-educated mechanic who'd fixed conveyer belts in the mines. They'd all get internship wages of $15 per hour while training, then a raise after they were actually doing work. Justice and his partner were investing their own money in the salaries, and it'd be up to them to scare up enough work to keep the team employed.

The miners began intense, daylong cram sessions on HTML, CSS, and eventually JavaScript and mobile-app languages.

"It was kind of like trying to drink from a firehouse," recalls William Stevens. Mining was physically exhausting, but "this was mentally draining, the most mentally intense work that I've ever done." Stevens had showed up at Bit Source after being laid off from a surface mine and hustling up a new mining job in a town three hours away. He left his wife and three daughters in Pikeville and would sleep in his car while off at his job, returning only on the weekends. He was, he says, insanely motivated to make coding work—he couldn't handle that sort of brutal commute, chasing vanishing coal jobs around the state. The miners were so intent on grasping code that Couch didn't learn his seatmate's last name for weeks.

Slowly, the work began to make sense. They'd end each week by making a small project with what they'd learned—a simple web page at first, then an interactive one, then ones that stored data in databases. Justice was right: They were rapid, attentive learners. One hard part was adjusting to the faster, looser ethics of development. In mining, moving slowly and meticulously was key. If you made a single mistake something could literally blow up, endangering lives and costing a company millions. With code, in contrast, the point was to iterate quickly and fix things as you went. Mistakes were fine; they were a given.

Within months, they were actually shipping code. Much of it was, at first, simple—a website for Pikeville's city council, sites for crane and earthmoving companies. But then it got steadily more complex, including an app that let patients redeem vouchers at farmers markets, an augmented-reality app, and—in a bit of pro bono work deeply relevant to Kentucky—an app to help communities struggling with opioid addiction.

Three years in, Justice's investment was close to breaking even. The hard part is sales, he says, getting clients to look past their coastal snobbery about hiring a bunch of *hillbillies*—a word Justice uses proudly—with thick Appalachian accents. But his programmers were now full-fledged developers, hobnobbing with MIT computer science professors at meetups and impressing Linux experts with their meticulously documented code. "They didn't have any bad habits to unlearn," Justice notes. Stevens had discovered he loved front-end design, tweaking the CSS and fonts on a site for hours. "That whole experience of when you see someone's face light up, that's just awesome," he gushes.

And Justice had become a minor celebrity in the world of economic development. He was fielding calls from all over the world, with civic leaders asking how they, too, could "do a Bit Source." It wasn't easy to give advice. You had to be pushed to the edge, confronted with a totally collapsing economy.

"It was motivation through starvation," he notes. "My daddy used to say, *Life ain't fair, so wear a helmet.*"

But he's convinced that Bit Source is, in its own small way, a glimpse of the next phase of coding. The first mass wave were the personal-computer pioneers, the nerdy kids who started with Commodore 64s or early HTML and parlayed it into millions. But now coding is mature. It's become more of a ticket to the middle class; something that the great mass of people can see as a route to reasonably stable, enjoyable employment. It's like, in other words, pretty much what mining used to be around Kentucky.

"These are blue-collar workers," Justice says of his programmers. "And this is blue-collar work."

Justice is right. The entire world of programming is now growing so quickly that it's changing the nature of who becomes a coder, and why.

We're likely to see the mainstreaming of coding—its rapid growth into a skill that many more educated adults, and possibly a majority of them, need to possess to some degree. The reigning image has been of the hoodied young Zuckerbergian coders, cribbing together apps they boast will "change the world." Now the new shift is toward programmers aiming instead for stable, high-quality jobs—a replacement for the well-paying trade work that sustained the middle classes in the latter half of the twentieth century. It is, as my friend the technology entrepreneur and thinker Anil Dash (whom you heard from in the last chapter) dubs it, the emergence of "the Blue-collar Coder."

The demand for programmers is exploding worldwide. In the US, the Bureau of Labor Statistics predicts that jobs for computer and IT workers will grow by 13 percent from 2016 to 2026, faster than the average for other occupations. The pay is likely to be high—it was already an average of $84,580 in May 2017, more than double the median annual wage for all job types. And there aren't yet enough people moving into coding to satisfy the demand. By some predictions, there will be 1 million more programming jobs than trained coders by 2020 alone. What's more, many jobs that aren't officially "software development" are increasingly requiring at least a bit of coding. A report by Burning Glass Technologies, a job-market analysis firm, found that knowing some amount of programming was an advantage in 20 percent of "career track" jobs in 2015. And all these coding jobs that are currently open, or are opening up in years to come, aren't just clustered in Silicon Valley. Indeed, barely a tenth of the programming jobs in the US are located in Silicon Valley; the rest are spread across the country, in every city and town of any size.

Where will all these future programmers come from?

The most straightforward route will be the same one people take today: going to college or university for a four-year computer science degree. Assuming a student performs well and it's a well-connected university, high-profile employers fight over hiring top computer science graduates. Indeed, those universities are so well connected to internships at major firms that many grads have offers in hand before they're finished with their coursework. When I hung out with two students who'd graduated the year before from Columbia University's computer science program, one had been scouted by Facebook, Microsoft, and a database firm in New York, and the other had juggled offers from Lyft and Twitter. "This is what it's like for basically everyone who came out of our program," they told me. And founders of high-tech start-ups tend to hire the friends they met at college; get into the right class and it's a ticket to a well-financed tech job.

The upshot is a growing stampede of students eager to study computer science. The number of US university undergraduates declaring it as their major doubled from 2011 to 2015—a mere four years.

How crazed are students to be coders? Consider that in 2015, fully 20 percent of all Stanford undergrads were pursuing computer science majors; it is now the most popular major on campus. Professors in other fields at Stanford—particularly the much-benighted humanities, left aside in the crush for STEM teaching—are increasingly left twiddling their thumbs as they watch students abandon fields like English, history, and anthropology. But the students know that the demand for coders is almost limitless. "If every Stanford student, heaven forbid, were to major in computer science, none of them even would have to leave *Santa Clara county* to get a job," says Eric Roberts, the former director of undergraduate studies for computer science there (and the guy you met in the chapter on women in coding).

To be sure, there were previous booms in computer science—in the early '80s, then the late '90s. Each one cooled down; they were a temporary vogue. But this time, observers argue, is different. The demand is likely to stay high for years, because of the sheer spread of coding into every industry. It's needed not just in start-ups but in insurance firms, local banks, the entertainment industry.

The kids flooding into CS have concluded precisely the same thing as Rusty Justice down in Kentucky: that programming is a reliable route to a steady job. In the world of these students—increasingly defined by precarity and college debt more mountainous than any generation before—coding promises a middle-class stability of the sort that has increasingly vanished from the American economic landscape. Previous surges into computer science were driven by boom-time mentality, kids looking to become Bill Gates or Mark Zuckerberg. Now they just want to become a well-paid employee.

"It's less of the, 'Wow, I could become a billionaire,' like we saw in the '90s," Roberts says.

The rub is, there isn't enough room in computer science schools to create that middle class. There aren't enough professors or instructors. Only about 2 percent of Stanford faculty are computer science professors, for example, which means that tiny slice of profs is already stretched thin as they try to teach 20 percent of all undergrads. Other colleges are in the same boat. They can't hire enough computer science profs. They don't have infinite budgets, and even if they did, they're competing for talent with high-tech firms, which frequently lure away senior faculty with high-six-figure job offers and open-ended research opportunities, which no college can match. PhD students, too, graduate and head to Uber or Google instead of coming back to teach. The upshot is that many computer science departments—as their own professors worry—are rapidly becoming overcrowded sardine tins. One single machine-learning class at Stanford had 760 students (and this was a graduate class). The only other alternative is to simply limit access: to tell most kids, *sorry, you can't study computer science here*; to admit only students who have absolutely top-flight grades.

"The alternative is to make our classes gigantic," muses Ed Lazowska, a computer science professor at the University of Washington. And that's not something they're willing to do; it would degrade the education too dramatically. The state has been fairly generous about giving them more money to increase in size, but even so, they haven't been able to meet the demand. "What this means is that we've been jerks about only expanding when we've got funding to expand."

But if computer science education can't expand fast enough, there will need to be many other ways for people to horn their way into the industry.

Avi Flombaum is trying to figure out one of these new routes. He's the founder of the Flatiron School, a boot camp that takes people and, for about $15,000 in tuition, puts them through an intense 15-week training curriculum. When I visit their campus in the Wall Street district of Manhattan, about 200 students sit at long tables, working in pairs as they puzzle through the nuances of Ruby. One student is sketching out a snippet of code for his partner with a dry-erase marker, writing right on the table itself. "They're whiteboards," Flombaum notes. "It took us time to figure it out like that, because we didn't want students moving around whiteboards and creating a big mess." But he wants them all talking and sharing their work nonstop, because "that's how you learn." On the far wall is a mural of nerdly dimensions: a huge looming head of Grace Hopper, an '80s-era PC, the phrase "LEARN LOVE CODE" in a spray of curved graffiti.

Flombaum is a 34-year-old who wears his hair swept in a thick mop back across his head; on his right shoulder is a tattoo of the logo of the Flatiron School, and on his chest is one for WeWork, the firm that bought his company. "I'm like a race car," he jokes. Flombaum is a self-taught coder who, after working for a hedge fund and running his own start-up for four years, started teaching five-week programming classes "for fun." He wound up getting a bunch of his students jobs, and thought, hmm, maybe he could scale this up. He and a partner launched the Flatiron School in 2012, and since then it has graduated almost 2,000 students.

Flatiron is, like many boot camps, renowned for being an absolute cram of knowledge. Before admission, students are encouraged to complete a free 15-week online course that introduces them to the basics of Ruby or JavaScript. While they're in session, many stay late into the evening, working on projects with colleagues. About half of the students are women, and most are young, including students who finished college but decided coding was a better bet for employment than the subject they majored in; others had been in the workforce but didn't like their job and wanted to switch careers. One recent student came from a pig farm in Texas.

Victoria Huang, a 25-year-old from New Jersey, is a neophyte who's only in her third week when I visit. She tells me she studied pharmacy at her parents' urging and worked for one year in Manhattan at a hospital, but she never felt passionate about it. What she dreamed of was making software: She'd spent her preteen years making anime websites, "and I just loved that feeling of making something from scratch." She thinks that her pharmacy training, in many ways, prepared her for the pace of the boot camp. "This form of super-intensive training is exactly like what it was like at the hospital—you're just absorbing all this practical knowledge all day long, and it works," she says. Nearly all her fellow students are like her, having worked for some years—lawyers, marketers—listlessly, feeling they had an itch to create things. "When I finished the first project at the end of our first session, I've never been more proud," she says.

Nobody is going to become an elite programmer in merely 15 weeks, as Flombaum admits. They're not going to learn the material that gets covered in a four-year computer science degree. College computer science courses focus heavily on the abstract theory of how computers do what they do, as well as on computer architecture and design. Those students graduate having achieved a nuanced sense, for example, of efficiency in crafting algorithms. They'll know the big O theory that tells you whether a particular sorting algorithm is maximally efficient—whether it'll sift through a 10-million-item data set in a few minutes or whether it'll take a few hours. It's these computer science grads, mostly, who'll create entirely new innovations in code, new forms of cryptography or AI or innovations like the Bitcoin blockchain. That's why the elite tech companies prefer computer science college grads and can be uninterested in boot-camp folks. "Our experience has found that most graduates from these programs are not quite prepared for software engineering roles at Google," Maggie Johnson, Google's vice president of education and university programs, has said.

A boot camp is more like traditional vocational training, akin to an old-fashioned trade school from the '60s or '70s. They acknowledge that boot-camp grads aren't going to learn how to innovate anything radically new. The students will focus just on learning how to do some of the most common things coders do in everyday jobs: how to design a web service that stores stuff in a database, retrieves it, and displays it for others to see.

But the thing is, that dull and uninnovative stuff? That's precisely what most coding work is actually like, once you step outside the empyrean of the elite tech firms like Amazon, Google, Baidu, or Alibaba. "A very tiny percentage of our industry works on genuinely hard algorithmic problems & on research," C. J. Silverio, the chief technological officer of the tech firm npm, noted on Twitter. "Most of us concatenate strings all day." (*Concatenating* means, say, appending the text-string "Clive" with another text string like "feels good!" or "feels lousy!" Silverio isn't joking; that's a healthy chunk of what coders do all day.) Every corporate website you see online has someone whose job it is to sling the JavaScript that manages all the interactive bits you click on. They're not constantly creating new gewgaws on the site; they're not building new stuff. No, they're often just maintaining the things that exist, in a job that's rather like being a digital plumber. When Google updates the Chrome browser? Or Microsoft updates its Edge browser? Somebody's gotta tweak the JavaScript to make sure it stays current with the latest browser specs, or things can break. They are like building superintendents, checking to make sure stuff isn't leaking.

The bootcamp phenomenon is global, with schools opening up in every country where coding jobs eclipse the supply of coders—which is to say, all over the world. In India, the veteran coder Santosh Rajan founded Geekskool, to try to address the fact that traditional Indian computer science schools are "ten years behind the times," as he says. Those schools to produce programmers who are well trained in old languages and databases, so they can go to work for India's huge firms that do back-end work cheaply for massive corporations worldwide. It's dull and reliable work. But let's say you wanted to study a new and rising computer language, like JavaScript or Ruby? You need an Indian boot camp, he says. Rajan's students have ranged from physicians to computer science grads who realized their degrees only qualified them to be spiritless database wranglers, and who crave a bit more. They're all hungry and driven, which is, he avers, all one needs to become a useful coder.

"There is no difference between becoming a great programmer and a great musician or great football player," he tells me. "It's the same thing—it's just that determination: *I want this.*" The promise of Geekskool is to align graduates with what the market actually wants—and to do so quickly, for far less money than a four-year college.

Fueled by this promise, boot camps have exploded in recent years. In 2013, boot camps in the US graduated 2,178 people. Only five years later that had risen by an astonishing factor of almost 10X, and more than 20,000 students were pouring out doors of boot camps, looking for jobs.

Does this quick-and-dirty route to coding work? How many people truly become software developers that way?

It's difficult to know for sure, because the quality of boot camps varies wildly. Some are well established, in operation for years, and offer audited reports on how often their grads get work. Others are sketchy and under-regulated. One survey by Course Report, a firm that studies the field, suggests that about two-thirds to three-fourths of those who go through boot camps find work in the field. Flatiron is comparatively transparent about their hiring rates; in an audited report, they found that of the students who graduated between November 2015 and December 2016 and looked for work, 97 percent found some sort of software engineering work in a six-month window around graduation—about 40 percent getting full-time work (with an average salary of $67,607 a year), half winding up with contract work, paid internships, or paid apprenticeship work that averaged $27 an hour, and the rest freelancing.

One student who'd enjoyed some success is Luis De Castro, a 29-year-old who lives in San Francisco. When I saw him at a GitHub conference in the fall of 2016, he was hunched over his laptop, his thick dark dreads spilling out of his hoodie as he frantically patched together a demo that used IBM's Watson AI to take text messages and identify their emotional state. He'd recently finished a stint at Dev Bootcamp, a since-shuttered school nearby, and was trying to get the demo working to show some fellow students at an event later that evening. It wasn't working, at least not yet. "Oh man, I have no idea what's going on here," he said, and laughed, as the code coughed up error after error, before he finally quashed the bug with minutes to go.

De Castro was born in the Philippines and raised in California from childhood onward. He loved automobiles, and after high school he got a car-washing job for BMW; he gradually bumped up the ladder until he landed in paperwork jobs that he found enervating. He quit and enrolled in college, but found that, too, couldn't keep his interest. He dropped out.

"I didn't know what the hell I was going to do," he told me. While pondering his options he spoke to a friend enrolled at Dev Bootcamp.

De Castro was intrigued and tried out its free online course. "I find this fun, and I understand it, and I'm pretty good at it," he realized. It took him a while to scrounge together the $15,000 in tuition—"That was rough; I was unemployed!"—but he finally enrolled. The boot camp was exhilarating, if somewhat terrifying in its intensity. He showed up early and stayed late, and he discovered he loved the Zen-like state of coding. "I love the time I have in the morning: I have my code, I have my laptop, and I have my headphones. It's like my peace time. I'm coding and nothing is bothering me."

Finding a job after graduation was not a foregone conclusion. De Castro figures only a third of his peers got jobs as programmers, while another third got something adjacent (a marketing job in a tech firm, say), and the final third abandoned the area entirely. He set up a regimented pace, applying to 15 or 20 jobs a day—"It was like spray and pray, you apply to everything without even looking." At first, auto-emailed rejections piled up in his in-box. Then he began inquiring with prospects: Why didn't you hire me? *We need more experience*, they'd say. Finally one application he'd sent via a job posting on LinkedIn resulted in an interview, his first, with Funding Circle, a company that helps small businesses find investment. De Castro figured this would be a rejection, too—almost no one got a job off their first interview. But, amazingly, they hired him.

Actually working as a coder was like treading water: He didn't feel prepared and was googling everything like mad on the job. Then one day he realized, hey: *This is actually fine.* He felt a burst of confidence for the first time. When I spoke to De Castro after a year on the job, he'd experimentally put his résumé on a job site online to see what he was now worth. The site asked for a minimum salary; just for the heck of it, he put $140,000. "And I'm getting emails from people saying 'senior level position, we want to interview you'!" He laughs, astonished.

Not everyone has that success, of course. Mitch Pronschinske is a tech-minded writer and former editor at *TechBeacon*, a website that covers coding. He got sufficiently interested in the field that he decided to give it a try himself. Since he couldn't afford to stop work entirely, he picked an online-only boot camp—Bloc—that cost him $5,000 and was something he could do in the evenings and on weekends. He beavered away at the classes, and the mentor he'd been assigned was encouraging, telling him, "Oh you could totally work for a start-up."

But when Pronschinske finished the program and looked for work, he quickly realized he wasn't anywhere near qualified. Employers were looking for someone with a GitHub repository filled with code projects they'd already built in their spare time. All he had were the "two or three" demos he'd done while at Bloc.io. The firms, in essence, wanted what they've always wanted—a passionate hobbyist who loves code for its own sake, not someone with a just-enough amount of vocational training. And he, Pronschinske accepted, wasn't that person. He didn't like coding enough to do it in his spare time, just for fun.

"I accept some of the responsibility," he says. "It wasn't fun for me to code night after night." But he also thinks the boot camps should be clearer about winnowing out those who are in it solely for good work—because he doubts the industry wants anyone like that. "Don't think of this just as a money grab," he says.

The boot camp world has a spotty record of being honest about its success rates. Many hype their hiring rates with dodgy stats; the California government issued cease-and-desist letters to several schools after finding they were operating without licenses. Subsequent complaints found that one, Coding House, boasted that its students had been hired at 21 different companies, but the government found that only 2 students had in reality ever been hired by any of the firms. (California levied a $50,000 fine on the founder and ordered the school to be shut down.) Even Flatiron was fined by New York State in a dispute over its hiring figures. (It paid a $375,000 settlement in the fall of

2017, to be disbursed to any eligible student who filed a claim.) In essence, the world of boot camps overlaps uneasily with the world of for-profit postsecondary schools in the US—which critics accuse of targeting kids who can't afford a four-year college, convincing them to take on big student debt, and then not giving them enough support to graduate or find jobs.

As Tressie McMillan Cottom, a scholar who has studied for-profit institutions, argues, boot camps seem poised halfway between the elite colleges like Stanford and low-graduation-rate, for-profit schools. They're self-selecting—at the moment, they mostly attract students who are more educationally ready, and less financially desperate, than the working-class kids drawn to for-profit colleges. "There's a very small pool of academically well-prepared, financially well-to-do students who can afford to pay for this training and who can benefit from it," she told *Logic* magazine.

If you wanted to actually use coding as a force for social mobility, you'd want to get more of those poor and working-class kids into programming. Boot camps are too expensive for them, or simply not culturally on their radar. Community colleges, by contrast, are a much cheaper and better-regulated route. The problem community colleges face is the "capacity crisis": With their relatively low salaries, they can't compete with elite universities—and frankly, the elite universities are already losing professors to tech firms. "If we can't hire, the community colleges sure can't hire," as Stanford's Roberts notes.

One of the things that makes coding weird, as an industry, is that people can teach themselves how to do it.

There aren't very many technical professions that work this way. Boeing does not employ engineers designing planes who taught themselves—who just sort of *fiddled around* building Dreamliner jet wings in their backyard, until they, you know, mastered the craft. You or I would be horrified at the prospect of having eye surgery done by someone who'd never been to med school and instead just practiced on friends. Yet we routinely use software that was written by people who never formally studied code at all, either at college or even at boot camp. They just started messing around with coding and learned it. Coding is, as several programmers pointed out to me, a curious technical trade because the tools for creation are as easily accessible to the amateur as to the pro. If you wanted to design a new mobile phone, you'd need to have some seriously elite tools at your disposal: a microchip-fabrication plant, a precision soldering station, maybe a 3-D printer to prototype the case. But if you want to write software, all you need is a basic laptop. Many "code editors," the tools used to write code itself, are free. These relatively low barriers to entry are partly why, according to surveys by Stack Overflow, 69 percent of coders are "at least partially self-taught," 56 percent never received a computer science degree, and 13 percent said they were "entirely" self-taught. It is an astonishing number for a well-paid technical field.

Quincy Larson is someone who learned this latter way. In his 20s, he ran English schools in China and then in Santa Barbara and knew nothing about tech. "My wife configured the wi-fi router," he says. But while directing his Santa Barbara school, he got annoyed at the many repetitive tasks he and the other employees had to engage in: cutting and pasting info into reports, formatting documents. In a move that by now (in this book) ought to be immediately visible as a sign of a new programmer being born, Larson wanted to automate it. So he read online how-tos on AutoHotkey macros, and soon he had crafted little shortcuts that saved staff hours each day.

"Our school became more efficient—and then more popular, because our teachers were able to hang out with their students more," he tells me.

Seized with curiosity now, Jones began sampling any learn-to-code materials he could get his hands on. He watched the online MOOC coding lectures from major universities that, back in the mid-'00s, had started to come online; he joined a local hackerspace and went to programming meetups. After seven months of this, he wound up being interviewed as a developer by a firm that mistakenly thought he knew more than he did—but decided to hire him anyway, figuring he knew enough already to learn on the job. In 2014, Larson began to pull together online resources and develop a curriculum; then over three days, he turned it into a free and nonprofit website called freeCodeCamp. It rapidly became one of the most popular spots for newbies to teach themselves; users praised it for carefully stepping neophytes through the basics of making a crude web page all the way up to running servers and databases. Doing the whole course would take 1,800 hours. Larson knew that people couldn't learn unless they had live people they could ask questions of, so he set up an online forum for those using the course, and he urged users to create face-to-face meetups. By 2018, millions were using it online, and users had met up in 2,000 locations offline to talk.

Who are these people? Who's actually trying to teach themselves online? When Larson surveyed almost 20,000 freeCodeCamp users, he found that they were mostly late-young adults trying to change their careers. The average age was 29, with perhaps a fifth of all users older than 35. They were mostly men—only one in five were women—and about half had a bachelor's degree or higher, though often not in anything technological. Almost two-thirds were outside the US, and about half didn't speak English as their first language. A significant minority had kids and were learning to code in their scant spare time. Perhaps most remarkably, 25 percent had actually gotten a programming job after teaching themselves to code.

Larson has become one of the most vocal advocates for the idea that virtually anyone can learn programming. "I think that anyone who's sufficiently motivated can learn enough to be employed," he tells me. He's particularly intrigued by the potential of free online courses to help "the huge volume of people who are living on less than $2 a day." India is the biggest market for freeCodeCamp outside the US; Brazil, Vietnam, and Nigeria are all in the top ten. Many users are, he says, clearly extraordinarily self-motivated by sheer need: There's no chance they could afford even a boot camp.

It can strain credulity to think that more than a tiny minority of people could truly teach themselves into a job. I spoke to several, though, who had indeed started with freeCodeCamp and wound up with jobs. One was Andrew Charlebois, a 29-year-old man in Montreal, Canada; he'd spent his twenties working as a carpenter and took the online courses during a winter lull, after he'd been laid off from a job. He discovered building interactive websites was oddly like carpentry—the finish and fit of components on-screen wasn't far from making things neatly square with wood. Four months after starting the online work, Charlebois began applying for developer jobs; after 78 applications he, to no one's surprise more than himself, landed one. "At the time I was googling stuff like, 'Is twenty-seven too late to do a career transition?' I was worried," he said. When I spoke to him, he'd been in the job for almost two years, and was now tutoring other newbie coders.

Could online self-education and boot camps help widen the doors to the industry, given how low its hiring rates of women and minorities have been? It clearly works for some, though when it comes to boot camps, critics like Tressie McMillan Cottom are dubious; she argues such routes tend to work only for "the cream of the crop," the ones with titanium-solid motivation or enough socio-economic backing. Self-teaching would not, to put it in nerd terms, scale.

What perhaps works even better, arguably, is employers actively hiring and training from nontraditional groups—offering paid on-the-job training, like what Bit Source did. And given the ravening need for developers, this, too, is emerging as a trend.

One software firm in Baltimore, Catalyte, uses an online aptitude test to identify neophytes who'd make good coders, then puts them through five months of intensive training. The online test was originally the brainchild of a Harvard teaching fellow who argued that traditional methods of hiring coders—sifting through résumés and doing interviews—too often failed to produce high-performing employees. "What he also found is that it reinforces social bias around class, when it comes down to it," says Jake Hsu, Catalyte's CEO: Well-off white kids had better-groomed CVs and educational specs, which meant they were getting more tech jobs regardless of whether they'd actually make good coders. So Catalyte instead focused on advertising its aptitude test broadly in areas that traditionally didn't produce coders and automatically offering a training opportunity, with a job guaranteed upon completion, to anyone who scored highly on it—without even asking for a résumé or doing a formal interview. Their first test city was Baltimore.

Aptitude testing is, of course, a great echo of the coder past. In the early days of software—back in the '50s and '60s—firms had no way of knowing who would make a good programmer, so they relied heavily on puzzles and pattern-recognition tests to figure out who could think logically. This opened the door to many different demographics, since it turned out lots of people who were perfectly adept at doing these sorts of puzzles. As we saw in the chapter on "The ENIAC Girls," it took a few decades for coding employers to pivot away from the raw aptitude test and begin to prioritize "culture fit," a shift that greatly favored men.

In the Catalyte test, applicants work on math questions, puzzles, and a writing section, with the system monitoring the flow of how the test-taker works. It verifies, as Hsu notes, "a person's ability to synthesize complexity—people who are able to take very complex things and break them into small slices."

Blind testing, deployed in a less-affluent city, appears to work quite well, judging by Catalyte's results. The firm currently employs around 600 people, and their demographics are remarkably more varied than the software industry at large. In their Baltimore office, 29 percent of the developers are African American, around three times higher than estimates of the industry's average. "Most are from working-class backgrounds," Hsu adds, and 44 percent didn't do a four-year degree. The average age is 33 years old, with most of their hires seeking a second career, and a better one.

"Talent truly is evenly distributed, or at least the aptitude to do this work is evenly distributed," he notes. One woman I spoke to who'd been working at Catalyte for two years, Carolina Erickson, was 35 and had studied music performance in college, trying for years to make a living as a flute player before concluding that she wasn't going to make it. Working part-time in tech call centers revived an old interest in web development, so she took a college course in it; then she came across Catalyte, she passed the test, and was hired. She's worked on teams there that helped revamp parts of the ticketing system and the iOS app for StubHub. Erickson suspected that second-career coders had some advantages that fresh-out-of-school computer science grads didn't, including a lot of experience working on teams.

"There's some overlap with playing in an orchestra," she said. Playing flute also required her to rapidly learn and think on her feet—as when sight-reading music—and to synch closely with workmates.

One benefit of hiring employees from nontraditional fields is that you gain their broader perspective. Eager computer science kids fresh out of college have little life experience. They easily fall into the naive, cocky belief that they can solve any problem—and often fail to even notice the real problems in the world that might be tackled with clever software, because they've not encountered them. And if they've studied very little of the humanities—history, sociology, literature—they often have what Northrop Frye might have called an "uneducated imagination": They have little ability to envision what motivates the users of their software. They're great at grasping the binary soul of the machine but not the quantum weirdness of human psychology.

"People with liberal arts degrees, they're critical thinkers," says David Kalt, CEO of Reverb, an online marketplace for selling music gear. Kalt is himself a musician and former music producer who retrained as a software engineer in the '90s, creating and selling a tech start-up. In 2013, having noticed that musicians complained about terrible experiences buying and selling instruments on eBay or Craigslist, he created Reverb; it offers a bundle of useful wares for gear selling, including a tool for figuring out what your drum set or rare old guitar pedal is worth. He used to advocate for more kids to study computer science—but after building two software companies, he noticed that his most productive developers had all studied humanities, like philosophy or political science (the latter of which is his own bachelor's degree). Now he argues the opposite: The software industry should be hiring those who studied broadly, and then either taught themselves to code on the side or retrained later.

"We've hired a lot of people out of boot camps," he tells me. "And you can tell the ones that will be senior developers in two years. Honestly, it goes back to their liberal arts degree. Critical thinkers can absorb new languages in a more dynamic way than more linear thinkers." Reverb also inherently winds up hiring engineers who are musicians, because they deeply intuit what musicians are looking for when buying online: large original pictures of the gear, videos of it in use, insurance policies covering expensive gear being shipped around the world by FedEx. And many musicians are also, Kalt points out, technically adept at their instrument, yet completely self-taught.

They know how to learn, how to practice until they can do something off the top of their head, he adds.

There are, of course, lots of subtle problems that emerge as more people crowd into coding. One is that the jobs appear to be bifurcating by gender. As women and minorities move into the field, the areas in which they work decline in prestige.

This is currently happening, it seems, to front-end coding. One common way for neophytes to learn coding is by hacking HTML and CSS to make simple websites, then adding JavaScript to make the sites interactive. A lot of online pop culture in the '90s and mid-'00s encouraged kids to do so, like the Neopets site, where kids could tweak the code to customize their digital pets and their homes; the old MySpace worked the same way, with teens terraforming their pages into intentionally aggro displays of glitched-out design. (I've heard many front-end engineers coo about the hundreds of hours they spent on Neopets.) Ditto for lavishly tricked-out tribute sites to a teenager's defiantly obscure favorite TV show or band. Video games were the door that got teenage boys into coding back in the Commodore 64 1980s—but it was pop-cultural web fandom that proved the onramp for a more diverse generation in the next two decades. These fans honed their web skills and walked into the exploding job market for web designers.

This is likely part of why the most common coding jobs for women are in front-end work. When Miriam Posner—a coder who teaches Information Studies at UCLA—looked at Stack Overflow coder-job data, two top titles listed by women were "Designer" and "Front-End Developer" jobs that fit in that basket. There were considerably fewer women working in back-end jobs that involve wrangling servers and databases, or in newly hot areas like blockchain or AI. In those areas, men rule. And the men are paid more for it: Front-end jobs, she found, pay on average about $30,000 less than back-end work.

The upshot, Posner notes, is that when women move into an area of coding, it gets devalued. The men leave that area, looking for new cutting-edge areas where they can reestablish artificial scarcity and a tacit no-girls-allowed culture, or at least one where girls are regarded as foreign interlopers. These days, that appears to be Bitcoin—or blockchain tech in general—and AI, where, whenever I go to events, it's a sea of men, far more than most other fields of coding.

What is actually going on, Posner argues, is the creation of a "pink-collar ghetto" in coding. The world of programmers is willing to admit that, sure, women and minorities may be expert at JavaScript and all the things that go into making a website or mobile app highly functional, and to respond precisely the way a user wants. But that gets denigrated as aesthetic, fuzzy stuff. It's not *real* coding. This is a preposterous idea; front-end coding is enormously complex, and changing rapidly, such that most of those doing it are continually learning new techniques at a ferocious rate. But Posner herself felt this sniffy disdain for front-end web work: She started out by teaching herself HTML and CSS so she could make websites, only to be told "that's not really coding"; so she learned PHP and Drupal, to work with servers, only to be told those were too simple; and so she learned JavaScript, only to watch the pattern of criticism repeat.

"Feminized jobs are less prestigious," she tells me. It's a pattern that economists have noted in nearly every industry: Once a specialty—nursing, elementary school teaching—becomes dominated by women, it's looked down upon (subtly or openly) and confers lower status and lower pay.

The snooty attitude toward front-end coding is also the latest manifestation of a long, historical fight over what's considered "real" programming. For decades, self-appointed elite coders have insisted that any language that's harder, more abstract, and more hellish to master is thus inherently more valuable. (Typically, these languages are those the elite guys themselves prefer to use.) Anything that makes coding easier and more accessible? That's a foul corruption of a priestly, necessarily grueling craft. Back in 1975, Edsger W. Dijkstra—a pioneering computer scientist—beheld the rise of computer languages like BASIC and COBOL, which were designed to read much like regular English and thus to be easier to learn for newbies. These languages appalled him. They seemed loose, baggy, and badly thought out; they encouraged coders to make terrible design decisions that resulted in inelegant code, he hissed. "The use of COBOL cripples the mind; its teaching should, therefore, be regarded as a criminal offence," Dijkstra wrote. "It is practically impossible to teach good programming to students that have had a prior exposure to BASIC: as potential programmers they are mentally mutilated beyond hope of regeneration."

Dijkstra's vitriol isn't entirely misplaced. COBOL really can be a hairball. When you design a language to be "readable" by managers who can't code, it's no wonder you produce something that's clumsy and inefficient. And as for BASIC, it did indeed have gruesome aspects. (One particularly terrible command—you can instruct a program to abruptly "GOTO" any other line in the program—encouraged spaghetti code that jumps all over the place.) But Dijkstra's disdain was also pure snobbishness, a sense that you shouldn't even *try* to make coding easier for the beginner. For him, a central glory of coding seemed to lie precisely in its difficulty, that it chased people away, leaving only the monk-like devotees behind, to ponder their ASCII printouts in blissful silence. COBOL and BASIC were the targets back then—but today, that snobbishness is leveled at the web languages so many newbies and underrepresented coders use as their on-ramp: JavaScript, HTML, CSS.

As the pink-collar ghetto emerges in the front end, the self-appointed alpha nerds flee it. These days, they're less and less interested in web or app development and are moving into emerging areas like blockchain—Bitcoin and other cryptocurrencies—or machine learning. They're fields that are newly technically challenging; serious machine-learning work requires some genuinely mathematical thinking (and, if you practice it at a high level, formal computer science education). And these coders know that these skills are the most lucrative, because they're necessary for the hot venture-capital-soaked fields like robotics and self-driving cars.

This is why thinkers like Marie Hicks, the academic who's closely studied the history of women in coding, are unenthusiastic about the idea that boot camps will bring more women deeply into the field. Sure, they'll get some work—but they'll be limited as to how high they can rise by the industry's belief that women innately aren't suited to the "harder" coding jobs; hey, if they were, wouldn't they already be getting hired for those top jobs? "I think these [boot camp] initiatives are well meaning," Hicks told Posner, "but they totally misunderstand the problem. The pipeline is not the problem; the meritocracy is the problem. The idea that we'll just stuff people into the pipeline assumes a meritocracy that does not exist."

One can thus perceive the emerging caste system of coding—who'll do what, and which jobs will offer the most pay and prestige. Boot camps and self-education and job training will indeed open the door for more people to be software developers, including many women, minorities, and working-class folks from all walks of life and neglected regions. These jobs will probably pay a lot better than the wages these workers could command in the service sector. But the top money? That'll go to the new, emerging fields, and it'll be dominated by the existing coders of today, as well as the computer science grads of tomorrow.

Blue-collar code will emerge, it seems; but so will pink- and white-collar.

... ...

Given the increasing demand for programming talent, should *everyone* in the next generation learn how to do it?

This turns out to be a divisive topic. Some argue hotly in favor, including Quincy Larson of freeCodeCamp.

"I think that everyone should learn to code, even if they're not going to work in it, because code is seeping into so many jobs," he points out. In the world of business, to pick just one example, you'd be much more employable if you could analyze sales data using Python or R to tease out useful insights. If you knew how to automate routine tasks, you'd punch above your weight in just about any office setting. "If you look at society, so much of the work's being done by machines, and they're just carrying out instructions," Larson adds. "To quote the computer scientist John McCarthy, he said, 'Everybody needs computer programming. It will be the way we speak to the servants.'"

Worldwide, propelled by their general STEM mania, many policy-makers are embracing the idea that coding ought to be treated like one of the three Rs, and taught at the elementary school level. The idea has some vintage. Back in the '80s, MIT's pioneering educational theorist Seymour Papert argued that computer programming constituted a mode of thought deeply useful to children. Much as you learn French best by living in a place where it's spoken daily (a French land, like Paris), you learn logic and math and systematic thinking by living in a "Mathland"—which is, essentially, computer programming. His descendants today argue that kids are too often taught only the weakest computer literacy, such as how to make PowerPoint presentations or catchy videos. But the real power of wielding computers comes from knowing how to coax them to do something new, something you uniquely want done. The result has been, in recent years, a welter of computer languages aimed at children—like MIT's popular Scratch coding language—and initiatives like Hour of Code and robotics competitions. School systems are scrambling to shoehorn bits of coding into their already-stuffed curriculum; England has made computer science classes mandatory for students between the ages of 5 through 16, and the rapidly growing middle-class parents of China are demanding their school systems follow suit.

But many coders themselves are unconvinced that programming is really such a core skill, that it ranks alongside writing or reading. Jeff Atwood, the cofounder of Stack Overflow, wrote an essay arguing that saying "everyone should learn coding" is as nonsensical as "everyone should learn plumbing." Society certainly needs many qualified plumbers, and a few genius ones; and it probably needs everyone to know what plumbing essentially *is*. But the world is better served by having people specialize in the things they're passionate about and letting the people who are passionate about plumbing (or coding) focus on that.

Indeed, one problem with assuming coding is crucial to navigating the world's future problems is that this drifts dangerously toward Silicon Valley–style solutionism—assuming that, hey, every problem in the world should be solved with some software! "It puts the method before the problem," Atwood noted. "Before you go rushing out to learn to code, *figure out what your problem actually is.* Do you even have a problem? Can you explain it to others in a way they can understand? Have you researched the problem, and its possible solutions, deeply? Does coding solve that problem? Are you *sure*?" The whole reason the humanities are still a crucial field of study is that they help us understand the grayscale, maddening complexities of human behavior. They help us define our vision for society and the human spirit, which is the crucial first step before grappling with which tools—coding? plumbing? urban planning?—can help us achieve it.

So anyone who's got young children likely beheld the florid rise of STEM mania and wondered, *Should my kid learn to code?* As with most things, the real answer is the dull middle ground. As several coders and educators put it to me: Sure, all elementary schools should introduce kids to at least *some* coding at some point, to help them discover themselves whether they're drawn to it. It may not be worth carving out scarce time from basic math, history, and literature to constantly drill all kids in coding. But unless schools allow kids to taste it, they won't know whether they like it. Or rather, the only kids who'll decide they want to be coders will be the well-off nerdy ones whose parents can offer technical support.

Even more important, really, is cultivating the odd cultural on-ramps that give the next generation of kids a way to tinker with code. Teaching kids in classrooms is fair enough, but often pretty leaden. The little abstract "how to write an algorithm" lessons do not capture the imagination. No, the powerful on-ramps tend to be in the after-school world of pop culture. Kids who get interested in coding usually do so because they want to make something fun that will impress their peers. That's why so many developers talk about designing video games or building fan websites for their beloved TV shows and bands: Those were projects that actually meant something, that let them show off to their friends. They learned coding not to "learn coding" but to achieve something culturally powerful—a game of their own, a website broadcasting to their people their nuanced unpacking of last night's episode of *Jane the Virgin*.

For example, in the last decade, one unexpectedly powerful vector for coding was the game Minecraft. Superficially, the game is like a digital form of Legos: The players "mine" blocks of material—chopping trees for wood, digging into the ground for dirt or iron or gold—then combine them to create hundreds of new blocks, using them to create structures from houses to entire cities. Most kids enjoy it merely at that level.

But a subset of players discover the game is more or less an introduction to programming. That's because Minecraft includes a sort of electrical wiring, "redstone," which lets you craft logic circuits that resemble the language of software. If you click this switch *and* that switch, a light turns on; turn that lever *or* that lever, and a door opens. (These are AND gates and OR gates, and in fact you can use Minecraft to build many of the main forms of logic you see in coding and microchips.) So these kids start building mechanisms of ridiculous complexity—complex doors, little traps that get triggered when you walk by—then showing them off to friends and posting videos of them online. Minecraft, these kids found, was a world where being able to build cool things out of logic was fun and made them impressive to the outside world. For these kids, Minecraft wasn't just a game. It was this generation's "personal computer," their Commodore 64, as my friend the philosopher and game designer Ian Bogost once noted. It was the machine that let them peel back the curtain, see how digital stuff was really made, and start making it themselves. And since redstone creations often don't work right the first time, you wind up learning the pain and pleasure of debugging, too.

I've never seen any data on how many kids made the transition from redstone to actual programming. It's likely only a small chunk of all Minecraft players, much as only a small chunk of kids who touched a Commodore 64—or "viewed source" on a website in 1999—went on to become programmers. But they certainly exist. One teenager in the UK, Oliver Brotherhood, became fascinated with redstone wiring and began making increasingly complex machines with it. He began putting videos of those creations online, along with step-by-step instructions of how he made them, and soon he had so many subscribers that he was making a decent part-time wage off it. But he also got interested in coding and discovered that after having spent so much time devising and untangling the logic gates of his Minecraft contraptions, he got it pretty quickly. He applied to study computer science at a college and was accepted.

"In the redstone community," he says, "a lot of people around me are programmers." It turned out that when he posted about redstone on discussion boards, he'd discover that while some of his fans were 10-year-old kids, others were adults who coded for a living.

It's a lesson for the overall question of "how to help young people get into coding": Cultural fun is often the most motivating route. People learn to wield logic and think like a machine not merely because that activity itself is fun but because they're trying—as Papert realized—to create something other people will find awesome. So while Minecraft may be a surprisingly powerful on-ramp for coding, it only has that power because the game is fun and cool. (The creators of Minecraft itself never intended for it to be remotely educational, nor did they intend to make it a way to learn logic. "We have never done things with that sort of intent," as Jens Bergensten, the lead Minecraft developer, told me. "We always made the game for ourselves.") The same is likely true of Neopets: Its creators didn't think, *Hey, we're making something that will give rise to a whole generation of front-end developers.* They made a virtual world that would be awesome to play with, and people used it to learn.

When it comes to luring people into coding, culture is a powerful force.

So is sheer cussedness.

Part of what made Rusty Justice decide to bring coding to rural Kentucky? He was told it wasn't possible, by a rich guy from New York City.

In 2011, Michael Bloomberg, the billionaire who was then mayor of

New York, donated $50 million to the Sierra Club to help pay for "Beyond Coal," a campaign designed to agitate for policy that promoted the rise of renewables and the end of coal-burning energy plants. That itself annoyed Justice, since he's a big defender of coal as a font of usefully cheap energy, and lucrative Kentucky jobs.

But in 2014, Bloomberg went even further by actively dismissing the idea that miners could learn to make software.

"You're not going to teach a coal miner to code," Bloomberg said at the Bloomberg New Energy Finance summit. "Mark Zuckerberg says you can teach them to code and everything will be great. I don't know how to break it to you . . . but no."

That pissed off Justice. "It touched every button of every stereotype you can put on us, that we're not smart and can't do things and are pitiful and all that," as Justice told *Backchannel.* "It was like waving a red flag in front of a bull's face." You think miners can't figure out how to write JavaScript? Think again.

Acknowledgments

This book wouldn't exist if it weren't for the crackling enthusiasm and insight of my agent, Suzanne Gluck, and the encouragement, brilliant pen, and big-picture vision of my editor, Scott Moyers.

I'm deeply grateful to everyone who took the time to be interviewed by me for this book. There are hundreds of them, and they were all enormously generous with their time and knowledge. Not all of them wound up in these pages, but each of their thoughts, observations, and stories helped inform all of my writing here.

Along the way to writing *Coders*, I've been fortunate to talk to many brilliant folks who offered invaluable feedback and conversation. That includes Max Whitney, Fred Benenson, Tom Igoe, Michelle Tepper, Saron Yitbarek, Katrina Owens, Cathy Pearl, Tim O'Reilly, Caroline Sinders, Heather Gold, Ian Bogost, Marie Hicks, Anil Dash, Robin Sloan, danah boyd, Bret Dawson, Evan Selinger, Gary Marcus, Gabriella Coleman, Greg Baugues, Holden Karau, Jessica Lam, Karla Starr, Mike Matas, Paul Ford, Ray Ozzie, Ross Goodwin, Scott Goodson, Zeynep Tufekci, Steve Silberman, Tim Omernick, Emily Pakulski, Darius Kazemi, Cyan

Banister, Craig Silverman, Chris Coyier, Chet Murthy, Chad Folwer, Brendan Eich, Lauren McCarthy, Annette Bowman, Allison Parrish, Dan Sullivan, Grant Paul, Guido van Rossum, Jens Bergensten, Mark Otto, Mitch Altman, Peter Skomoroch, Jimoh Ovbiagele and all the hackers at Ross Intelligence, Rob Graham, Steve Klabnik, Rob Liguori, Adam D'Angelo, Belle Cooper, Dug Song, Kim Zetter, David Silva, Sam Lang, Ron Jeffries, Susan Tan, and John Reisig. This is a very incomplete list, alas, given the frailties of human memory.

I've been lucky to have superb magazine editors who've encouraged and shaped my writing on technology. In recent years that has included Dean Robinson, Bill Wasik, Jessica Lustig, and Jake Silverstein at the *New York Times Magazine*; Adam Rogers, Vera Titunik, and Nick Thompson at *Wired*; Debra Rosenberg and Michael Caruso at *Smithsonian*; Clara Jeffery and Mike Mechanic at *Mother Jones*; and Erica Lenti of *This Magazine*, to name just a few.

All the factual errors in this book are my responsibility. But I had some excellent help in minimizing them from my fact-checking team, which included Sharmila Venkatasubban, Lukas Vrbka, Benji Jones, James Gaines, Carla Murphy, Annie Ma, Rowan Walrath, Karen Font, and Fergus McIntosh.

I'm particularly grateful for Mia Council at Penguin, who did a ton of blocking and tackling to make this book happen.

I owe a large debt to a few people from my deep past—including Hal, a friend of my father's who lent me his VIC-20 for a summer back in the early '80s; he kickstarted my first exhilarating dive into nonstop daily coding in BASIC. I also have to thank someone whose name I don't know: The farsighted 1970s librarian at Toronto's Summit Heights Public School, who decided to acquire a book explaining how to build logic gates using electromechanical relays. I read that book at age 11 and it probably changed the arc of my life. It certainly led to *this* book!

I've been very lucky to have friends who've been enormously encouraging while I did this work. That includes all the Wack discussion-board diaspora, my Toronto and Brooklyn crews, and The Delorean Sisters; thanks to all of you.

Most indispensable of all was my family. My wife, Emily (a nascent coder herself back in the '80s!), engaged in many late-night brainstorming sessions, and gave me crucial first feedback on these pages. My kids, Gabriel and Zev, patiently listened to me jabber about this subject for three years solid. I couldn't have done it without you.

Notes

CHAPTER 1: THE SOFTWARE UPDATE THAT CHANGED REALITY

3 **"riding a bull dog":** Adam Fisher, *Valley of Genius: The Uncensored History of Silicon Valley (As Told by the Hackers, Founders, and Freaks Who Made It Boom)* (New York: Twelve, 2017), 357.

3 **new feature working:** Fisher, *Valley of Genius*, 361.

3 **"Move fast and break things":** This section draws from my interview with Sanghvi, as well as several books, articles, and videos about the early days of Facebook, including: Daniela Hernandez, "Facebook's First Female Engineer Speaks Out on Tech's Gender Gap," *Wired*, December 12, 2014, https://www.wired.com/2014/12/ruchi-qa/; Mark Zuckerberg, "Live with the Original News Feed Team," Facebook video, 25:36, September 6, 2016, https://www.facebook.com/zuck/videos/10103087013971051; David Kirkpatrick, *The Facebook Effect: The Inside Story of the Company That Is Connecting the World* (New York: Simon & Schuster, 2011); INKtalksDirector, *Ruchi Sanghvi: From Facebook to Facing the Unknown*, YouTube, 11:50, March 20, 2012, https://www.youtube.com/watch?v=64AaXC00bkQ; TechCrunch, *TechFellow Awards: Ruchi Sanghvi*, TechCrunch video, 4:40, March 4, 2012, https://techcrunch.com/video/techfellow-awards-ruchi-sanghvi/517287387/; FWDus2, *Ruchi's Story*, YouTube, 1:24, May 10, 2013, https://www.youtube.com/watch?v=i86ibVt1OMM.; all videos accessed August 16, 2018.

5 **did a keg stand:** Clare O'Connor, "Video: Mark Zuckerberg in 2005, Talking Facebook (While Dustin Moskovitz Does a Keg Stand)," *Forbes*, August 15, 2011, accessed October 7, 2018, https://www.forbes.com/sites/clareoconnor/2011/08/15/video-mark-zuckerberg-in-2005-talking-facebook-while-dustin-moskovitz-does-a-keg-stand/#629cb86571a5.

6 **"anything you can find on the web":** Ruchi Sanghvi, "Facebook Gets a Facelift," Facebook, September 5, 2006, accessed August 18, 2018, https://www.facebook.com/notes/facebook/facebook-gets-a-facelift/2207967130/.

6 **"just too creepy, too stalker-esque":** Brenton Thornicroft, "Something to Consider before You Complain about Facebook's News Feed Updates," *Forbes*, April 2, 2013, accessed August 18, 2018, https://www.forbes.com/sites/quora/2013/04/02/something-to-consider-before-you-complain-about-facebooks-news-feed-updates/#7154da847938.

8 **"you control of them":** Mark Zuckerberg, "An Open Letter from Mark Zuckerberg," Facebook, September 8, 2006, accessed September 18, 2018, https://www.facebook.com/notes/facebook/an-open-letter-from-mark-zuckerberg/2208562130/.

9 **for each American:** Evan Asano, "How Much Time Do People Spend on Social Media?," *SocialMediaToday*, January 4, 2017, accessed August 18, 2018, https://www.socialmediatoday.com/marketing/how-much-time-do-people-spend-social-media-infographic.

9 **neatly with your preferences:** Facebook does not openly discuss the nuances of how it ranks items in its feed, but it discusses the general details occasionally, as in: Miles O'Brien, "How Does the Facebook News Feed Work? An Interview with Dan Zigmond, Head of Facebook News Feed Analytics," *Miles O'Brien Productions*, March 30, 2018, accessed August 18, 2018, https://milesobrien.com/how-does-the-facebook-news-feed-work-an-interview-with-dan-zigmond-head-of-facebook-news-feed-analytics/; Will Oremus, "Who Controls Your Facebook Feed," *Slate*, January 3, 2016, accessed August 18, 2018, http://www.slate.com/articles/technology/cover_story/2016/01/how_facebook_s_news_feed_algorithm_works.html.

9 **a year in advertising:** Janko Roettgers, "Facebook Says It's Cutting Down on Viral Videos as 2017 Revenue Tops $40 Billion," *Variety*, January 31, 2018, accessed August 18, 2018, https://variety.com/2018/digital/news/facebook-q4-2017-earnings-1202683184/.

9 **what you already "liked":** Clive Thompson, "Social Networks Must Face Up to Their Political Impact," *Wired*, February 5, 2017, accessed August 18, 2018, https://www.wired.com/2017/01/social-networks-must-face-political-impact.

9 **"divisiveness and isolation":** Mark Zuckerberg, "Building Global Community," Facebook, February 16, 2017, accessed August 18, 2018, https://www.facebook.com/notes/mark-zuckerberg/building-global-community/10154544292806634/.

10 **"eating the world":** Marc Andreessen, "Why Software Is Eating the World," *Wall Street Journal*, August 20, 2011, accessed August 18, 2018, https://www.wsj.com/articles/SB10001424053111903480904576512250915629460.

10 **trained machine learning:** John Morris, "How Facebook Scales AI," *ZDNet*, June 6, 2018, accessed August 18, 2018, https://www.zdnet.com/article/how-facebook-scales-ai/.

12 **operating system of its democracy:** Erwin C. Surrency, "The Lawyer and the Revolution," *American Journal of Legal History* 8, no. 2 (April 1964): 125–35.

13 **move fast and break things:** My description of Moses's work here draws from: Robert A. Caro, *The Power Broker: Robert Moses and the Fall of New York* (New York: Knopf, 1974), 850–84; David W. Dunlap, "Why Robert Moses Keeps Rising from an

Unquiet Grave," *New York Times*, March 21, 2017, accessed August 18, 2018, https://www.nytimes.com/2017/03/21/nyregion/robert-moses-andrew-cuomo-and-the-saga-of-a-bronx-expressway.html; Sydney Sarachan, "The Legacy of Robert Moses," *PBS*, January 17, 2013, accessed August 18, 2018, http://www.pbs.org/wnet/need-to-know/environment/the-legacy-of-robert-moses/16018/.

14 **there are over 250:** There may be far more than 250; it's hard to count, in part because companies and coders invent new special-purpose ones all the time. Two attempts to count them include Robert Diana, "The Big List of 256 Programming Languages," *DZone*, May 16, 2013, accessed August 18, 2018, https://dzone.com/articles/big-list-256-programming; "How Many Programming Languages Are There in the World?," *CodeLani*, November 17, 2017, accessed August 18, 2018, http://codelani.com/posts/how-many-programming-languages-are-there-in-the-world.html.

15 **"never were nor could be":** Fred Brooks, *The Mythical Man-Month: Essays on Software Engineering* (New York: Addison-Wesley, 1975), 7–8.

15 **"It is good":** Jon Carroll, "Guerrillas in the Myst," *Wired*, August 1, 1994, accessed August 18, 2018, https://www.wired.com/1994/08/myst/.

16 **"burned at the stake":** Daniel Hillis, *The Pattern on the Stone: The Simple Ideas That Make Computers Work*, 2nd ed. (New York: Basic Books, 2014), location 112 of 2741, Kindle.

16 **"promotes a dangerous confidence":** Maciej Cegłowski, "The Moral Economy of Tech," *Idle Words* (blog), accessed August 18, 2018, http://idlewords.com/talks/sase_panel.htm.

16 **"so easily achievable":** Joseph Weizenbaum, *Computer Power and Human Reason: From Judgment to Calculation* (New York: W. H. Freeman, 1976), 111.

17 **back in 1980:** Seymour Papert, *Mindstorms: Children, Computers, and Powerful Ideas* (New York: Basic Books, 1980), 23.

18 **"errors in my own programs":** Daniel Kohanski, *The Philosophical Programmer: Reflections on the Moth in the Machine* (New York: St. Martin's, 1998), 160.

21 **"higher probability of succeeding":** INKtalksDirector, *Ruchi Sanghvi: From Facebook to Facing the Unknown.*

22 **"than to their profession":** Barry A. Stevens, "Probing the DP Psyche," *Computerworld*, July 21, 1980, 28.

23 **less than half that, around 17 percent:** "Degrees in Computer and Information Sciences Conferred by Degree-granting Institutions, by Level of Degree and Sex of Student: 1970–71 through 2010–11," National Center for Education Statistics, July 2012, accessed August 16, 2018, https://nces.ed.gov/programs/digest/d12/tables/dt12_349.asp.

23 **high teens to around twenty:** Roger Cheng, "Women in Tech: The Numbers Don't Add Up," *CNET*, May 6, 2015, accessed August 16, 2018, https://www.cnet.com/news/women-in-tech-the-numbers-dont-add-up/.

23 **across the country:** "Employed Persons by Detailed Occupation, Sex, Race, and Hispanic or Latino Ethnicity," Bureau of Labor Statistics, accessed August 16, 2018, https://www.bls.gov/cps/cpsaat11.pdf.

23 **at top Silicon Valley firms:** Rani Molla, "It's Not Just Google—Many Major Tech Companies Are Struggling with Diversity," *Recode*, August 7, 2017, accessed August 16,

2018, https://www.recode.net/2017/8/7/16108122/major-tech-companies-silicon-valley-diversity-women-tech-engineer.

23 **alongside people with PhDs:** Alyssa Mazzina, "Do Developers Need College Degrees?," *Stack Overflow* (blog), October 7, 2016, accessed August 16, 2018, https://stackoverflow.blog/2016/10/07/do-developers-need-college-degrees.

24 **number plummeted to 5 percent:** Erin Carson, "When Tech Firms Judge on Skills Alone, Women Land More Job Interviews," *CNET*, August 27, 2016, accessed August 16, 2018, https://www.cnet.com/news/when-tech-firms-judge-on-skills-alone-women-land-more-job-interviews.

26 **a "high leverage point":** Douglas Rushkoff, *Program or Be Programmed: Ten Commands for a Digital Age* (New York: OR Books, 2010), 133.

CHAPTER 2: THE FOUR WAVES OF CODERS

28 **a computer science department until 1965:** Andrew Myers, "Period of Transition: Stanford Computer Science Rethinks Core Curriculum," Stanford Engineering, June 14, 2012, accessed August 16, 2018, https://engineering.stanford.edu/news/period-transition-stanford-computer-science-rethinks-core-curriculum.

29 **and Gottfried Leibniz:** Chris Dixon, "How Aristotle Created the Computer," *The Atlantic*, March 20, 2017, accessed August 16, 2018, https://www.theatlantic.com/technology/archive/2017/03/aristotle-computer/518697.

31 **were staffed by women:** Bryony Norburn, "The Female Enigmas of Bletchley Park in the 1940s Should Encourage Those of Tomorrow," *The Conversation*, January 26, 2015, accessed August 16, 2018, https://theconversation.com/the-female-enigmas-of-bletchley-park-in-the-1940s-should-encourage-those-of-tomorrow-36640; Sarah Rainey, "The Extraordinary Female Codebreakers of Bletchley Park," *The Telegraph*, January 4, 2015, accessed August 16, 2018, https://www.telegraph.co.uk/history/world-war-two/11308744/The-extraordinary-female-codebreakers-of-Bletchley-Park.html.

31 **calculating ballistics trajectories, were women:** Jennifer S. Light, "When Computers Were Women," *Technology and Culture* 40, no. 3 (July 1999): 455–83.

32 **hired back both times:** Charles E. Molnar and Wesley A. Clark, "Development of the LINC," in *A History of Medical Informatics*, eds. Bruce I. Blum and Karen A. Duncan (New York: ACM Press, 1990), 119–38.

32 **or laboratory room:** John Markoff, "Wesley A. Clark, Who Designed First Personal Computer, Dies at 88," *New York Times*, February 27, 2016, accessed August 16, 2018, https://www.nytimes.com/2016/02/28/business/wesley-a-clark-made-computing-personal-dies-at-88.html.

32 **"conversational access" to the LINC:** Mary Allen Wilkes, "Conversational Access to a 2048-Word Machine," *Communications of the ACM* 13, no. 7 (July 1970): 407–14.

33 **"terrible food," she recalls:** This specific comment is from Wilkes's interview in this video: Dr. Bruce Damer *DigiBarn TV: Mary Allen Wilkes Programming the LINC Computer in the mid-1960s*, YouTube, 15:41, April 25, 2011, accessed August 16, 2018, https://www.youtube.com/watch?v=Cmv6p8hN0xQ.

33 **"a jig right around the equipment":** Joe November, "LINC: Biology's Revolutionary Little Computer," *Endeavour* 28, no. 3 (September 2004): 125–31.

35 **began to cluster around the lab:** This section is drawn from Steven Levy's superb

book, particularly chapters 3 ("Spacewar") and 4 ("Greenblatt and Gosper"): Steven Levy, *Hackers: Heroes of the Computer Revolution—25th Anniversary Edition* (Sebastopol, CA: O'Reilly Media, 2010).

36 **"in milliseconds to what you were doing":** Levy, *Hackers*, 67.

36 **"of the sun or moon it was":** Levy, *Hackers*, 139.

36 **a $120,000 machine:** Russell Brandom, "'Spacewar!': The Story of the World's First Digital Video Game," *The Verge*, February 4, 2013, accessed August 16, 2018, https://www.theverge.com/2013/2/4/3949524/the-story-of-the-worlds-first-digital-video-game.

36 **a "hacker ethic":** Levy, *Hackers*, 26–37.

37 **"hacking in general":** Levy, *Hackers*, 107.

38 **tinker with the code:** "GNU General Public License," Free Software Foundation, June 29, 2007, accessed August 16, 2018, https://www.gnu.org/licenses/gpl-3.0.en.html.

39 **"bunch of other robots":** Levy, *Hackers*, 129.

39 **like the MIT hackers:** Clive Thompson, "Steve Wozniak's Apple I Booted Up a Tech Revolution," *Smithsonian*, March 2016, accessed August 18, 2018, https://www.smithsonianmag.com/smithsonian-institution/steve-wozniaks-apple-i-booted-up-tech-revolution-180958112/.

39 **machine for $300:** Philip H. Dougherty, "Commodore Computers Plans Big Campaign," *New York Times*, February 18, 1982, accessed August 18, 2018, https://www.nytimes.com/1982/02/18/business/advertising-commodore-computers-plans-big-campaign.html.

40 **grasp and wield:** Harry McCracken, "Fifty Years of BASIC, the Programming Language That Made Computers Personal," *Time*, April 29, 2014, accessed August 18, 2018, http://time.com/69316/basic/; "BASIC Begins at Dartmouth," Dartmouth, accessed August 18, 2018, https://www.dartmouth.edu/basicfifty/basic.html; Jimmy Maher, "In Defense of BASIC," *The Digital Antiquarian* (blog), May 2, 2011, accessed August 18, 2018, https://www.filfre.net/2011/05/in-defense-of-basic.

42 **varieties of games:** Nate Anderson, "First Encounter: *COMPUTE!* Magazine and Its Glorious, Tedious Type-in Code," *Ars Technica*, December 28, 2012, accessed August 18, 2018, https:// arstechnica.com/staff/2012/12/first-encounter-compute-magazine-and-its-glorious-tedious-type-in-code; Shelby Goldstein, "Making Music with Your Vic," *Creative Computing* 9, no. 7 (July 1983): 43; Marek Karcz, "Conway's Game of Life on a Commodore 64," *Commodore and Retro Computing*, September 15, 2013, accessed August 18, 2018, http://c64retr.blogspot.com/2013/09/conways-game-of-life-on-commodore-64.html.

43 **calling the trick "war dialing":** Patrick S. Ryan, "War, Peace, or Stalemate: Wargames, Wardialing, Wardriving, and the Emerging Market for Hacker Ethics," *Virginia Journal of Law & Technology* 9, no. 7 (Summer 2004): 1–57, accessed August 18, 2018, https://papers.ssrn.com/sol3/papers.cfm?abstract_id=585867.

44 **primarily one of boys:** I discuss the boy-centric nature of the home-computer coding scene in greater length in chapter 7, "The ENIAC Girls Vanish," but some documents of this phenomenon include Sara Kiesler, Lee Sproull, and Jacquelynne Eccles, "Pool Halls, Chips, and War Games: Women in the Culture of Computing," *Psychology of Women Quarterly* 9, no. 4 (December 1985): 451–62; Jane Margolis and Allan Fisher, *Unlocking the Clubhouse: Women in Computing* (Cambridge, MA: MIT Press, 2003).

46 **down onto the coders:** Janet Abbate, "Oral-History: Judy Clapp," Engineering and Technology History Wiki, February 11, 2001, accessed August 18, 2018, https://ethw.org/Oral-History:Judy_Clapp.

46 **"Worse is better":** Tom Steinert-Threlkeld, "Can You Work in Netscape Time?," *Fast Company*, October 31, 1995, accessed September 27, 2018, https://www.fastcompany.com/26443/can-you-work-netscape-time.

47 **in a single year:** "Netscape Navigator," Blooberry, accessed August 18, 2018, http://www.blooberry.com/indexdot/history/netscape.htm.

49 **weird typography and graphics:** Anil Dash, "The Missing Building Blocks of the Web," *Medium*, March 21, 2018, accessed August 18, 2018, https://medium.com/@anildash/the-missing-building-blocks-of-the-web-3fa490ae5cbc; Amélie Lamont, "From Designing Neopets Pages to Becoming a Professional Web Developer," *Superyesmore*, June 20, 2016, accessed August 18, 2018, https://superyesmore.com/d6121b7fe42324e456deb8988d481ec8; Brittney Lopez, "How I Became a Web Designer," *Branded by Britt*, accessed August 18, 2018, https://www.brandedbybritt.co/how-i-became-a-web-designer.

51 **Send the Sunshine:** Ian Leslie, "The Scientists Who Make Apps Addictive," *1843*, October/November 2016, accessed August 18, 2018, https://www.1843magazine.com/features/the-scientists-who-make-apps-addictive.

52 **"get the psychology right":** Stephen Wendel, *Designing for Behavior Change: Applying Psychology and Behavioral Economics* (Sebastapol, CA: O'Reilly Media, 2013), location 189 of 7988, Kindle.

54 **uploaded every second:** Alyson Shontell, "Meet the 13 Lucky Employees and 9 Investors Behind $1 Billion Instagram," *Business Insider*, April 9, 2012, accessed August 18, 2018, https://www.businessinsider.com/instagram-employees-and-investors-2012-4; Marty Swant, "This Instagram Timeline Shows the App's Rapid Growth to 600 Million," *AdWeek*, December 15, 2016, accessed August 18, 2018, https://www.adweek.com/digital/instagram-gained-100-million-users-6-months-now-has-600-million-accounts-175126/; Nancy Messieh, "Instagram Could Hit 1bn Photos by April, Twice as Fast as Flickr Managed," *The Next Web*, January 19, 2012, accessed August 18, 2018, https://thenextweb.com/socialmedia/2012/01/19/instagram-could-hit-1bn-photos-by-april-twice-as-fast-as-flickr-managed/.

56 **particularly young women:** Olivia Fleming, "'Why Don't I Look Like Her?': How Instagram Is Ruining Our Self Esteem," *Cosmopolitan*, January 15, 2017, accessed August 18, 2018, https://www.cosmopolitan.com/health-fitness/a8601466/why-dont-i-look-like-her-how-instagram-is-ruining-our-self-esteem/; Amanda MacMillan, "Why Instagram Is the Worst Social Media for Mental Health," *Time*, May 25, 2017, accessed August 18, 2018, http://time.com/4793331/instagram-social-media-mental-health/; Mahita Gajanan, "Young Women on Instagram and Self-esteem: 'I Absolutely Feel Insecure,'" *Guardian*, November 4, 2015, accessed August 18, 2018, https://www.theguardian.com/media/2015/nov/04/instagram-young-women-self-esteem-essena-oneill.

56 **"negative mood and body dissatisfaction":** Zoe Brown and Marika Tiggemann, "Attractive Celebrity and Peer Images on Instagram: Effect on Women's Mood and Body Image," *Body Image* 19 (December 2016): 37–43.

56 **like #thygap or #thynspo:** Lily Herman, "Pro Eating Disorder Content Continues to Spread Online, Researchers Say," *Allure*, October 17, 2017, accessed August 18, 2018, https://www.allure.com/story/bonespiration-thinspiration-instagram-hashtag; Stevie Chancellor et al., "#thyghgapp: Instagram Content Moderation and Lexical Variation in Pro–Eating Disorder Communities," ACM Conference on Computer-Supported Cooperative Work and Social Computing, February 27–March 2, 2016, accessed August 18, 2018, http://www.munmund.net/pubs/cscw16_thyghgapp.pdf; Emily Reynolds, "Instagram's Pro-anorexia Ban Made the Problem Worse," *Wired UK*, March 14, 2016, accessed August 18, 2018, https://www.wired.co.uk/article/instagram-pro-anorexia-search-terms.

57 **"our psychological vulnerabilities":** Leslie, "The Scientists Who Make Apps Addictive."

57 **"this attention city":** *Recode* Staff, "Full Transcript: Time Well Spent Founder Tristan Harris on *Recode* Decode," *Recode*, February 7, 2017, accessed August 18, 2018, https://www.recode.net/2017/2/7/14542504/recode-decode-transcript-time-well-spent-founder-tristan-harris.

58 **exceeded that amount:** Ameet Ranadive, "New Tools to Manage Your Time on Facebook and Instagram," Facebook Newsroom, August 1, 2018, accessed October 2, 2018, https://newsroom.fb.com/news/2018/08/manage-your-time/.

CHAPTER 3: CONSTANT FRUSTRATION AND BURSTS OF JOY

63 **to dismantle and rebuild:** Elizabeth Dickason, "Looking Back: Grace Murray Hopper's Younger Years," *Chips Ahoy*, July 1986, posted online June 27, 2011, accessed August 18, 2018, http://www.doncio.navy.mil/chips/ArticleDetails.aspx?ID=2388.

64 **single mistyped command:** Casey Newton, "How a Typo Took Down S3, the Backbone of the Internet," *The Verge*, March 2, 2017, accessed August 18, 2018, https://www.theverge.com/2017/3/2/14792442/amazon-s3-outage-cause-typo-internet-server.

64–65 **as he wrote in his notebook:** Alexander B. Magoun and Paul Israel, "Did You Know? Edison Coined the Term 'Bug,'" *The Institute*, August 23, 2013, accessed August 18, 2018, http://theinstitute.ieee.org/tech-history/technology-history/did-you-know-edison-coined-the-term-bug.

65 **next to it:** "Log Book with Computer Bug," National Museum of American History, accessed August 18, 2018, http://americanhistory.si.edu/collections/search/object/nmah_334663.

66 **"the less obvious ones":** Michael Lopp, "Please Learn to Write," *Rands in Repose* (blog), May 16, 2012, accessed August 18, 2018, http://randsinrepose.com/archives/please-learn-to-write/.

68 **to a newbie's question:** "Angular 2 Passing Data into a For Loop," *Stack Overflow*, June 17, 2017, accessed August 18, 2018, https://stackoverflow.com/questions/44610183/angular-2-passing-data-into-a-for-loop.

68 **in a populated area:** Alex Pasternack, "Sometimes a Typo Means You Need to Blow Up Your Own Spacecraft," *Motherboard*, July 26, 2014, accessed August 18, 2018, https://motherboard.vice.com/en_us/article/4x3n9b/sometimes-a-typo-means-you-need-to-blow-up-your-spacecraft.

71 **than they do writing them:** Robert C. Martin, *Clean Code: A Handbook of Agile Software Craftsmanship* (New York: Pearson Education, 2009), 14.

73 **"that gamblers call the zone":** Natasha Dow Schüll, *Addiction by Design: Machine Gambling in Las Vegas* (Princeton, NJ: Princeton University Press, 2012), 68.

73 **"addicted" to coding:** In addition to the many programmers who talked to me about their feelings of addiction—or feeling a "coder's high"—there's Joseph Weizenbaum's (rather gloomy) observation about the "compulsive programmer" type he saw at MIT, cited in Levy, *Hackers*, 107.

74 **"in a good way":** Jacob Thornton, "Isn't Our Code Just the *BEST* ☺," *Medium*, January 19, 2017, accessed August 18, 2018, https://medium.com/bumpers/isnt-our-code-just-the-best-f028a78f33a9.

74 **at length:** Blake Ross, "Mr. Fart's Favorite Colors," *Medium*, March 4, 2016, accessed August 18, 2018, https://medium.com/@blakeross/mr-fart-s-favorite-colors-3177a406c775.

76 **what's known as "agile" development:** Caroline Mimbs Nyce, "The Winter Getaway That Turned the Software World Upside Down," *The Atlantic*, December 8, 2017, accessed August 18, 2018, https://www.theatlantic.com/technology/archive/2017/12/agile-manifesto-a-history/547715.

80 **"the lights are on":** Matthew B. Crawford, *Shop Class as Soulcraft: An Inquiry into the Value of Work* (New York: Penguin, 2009), 14.

81 **"how foolish we were":** Chad Fowler, *The Passionate Programmer: Creating a Remarkable Career in Software Development* (Raleigh, NC: Pragmatic Bookshelf, 2009), location 929 of 2976, Kindle.

84–85 **regular, daily, visible progress:** Teresa M. Amabile and Steven J. Kramer, "The Power of Small Wins," *Harvard Business Review* 89, no. 5 (May 2011), accessed online August 18, 2018, https://hbr.org/2011/05/the-power-of-small-wins.

CHAPTER 4: AMONG THE INTJs

87 **"That was a bad move":** Parts of this section on Bram Cohen draw from my previous profile of him in *Wired* magazine: Clive Thompson, "The BitTorrent Effect," *Wired*, January 2005, accessed online August 18, 2018, https://www.wired.com/2005/01/bittorrent-2.

90 **about bootlegging, like Phish:** Sarah Kessler, "The Infinite Lives of BitTorrent," *Fast Company*, March 10, 2014, accessed August 18, 2018, https://www.fastcompany.com/3027441/the-infinite-lives-of-bittorrent.

93 ***The Organization Man*:** William H. Whyte, *The Organization Man*, rev. ed. (Philadelphia: University of Pennsylvania Press, 2013), 3.

93 **in a 1966 paper:** William M. Cannon and Dallis K. Perry, "A Vocational Interest Scale for Computer Programmers," *Proceedings of the Fourth SIGCPR Conference on Computer Personnel Research* (June 1966): 61–82.

93–94 **ought to be a coder:** Nathan Ensmenger discusses IBM's "Programmer Aptitude Test" in *The Computer Boys Take Over: Computers, Programmers, and the Politics of Technical Expertise* (Cambridge, MA: MIT Press, 2012), 64–67; the actual tests are quite fascinating to look at, and there's one scanned at this location: "IBM

Programmer Aptitude Test (Revised)," accessed August 18, 2018, http://ed-thelen.org/comp-hist/IBM-ProgApti-120-6762-2.html.

94 **"bridge or anagrams":** Ensmenger, *The Computer Boys*, 52.

94 **"responsibility for helping people":** Dallis Perry and William Cannon, "Vocational Interests of Female Computer Programmers," *Journal of Applied Psychology* 52, no. 1 (1968): 34.

95 **"this demographic group":** Nathan Ensmenger, "Making Programming Masculine," in *Gender Codes: Why Women Are Leaving Computing*, ed. Thomas J. Misa (New York: IEEE Computer Society, 2010), 128.

95 **a fretful 1971 report:** Ensmenger, *The Computer Boys*, 159.

95 **"introverted, logical, and analytical":** Herendira Garcia–de Galindo, "An Investigation of Factors Related to Preservice Secondary Mathematics Teachers' Computer Environment Preferences for Teaching High School Geometry" (PhD diss., Ohio State University, 1994), 56–57, accessed September 27, 2018, https://etd.ohiolink.edu/!etd.send_file?accession=osu1487856906259719&disposition=inline.

95 **"became a more desirable companion":** Mariko R. Pope, "Creativity and the Computer Professional: The Impact of Personality Perception on Innovation Approach Preferences in Terms of Creative Thinking and Behavior" (PhD diss., Colorado Technical University, 1997), 31.

95 **cut off from humanity:** Levy, *Hackers*, 107.

97 **professional programmers in America:** Nathan Ensmenger and William Aspray, "Software as a Labor Process," in *History of Computing: Software Issues*, eds. Ulf Hashagen, Reinhard Keil-Slawik, and Arthur L. Norberg (Berlin: Springer Science & Business Media, 2013), 142.

98 **4 million programmers today:** This analysis finds "between 3,357,626 and 4,185,114 people working in a role which required some software development": P. K., "How Many Developers Are There in America, and Where Do They Live?," Don't Quit Your Day Job, August 10, 2017, accessed August 18, 2018, https://dqydj.com/number-of-developers-in-america-and-per-state; the Bureau of Labor Statistics for 2017 reports that for "Computer and mathematical occupations" there were 4.8 million employed: Bureau of Labor Statistics, "Employed Persons."

103 **"ideas to their partners":** Jean Hollands, *The Silicon Syndrome: How to Survive a High-tech Relationship* (New York: Bantam Books, 1985), 1–2.

105 **only 37 percent of the time:** Matt Parker, "The Secretary Problem," *Slate*, December 17, 2014, accessed August 18, 2018, http://www.slate.com/articles/technology/technology/2014/12/the_secretary_problem_use_this_algorithm_to_determine_exactly_how_many_people.html.

106 **"the logic stuff doesn't work":** Scott Hanselman, "More Relationship Hacks with Scott's Wife," *Hanselminutes* (podcast), April 17, 2012, accessed August 18, 2018, https://hanselminutes.com/314/more-relationship-hacks-with-scotts-wife.

108 **"your skills to the utmost":** John Geirland, "Go with the Flow," *Wired*, September 1996, accessed online August 18, 2018, https://www.wired.com/1996/09/czik.

109 **superglued his Ethernet port shut:** Lev Grossman, "Jonathan Franzen: Great

American Novelist," *Time*, August 12, 2010, accessed August 18, 2018, http://content.time.com/time/magazine/article/0,9171,2010185-2,00.html.

109 **break his concentration:** Lauren Passell, "Stephen King's Top 20 Rules for Writers," *B&N Reads* (blog), March 22, 2013, accessed August 18, 2018, https://www.barnesandnoble.com/blog/stephen-kings-top-20-rules-for-writers.

110 **"his floating hair":** Samuel Taylor Coleridge, *Kubla Khan*, Poetry Foundation, accessed August 18, 2018, https://www.poetryfoundation.org/poems/43991/kubla-khan.

111 **"to kill them off":** Paul Graham, "Maker's Schedule, Manager's Schedule," *Paul Graham* (blog), July 2009, accessed August 18, 2018, http://www.paulgraham.com/makersschedule.html.

112 **self-medicates with morphine:** Matt Giles, "'Mr. Robot' Creator Explains What's Really Going on in Elliot's Mind," *Popular Science*, September 3, 2015, accessed August 18, 2018, https://www.popsci.com/mr-robot-creator-explains-whats-really-going-on-in-elliots-mind.

CHAPTER 5: THE CULT OF EFFICIENCY

123 **invented bifocal glasses:** Walter Isaacson, *Benjamin Franklin: An American Life* (New York: Simon & Schuster, 2004), 426.

123 **"Saving of Wood to the Inhabitants":** Isaacson, *Benjamin Franklin*, 130–32; Benjamin Franklin, *Memoirs of the Life and Writings of (the Same), Continued to the Time of His Death by William Temple Franklin, Vol. 1* (London: H. Colburn, 1818), 94.

123 **"simultaneously wholesome and insane":** Jennifer Brostrom, "The Time-management Gospel," in *Commodify Your Dissent: Salvos from the Baffler*, eds. Thomas Frank and Matt Weiland (New York: W. W. Norton, 2011), 116.

123 **Charles Hermany from 1904:** "Address of President Charles Hermany," *Transactions of the American Society of Civil Engineers* 53 (1904): 464.

124 **to ensure maximum output:** Frederick Winslow Taylor, *The Principles of Scientific Management* (New York: Harper & Brothers, 1919), 5; David A. Hounshell, "The Same Old Principles in the New Manufacturing," *Harvard Business Review* (November 1988), accessed online August 18, 2018, https://hbr.org/1988/11/the-same-old-principles-in-the-new-manufacturing.

124 **bricklaying to vest buttoning:** Jill Lepore, "Not So Fast," *New Yorker*, October 12, 2009, accessed August 18, 2018, https://www.newyorker.com/magazine/2009/10/12/not-so-fast; Dennis McLellan, "Ernestine Carey, 98; Wrote a Comical Look at Her Big Family in 'Cheaper by the Dozen,'" *Los Angeles Times*, November 7, 2006, accessed August 18, 2018, http://articles.latimes.com/2006/nov/07/local/me-carey7.

124 **as *Better Homes Manual* enthused:** Alexandra Lange, "The Woman Who Invented the Kitchen," *Slate*, October 25, 2012, http://www.slate.com/articles/life/design/2012/10/lillian_gilbreth_s_kitchen_practical_how_it_reinvented_the_modern_kitchen.html.

124 **psychologists call "prospective memory":** Clive Thompson, "We Need Technology to Help Us Remember the Future," *Wired*, January 22, 2013, accessed August 18, 2018, https://www.wired.com/2013/01/a-sense-of-place-ct.

124 **Greek word for "time":** Kah Seng Tay, "What Is the Etymology of 'Cron'?," Quora, December 22, 2015, accessed August 18, 2018, https://www.quora.com/What-is-the-etymology-of-cron.

126 **do it over and over:** Peter Seibel, *Coders at Work: Reflections on the Craft of Programming* (New York: Apress, 2009), 77.

126 **we take it:** Nicholas Carr, *The Glass Cage* (New York: W. W. Norton, 2014).

127 **"so many questions about it":** Tom Christiansen, Brian D. Foy, Larry Wall, and Jon Orwant, *Programming Perl: Unmatched Power for Text Processing and Scripting* (Sebastapol, CA: O'Reilly Media, 2012), 387, 1062.

128 **"when are you free?":** "As a programmer, what tasks have you automated to make your everyday life easier? How can one expect to improve life through automated programming?," Quora, June 4, 2017, accessed August 18, 2018, https://www.quora.com/As-a-programmer-what-tasks-have-you-automated-to-make-your-everyday-life-easier-How-can-one-expect-to-improve-life-through-automated-programming.

129 **his colleague marveled:** Alexander Yumashev, "Now That's What I Call a Hacker," JitBit, November 20, 2015, accessed August 18, 2018, https://www.jitbit.com/alexblog/249-now-thats-what-i-call-a-hacker/.

129 **"will become like computers":** Martin Campbell-Kelly, "OBITUARY: Konrad Zuse," *Independent*, December 21, 1995, accessed August 18, 2018, https://www.independent.co.uk/news/people/obituary-konrad-zuse-1526795.html; Paul A. Youngman, *We Are the Machine: The Computer, the Internet, and Information in Contemporary German Literature* (Rochester, NY: Camden House, 2009), 94.

131 **searches people type:** Jake Brutlag, "Speed Matters," Google AI Blog, June 23, 2009, accessed October 2, 2018, https://ai.googleblog.com/2009/06/speed-matters.html.

135 **Pascal once apologized:** Blaise Pascal, *The Provincial Letters of Blaise Pascal* (New York: Hurd and Houghton, 1866), 18.

135 **"Brevity is the soul of wit":** William Shakespeare, *Hamlet* (London: Claredon Press, 1998), 35.

136 **2 billion lines of code:** Cade Metz, "Google Is 2 Billion Lines of Code—and It's All in One Place," *Wired*, September 16, 2015, accessed August 18, 2018, https://www.wired.com/2015/09/google-2-billion-lines-codeand-one-place.

137 **"lines removed per hour":** Jinghao Yan, "How many lines of code do professional programmers write per hour?," Quora, July 6, 2014, accessed August 18, 2018, https://www.quora.com/How-many-lines-of-code-do-professional-programmers-write-per-hour.

137 **"Now, that's efficiency":** Matt Ward, "The Poetics of Coding," *Smashing*, May 5, 2010, accessed August 18, 2018, https://www.smashingmagazine.com/2010/05/the-poetics-of-coding/.

138 **"the better craftsman":** Mark Ford, "Ezra Pound and the Drafts of *The Waste Land*," British Library, December 13, 2016, accessed August 18, 2018, https://www.bl.uk/20th-century-literature/articles/ezra-pound-and-the-drafts-of-the-waste-land; Helen Vendler, "The Most Famous Modern Poem—What Was Left In and What Was Cut Out," *New York Times*, November 7, 1971, accessed online August 18, 2018, https://www.nytimes.com/1971/11/07/archives/review

-1-no-title-the-waste-land-a-facsimile-and-transcript-of-the.html; Charles McGrath, "Il Miglior Fabbro," *New York Times*, January 27, 2008, accessed online August 18, 2018, https://www.nytimes.com/2008/01/27/books/review/McGrath-t.html.

138 **"Bad Smells in Code":** Kent Beck and Martin Fowler, "Bad Smells in Code," in *Refactoring: Improving the Design of Existing Code*, Martin Fowler, Kent Beck, John Brant, William Opdyke, and Don Roberts (New York: Addison-Wesley, 2012), 75.

138 **"more prone to introducing *smell instances*":** Michele Tufano, Fabio Palomba, Gabriele Bavota, Rocco Oliveto, Massimiliano Di Penta, Andrea De Lucia, and Denys Poshyvanyk, "When and Why Your Code Starts to Smell Bad," presented at IEEE /ACM 37th IEEE International Conference on Software Engineering, May 2015, accessed online August 18, 2018, https://www.cs.wm.edu/~denys/pubs/ICSE'15-Bad-Smells-CRC.pdf.

139 **"gross," "disgusting," or "vile":** Bryan Cantrill, "A Spoonful of Sewage," in *Beautiful Code: Leading Programmers Explain How They Think*, eds. Greg Wilson and Andy Oram (Sebastapol, CA: O'Reilly Media, 2007), 367–68.

140 **problem of everyday life: food:** Lizzie Widdicombe, "The End of Food," *New Yorker*, May 12, 2014, accessed online August 18, 2018, https://www.newyorker.com /magazine/2014/05/12/the-end-of-food.

140 **"it's just a hassle, though":** Rob Rhinehart, "How I Stopped Eating Food," *Mostly Harmless* (blog), February 13, 2013, accessed August 18, 2018, via the Internet Archive, https://web.archive.org/web/20130517220351/http://robrhinehart .com:80/?p=298.

141 **time spent on eating:** "Who Are You and Why Do You Use Soylent?," Reddit, accessed August 18, 2018, https://www.reddit.com/r/soylent/comments/5j57i5/who _are_you_and_why_do_you_use_soylent.

141 **"or Amazon Go":** Ruhi Sarikaya, "Making Alexa More Friction-free," *Alexa Blogs*, April 25, 2018, accessed August 18, 2018, https://developer.amazon.com/blogs/alexa /author/Ruhi+Sarikaya.

141 **anything else for you:** Steven Overly, "Washio Picks Up Your Dirty Laundry, Dry Cleaning with the Tap of an App," *Washington Post*, January 30, 2014, https://www .washingtonpost.com/business/capitalbusiness/washio-picks-up-your-dirty-laundry -dry-cleaning-with-the-tap-of-an-app/2014/01/29/08509ae4-8865-11e3-833c -33098f9e5267_story.html; Steven Bertoni, "Handybook Wants to Be the Uber for Your Household Chores," *Forbes*, March 26, 2014, https://www.forbes.com/sites /stevenbertoni/2014/03/26/handybook-wants-to-be-the-uber-for-your-household -chores/#221628987fa9; Brittain Ladd, "The Trojan Horse: Will Instacart Become a Competitor of the Grocery Retailers It Serves?," *Forbes*, July 1, 2018, https://www .forbes.com/sites/brittainladd/2018/07/01/__trashed-2/#7cc74ef1e4d1; Ken Yeung, "TaskRabbit's App Update Focuses on Getting Tasks Done in under 90 Minutes," *VentureBeat*, March 1, 2016, https://venturebeat.com/2016/03/01/taskrabbits -app-update-focuses-on-getting-tasks-done-in-under-90-minutes; all accessed August 18, 2018.

142 **"wanting to replicate mom":** Clara Jeffery (@ClaraJeffery), "So many Silicon Valley startups," Twitter, September 13, 2017, accessed August 18, 2018, https://twitter.com /clarajeffery/status/907997677048045568?lang=en.

143 **for a predictable income:** Corky Siemaszko, "In the Shadow of Uber's Rise, Taxi

Driver Suicides Leave Cabbies Shaken," *NBC News*, June 7, 2018, accessed August 18, 2018, https://www.nbcnews.com/news/us-news/shadow-uber-s-rise-taxi-driver-suicides-leave-cabbies-shaken-n879281.

143 **she once joked:** TEDx Talks, *Do You Like Me? Do I? | Leah Pearlman | TED xBoulder*, YouTube, 12:21, October 31, 2016, accessed August 18, 2018, https://www.youtube.com/watch?v=5nwSjRA3kQA.

143 **"required was so low":** Stanford eCorner, *Justin Rosenstein: No Dislike Button on Facebook*, YouTube, 1:33, May 13, 2013, accessed August 18, 2018, https://www.youtube.com/watch?v=11WbGqALF_I.

143 **"certain kinds of interaction":** Victor Luckerson, "The Rise of the Like Economy," *The Ringer*, February 15, 2017, accessed August 18, 2018, https://www.theringer.com/2017/2/15/16038024/how-the-like-button-took-over-the-internet-ebe778be2459.

143–144 **neatly, cleanly, and quickly:** Andrew Bosworth, "What's the history of the Awesome Button (that eventually became the Like button) on Facebook?," Quora, October 16, 2014, accessed August 18, 2018, https://www.quora.com/Whats-the-history-of-the-Awesome-Button-that-eventually-became-the-Like-button-on-Facebook.

144 **that user's News Feed:** Will Oremus, "Who Controls Your Facebook Feed," *Slate*, January 3, 2016, accessed August 18, 2018, http://www.slate.com/articles/technology/cover_story/2016/01/how_facebook_s_news_feed_algorithm_works.html.

144 **launched on February 9, 2009:** Kathy H. Chan, "I Like This," Facebook, February 9, 2009, accessed August 18, 2018, https://www.facebook.com/notes/facebook/i-like-this/53024537130.

144 **a trillion times by now:** Aaron Souppouris, "One Billion People Now 'Actively Using' Facebook," *The Verge*, October 4, 2012, accessed August 18, 2018, https://www.theverge.com/2012/10/4/3453350/facebook-one-billion-monthly-users-announcement.

145 **"truly peculiar reactions":** Adam Alter, *Irresistible: The Rise of Addictive Technology and the Business of Keeping Us Hooked* (New York: Penguin, 2017), 128.

145 **known as "Campbell's Law":** Donald T. Campbell, "Assessing the Impact of Planned Social Change," *Evaluation and Program Planning* 2, no. 1 (1979): 67–90.

145 **explosive emotionality, clickbait:** James Somers, "The Like Button Ruined the Internet," *The Atlantic*, March 21, 2017, accessed August 18, 2018, https://www.theatlantic.com/technology/archive/2017/03/how-the-like-button-ruined-the-internet/519795/; M. J. Crockett, "Modern Outrage Is Making It Harder to Better Society," *Globe and Mail*, March 2, 2018, accessed August 18, 2018, http://www.theglobeandmail.com/opinion/modern-outrage-is-making-it-harder-to-bettersociety/article38179877.

145 **across the entire web:** Allen St. John, "How Facebook Tracks You, Even When You're Not on Facebook," *Consumer Reports*, April 11, 2018, accessed August 18, 2018, https://www.consumerreports.org/privacy/how-facebook-tracks-you-even-when-youre-not-on-facebook; Alex Kantrowitz, "Here's How Facebook Tracks You When You're Not on Facebook," *BuzzFeed News*, April 11, 2018, accessed August 18, 2018, https://www.buzzfeednews.com/article/alexkantrowitz/heres-how-facebook-tracks-you-when-youre-not-on-facebook#.elVbWNnav.

145 **as Rosenstein told *The Verge*:** Casey Newton, "The Person Behind the Like Button Says Software Is Wasting Our Time," *The Verge*, March 28, 2018, https://www.theverge.com/2018/3/28/17172404/justin-rosenstein-asana-social-media-facebook-timeline-gantt.

145–146 **2,617 times a day:** Paul Lewis, "'Our Minds Can Be Hijacked': The Tech Insiders Who Fear a Smartphone Dystopia," *Guardian*, October 6, 2017, accessed August 18, 2018, https://www.theguardian.com/technology/2017/oct/05/smartphone-addiction-silicon-valley-dystopia.

CHAPTER 6: 10X, ROCK STARS, AND THE MYTH OF MERITOCRACY

147 **a frenzy of programming:** This section on Max Levchin draws from several sources, including Adam Penenberg, *Viral Loop: From Facebook to Twitter: How Today's Smartest Businesses Grow Themselves* (New York: Hyperion, 2009), 158–275, Kindle; Sarah Lacy, *Once You're Lucky, Twice You're Good: The Rebirth of Silicon Valley and the Rise of Web 2.0* (New York: Penguin, 2008), 17–41, Kindle; Jessica Livingston, *Founders at Work: Stories of Startups' Early Days* (New York: Apress, 2008), locations 200–605 of 12266, Kindle; Krissy Clark, "What Does Meritocracy Really Mean in Silicon Valley?," *Marketplace*, October 4, 2013, accessed August 18, 2018, https://www.marketplace.org/2013/10/04/wealth-poverty/what-does-meritocracy-really-mean-silicon-valley; Peter Thiel and Blake Masters, *Zero to One: Notes on Startups, or How to Build the Future* (New York: Crown Publishing Group, 2014).

149 **"perfect validation of merit":** Emily Chang, *Brotopia: Breaking Up the Boys' Club of Silicon Valley* (New York: Penguin, 2018), 60.

150 **"People wouldn't talk to us":** Chang, *Brotopia*, 48.

150 **"it is in Silicon Valley":** Jodi Kantor, "A Brand New World in Which Men Ruled," *New York Times*, December 23, 2014, accessed August 18, 2018, www.nytimes.com/interactive/2014/12/23/us/gender-gaps-stanford-94.html.

151 **"Offline Programming Performance":** H. Sackman, W. J. Erikson, and E. E. Grant, "Exploratory Experimental Studies Comparing Online and Offline Programming Performance," *Communications of the ACM* 11, no. 1 (January 1968): 3–11.

152 **critics have noted:** Laurent Bossavit, *The Leprechauns of Software Engineering: How Folklore Turns into Fact and What to Do about It* (Leanpub, 2016), 36–47.

153 **"creative work in this regard":** Butler Lampson, "A Critique of 'An Exploratory Investigation of Programmer Performance Under On-line and Off-line Conditions,'" *IEEE Transactions on Human Factors in Electronics* 8, no. 1 (March 1967): 48–51.

153 **an elite virtuoso class:** This blog post describes several of the papers that historically claimed to track large deltas in programmer performance: Steve McConnell, "Origins of 10X—How Valid Is the Underlying Research?," *Construx*, January 9, 2011, accessed August 18, 2018, http://www.construx.com/blog/the-origins-of-10x-how-valid-is-the-underlying-research.

153 **factors of 8X to 13X:** Lampson, "A Critique of 'An Exploratory Investigation of Programmer Performance Under On-Line and Off-Line Conditions.'"

153 **"The Mongolian Hordes versus Superprogrammer":** Bossavit, *Leprechauns*, 47.

153 **get the work done**: Brooks, *The Mythical Man-Month*, 30.

153 ***makes it later***: Brooks, *The Mythical Man-Month*, 25.

154 **"an average software writer":** "The Other Side of Paradise," *The Economist*, January

14, 2016, accessed online August 18, 2018, https://www.economist.com/business /2016/01/14/the-other-side-of-paradise.

154 **by two brothers:** Len Shustek, "Adobe Photoshop Source Code," Computer History Museum, February 13, 2013, accessed August 18, 2018, http://www.computerhistory .org/atchm/adobe-photoshop-source-code.

155 **freshman Monte Davidoff:** Paul Allen, "Microsoft's Odd Couple," *Vanity Fair*, May 2011, accessed online August 18, 2018, https://www.vanityfair.com/news/2011/05 /paul-allen-201105.

155 **written by Brad Fitzpatrick:** "Frequently Asked Question #4. How Did LiveJournal Get Started? Who Runs It Now?," LiveJournal, last updated April 3, 2017, accessed August 18, 2018, https://www.livejournal.com/support/faq/4.html.

155 **Larry Page and Sergey Brin:** John Battelle, "The Birth of Google," *Wired*, August 1, 2005, accessed August 18, 2018, https://www.wired.com/2005/08/battelle.

155 **trio of coworkers:** Laura Fitzpatrick, "Brief History of YouTube," *Time*, May 31, 2010, accessed August 18, 2018, http://content.time.com/time/magazine/ article/0,9171,1990787,00.html.

155 **one person, Bobby Murphy:** Alex Hern, "Snapchat Boss Evan Spiegel on the App That Made Him One of the World's Youngest Billionaires," *Guardian*, December 5, 2017, https://www.theguardian.com/technology/2017/dec/05/snapchat-boss-evan -spiegel-on-the-app-that-made-him-one-of-the-worlds-youngest-billionaires.

155 **the pseudonymous "Satoshi Nakamoto":** Joshua Davis, "The Crypto-Currency," *New Yorker*, October 10, 2011, accessed August 18, 2018, https://www.newyorker.com /magazine/2011/10/10/the-crypto-currency.

155 **first-person shooter video games:** Chris Kohler, "Q&A: Doom's Creator Looks Back on 20 Years of Demonic Mayhem," *Wired*, December 10, 2013, accessed August 18, 2018, https://www.wired.com/2013/12/john-carmack-doom.

156 **"'QA team put together'":** Joel Spolsky, "Top Five (Wrong) Reasons You Don't Have Testers," *Joel on Software* (blog), April 30, 2000, accessed August 18, 2018, https:// www.joelonsoftware.com/2000/04/30/top-five-wrong-reasons-you-dont-have-testers.

156 **"if they work at it":** Mark Guzdial, "Anyone Can Learn Programming: Teaching > Genetics," *Blog@CACM*, October 14, 2014, accessed August 18, 2018, https://cacm. acm.org/blogs/blog-cacm/179347-anyone-can-learn-programming-teaching-genetics /fulltext.

160 **"not the other way around":** Meredith L. Patterson, "When Nerds Collide," *Medium*, March 24, 2014, accessed August 18, 2018, https://medium.com/@maradydd /when-nerds-collide-31895b01e68c.

160 **"manages the most people":** "Zuckerberg's Letter to Investors," *Sydney Morning Herald*, February 2, 2012, accessed August 18, 2018, https://www.smh.com.au /business/zuckerbergs-letter-to-investors-20120202-1qu9p.html.

162 **he first announced it online:** Michael Calore, "Aug. 25, 1991: Kid from Helsinki Foments Linux Revolution," *Wired*, August 25, 2009, accessed August 18, 2018, https://www.wired.com/2009/08/0825-torvalds-starts-linux.

163 **Intel, Red Hat, or Samsung:** Dawn Foster, "Who Contributes to the Linux Kernel?," *The New Stack*, January 18, 2017, accessed August 18, 2018, https://thenewstack.io /contributes-linux-kernel.

164 **with a lifesaving fix:** This opening story is from this blog post: Jonathan Solórzano-Hamilton, "We Fired Our Top Talent. Best Decision We Ever Made," freeCodeCamp, October 13, 2017, accessed August 18, 2018, https://medium.freecodecamp.org/we-fired-our-top-talent-best-decision-we-ever-made-4c0a99728fde.

168 **"not really a programmer":** Jake Edge, "The Programming Talent Myth," LWN, April 28, 2015, accessed August 18, 2018, https://lwn.net/Articles/641779; the original speech is here: PyCon 2015, "Keynote—Jacob Kaplan-Moss—Pycon 2015," YouTube, 35:50, April 12, 2015, accessed August 18, 2018, https://www.youtube.com/watch?v=hIJdFxYlEKE.

168 **claimed it to be:** Chang, *Brotopia*, 60–63.

169 **"to be equally obsessed":** Thiel, *Zero to One*, 122, Kindle.

169 **you have a safety net:** Ross Levine and Yona Rubinstein, "Smart and Illicit: Who Becomes an Entrepreneur and Do They Earn More?," National Bureau of Economic Research, issued August 2013, revised September 2015, accessed August 18, 2018, https://www.nber.org/papers/w19276.pdf.

169 **Stanford, Harvard, or MIT:** Sarah McBride, "Insight: In Silicon Valley Start-up World, Pedigree Counts," Reuters, September 12, 2013, accessed August 18, 2018, https://www.reuters.com/article/us-usa-startup-connections-insight/insight-in-silicon-valley-start-up-world-pedigree-counts-idUSBRE98B15U20130912.

169 **cascade of good fortune downstream:** Robert H. Frank and Philip J. Cook, *The Winner-Take-All Society: Why the Few at the Top Get So Much More Than the Rest of Us* (New York: Virgin Books, 2010).

170 **"you start acting like one":** Antonio García Martínez, *Chaos Monkeys: Obscene Fortune and Random Failure in Silicon Valley* (New York: HarperCollins, 2016), 490.

170 **"an awful lot like me":** Johnathan Nightingale, "Some Garbage I Used to Believe about Equality," *Co-Pour*, November 27, 2016, accessed August 18, 2018, https://mfbt.ca/some-garbage-i-used-to-believe-about-equality-e7c771784f26.

171 **its respondents identified as men:** Klint Finley, "Diversity in Open Source Is Even Worse Than in Tech Overall," *Wired*, June 2, 2017, accessed June 23, 2018, https://www.wired.com/2017/06/diversity-open-source-even-worse-tech-overall.

171 **10 percent or lower:** Gregorio Robles, Laura Arjona Reina, Jesús M. González-Barahona, and Santiago Dueñas Domínguez, "Women in Free/Libre/Open Source Software," *Proceedings of the 12th IFIP WG 2.13 International Conference OSS 2016*, Gothenburg, Sweden, May 30–June 2, 2016, 163–173, accessed October 7, 2018, https://flosshub.org/sites/flosshub.org/files/paper-pre.pdf; Breanden Beneschott, "Is Open Source Open to Women?," Toptal, accessed October 7, 2018, https://www.toptal.com/open-source/is-open-source-open-to-women.

171 **open source conferences:** Valeria Aurora, "The Dark Side of Open Source Conferences," LWN, December 1, 2010, accessed October 7, 2018, https://lwn.net/Articles/417952/.

171 **as he wrote:** Noam Cohen, "After Years of Abusive E-mails, the Creator of Linux Steps Aside," *New Yorker*, September 19, 2018, accessed October 7, 2018, https://www.newyorker.com/science/elements/after-years-of-abusive-e-mails-the-creator-of-linux-steps-aside.

172 **weary giants of flesh and steel:** Hettie O'Brien, "The Floating City, Long a Libertarian Dream, Faces Rough Seas," *CityLab*, April 27, 2018, accessed August 18, 2018, https://www.citylab.com/design/2018/04/the-unsinkable-dream-of-the-floating-city/559058/.

172 **"that freedom and democracy are compatible":** Peter Thiel, "The Education of a Libertarian," *Cato Unbound*, April 13, 2009, accessed August 18, 2018, https://www.cato-unbound.org/2009/04/13/peter-thiel/education-libertarian.

172 **"that offer poor services":** Paul Bradley Carr, "Travis Shrugged: The Creepy, Dangerous Ideology Behind Silicon Valley's Cult of Disruption," *Pando*, October 24, 2012, accessed August 18, 2018, https://pando.com/2012/10/24/travis-shrugged.

172 **probably have been stillborn:** Fred Kaplan, "When America First Met the Microchip," *Slate*, June 18, 2009, accessed August 18, 2018, http://www.slate.com/articles/arts/books/2009/06/when_america_first_met_the_microchip.html.

172 **the Internet Protocols themselves:** "Funding a Revolution: Government Support for Computing Research," National Research Council, 1999, accessed August 18, 2018, https://www.nap.edu/read/6323/chapter/1; specifically chapters 8, 9, 10, and 12.

172 **being pooh-poohed worldwide:** Katrina Onstad, "Mr. Robot," *Toronto Life*, January 29, 2018, accessed August 18, 2018, https://torontolife.com/tech/ai-superstars-google-facebook-apple-studied-guy.

172 **by federal research dollars:** Mariana Mazzucato, *The Entrepreneurial State: Debunking Public vs. Private Sector Myths* (London: Anthem Press, 2015), 70.

173 **the general population:** Peter Ryan, "Left, Right and Center: Crypto Isn't Just for Libertarians Anymore," *CoinDesk*, July 27, 2018, https://www.coindesk.com/no-crypto-isnt-just-for-libertarians-anymore; Nate Silver, "There Are Few Libertarians. But Many Americans Have Libertarian Views," *FiveThirtyEight*, April 9, 2015, https://fivethirtyeight.com/features/there-are-few-libertarians-but-many-americans-have-libertarian-views/; both accessed October 7, 2018.

174 **got interested in this question:** David Broockman, Greg F. Ferenstein, and Neil Malhotra, "The Political Behavior of Wealthy Americans: Evidence from Technology Entrepreneurs," Stanford Graduate School of Business, Working Paper No. 3581, December 9, 2017, accessed August 18, 2018, https://www.gsb.stanford.edu/faculty-research/working-papers/political-behavior-wealthy-americans-evidence-technology.

174 **than to Donald Trump:** Ari Levy, "Silicon Valley Donated 60 Times More to Clinton Than to Trump," *CNBC*, November 7, 2016, accessed August 18, 2018, https://www.nbcnews.com/storyline/2016-election-day/silicon-valley-donated-60-times-more-clinton-trump-n679156.

176 **philosopher and technologist Ian Bogost says:** Alexis C. Madrigal, "What Should We Call Silicon Valley's Unique Politics?," *The Atlantic*, September 7, 2017, accessed August 18, 2018, https://www.theatlantic.com/technology/archive/2017/09/what-to-call-silicon-valleys-anti-regulation-pro-redistribution-politics/539043.

176 **civic goods like public libraries:** Susan Stamberg, "How Andrew Carnegie Turned His Fortune into a Library Legacy," *NPR*, August 1, 2013, accessed August 18, 2018,

https://www.npr.org/2013/08/01/207272849/how-andrew-carnegie-turned-his-fortune-into-a-library-legacy.

CHAPTER 7: THE ENIAC GIRLS VANISH

188 **from 8 percent of chemists to 39 percent:** Christianne Corbett and Catherine Hill, "Solving the Equation: The Variables for Women's Success in Engineering and Computing," AAUW (2015), accessed August 18, 2018, https://www.aauw.org/research/solving-the-equation.

189 **sequence of numbers:** My description of Lovelace's life and work draws from James Essinger, *Ada's Algorithm: How Lord Byron's Daughter Ada Lovelace Launched the Digital Age* (New York: Melville House, 2014); Betsy Morais, "Ada Lovelace, the First Tech Visionary," *New Yorker*, October 15, 2013, accessed August 18, 2018, https://www.newyorker.com/tech/elements/ada-lovelace-the-first-tech-visionary; Amy Jollymore, "Ada Lovelace, An Indirect and Reciprocal Influence," *Forbes*, October 15, 2013, accessed August 18, 2018, https://www.forbes.com/sites/oreillymedia/2013/10/15/ada-lovelace-an-indirect-and-reciprocal-influence; Valerie Aurora, "Deleting Ada Lovelace from the History of Computing," Ada Initiative, August 24, 2013, accessed August 18, 2018, https://adainitiative.org/2013/08/24/deleting-ada-lovelace-from-the-history-of-computing.

189 **it contained a bug:** Eugene Eric Kim and Betty Alexandra Toole, "Ada and the First Computer," *Scientific American*, May 1999, 76–81.

189 **she wrote in a letter:** Essinger, *Ada's Algorithm*, 184.

189 **"to the sound of *Music*":** James Gleick, *The Information: A History, a Theory, a Flood* (New York: Pantheon, 2011), 118–19, 124.

189 **recounts in *Recoding Gender*:** Janet Abbate, *Recoding Gender: Women's Changing Participation in Computing* (Cambridge, MA: MIT Press, 2012), Kindle.

189 **and 70,000 resistors:** "ENIAC," Computer Hope, updated May 22, 2018, accessed August 18, 2018, https://www.computerhope.com/jargon/e/eniac.htm.

190 **menial, even secretarial:** Ensmenger, "Making Programming Masculine," 121.

190 **the "ENIAC Girls":** Light, "When Computers Were Women," 459.

190 **"if not better than, the engineer":** Abbate, *Recoding Gender*, 24, Kindle.

191 **a key part of debugging:** Abbate, 32.

191 **"can do awake," Jennings later said:** Abbate, 33.

191 **of the women, was considered:** Abbate, 36–37.

191 **"FLOW-MATIC" language:** Kurt W. Beyer, *Grace Hopper and the Invention of the Information Age* (Cambridge, MA: MIT Press, 2012), Kindle, particularly chapter 7, "The Education of a Computer"; "Grace Murray Hopper," Lemelson-MIT, accessed August 18, 2018, https://lemelson.mit.edu/resources/grace-murray-hopper.

192 **"communication with the computer":** Steve Lohr, "Jean Sammet, Co-Designer of a Pioneering Computer Language, Dies at 89," *New York Times*, June 4, 2017, accessed August 18, 2018, https://www.nytimes.com/2017/06/04/technology/obituary-jean-sammet-software-designer-cobol.html; Jean E. Sammet, "The Early History of

COBOL," in *History of Programming Languages*, ed. Richard L. Wexelblat (New York: ACM, 1981), 199–243.

192 **first female IBM fellow:** Janet Abbate, "Oral-History: Frances 'Fran' Allen," Engineering and Technology History Wiki, accessed August 18, 2018, https://ethw.org/Oral-History:Frances_"Fran"_Allen.

192 **make good programmers:** Abbate, *Recoding Gender*, 65.

192 **as one ad enthused:** Marie Hicks, "Meritocracy and Feminization in Conflict: Computerization in the British Government," in *Gender Codes*, 105.

192 ***My Fair Ladies*:** Abbate, *Recoding Gender*, 65.

192 **"as female as anything":** Abbate, *Recoding Gender*, 62.

193 **"had it much harder":** Reginald Braithwaite, "A Woman's Story," *braythwayt* (blog), March 29, 2012, accessed August 18, 2018, http://braythwayt.com/posterous/2012/03/29/a-womans-story.html.

193 **in today's money:** Ensmenger, "Making Programming Masculine"; the illustrations of the *Cosmopolitan* article appear in a version of this article published on Ensmenger's page at the University of Indiana, accessed August 18, 2018, http://homes.sice.indiana.edu/nensmeng/files/ensmenger-gender.pdf.

194 **"I thought it was women's work":** Janet Abbate, "Oral-History: Elsie Shutt," Engineering and Technology History Wiki, accessed August 18, 2018, https://ethw.org/Oral-History:Elsie_Shutt.

194 **"Mixing Math and Motherhood":** Abbate, *Recoding Gender*, 113–44.

195 **and unkempt grooming:** Ensmenger, "Making Programming Masculine," 128–29.

195 **"huge glass ceiling":** Abbate, "Oral-History: Frances 'Fran' Allen."

195 **as a career was equal:** Steven James Devlin, "Sex Differences among Computer Programmers, Computer Application Users and General Computer Users at the Secondary School Level: An Investigation of Sex Role Self-concept and Attitudes toward Computers" (PhD diss., Temple University, 1991), 2, accessed September 27, 2018, https://dl.acm.org/citation.cfm?id=918494.

196 **computer science programs were women:** "Degrees in Computer and Information Sciences Conferred by Degree-granting Institutions, by Level of Degree and Sex of Student: 1970–71 through 2010–11," National Center for Educational Statistics, accessed August 18, 2018, https://nces.ed.gov/programs/digest/d12/tables/dt12_349.asp.

197 **an ambitious study:** The results of Margolis and Fisher's research were written up in their book: Jane Margolis and Allan Fisher, *Unlocking the Clubhouse: Women in Computing* (Cambridge, MA: MIT Press, 2003).

201 **women were precisely the reverse:** Lilly Irani, "A Different Voice: Women Exploring Stanford Computer Science" (Honors thesis, Stanford University, 2003), 46, citeseerx.ist.psu.edu/viewdoc/download?doi=10.1.1.107.1406&rep=rep1&type=pdf.

201 **pleasure of tinkering:** Margaret Burnett, Scott D. Fleming, Shamsi Iqbal, Gina Venolia, Vidya Rajaram, Umer Farooq, Valentina Grigoreanu, and Mary Czerwinski, "Gender Differences and Programming Environments: Across Programming Populations," in ESEM '10 Proceedings of the 2010 ACM-IEEE International Symposium on Empirical Software Engineering and Measurement, accessed August

18, 2018, https://www.microsoft.com/en-us/research/wp-content/uploads/2016/02/a28-burnett.pdf.

202 **They loved it:** Irani, "A Different Voice," 61–64.

202 **the "capacity crisis":** Eric Roberts, "Conserving the Seed Corn: Reflections on the Academic Hiring Crisis," *ACM SIGCSE Bulletin* (December 1999), accessed August 18, 2018, https://www.researchgate.net/profile/Eric_Roberts2/publication/220612646_Conserving_the_seed_corn_Reflections_on_the_academic_hiring_crisis/links/00b4951cafd2900e86000000/Conserving-the-seed-corn-Reflections-on-the-academic-hiring-crisis.pdf.

202 **the computer science major:** Mark Guzdial, "NPR When Women Stopped Coding in 1980's: As We Repeat the Same Mistakes," *Computing Education Research Blog*, October 30, 2014, accessed August 18, 2018, https://computinged.wordpress.com/2014/10/30/npr-when-women-stopped-coding-in-1980s-are-we-about-to-repeat-the-past.

203 **taking in more students:** "Degrees in Computer and Information Sciences," National Center for Educational Statistics.

203 **experiences in programming classes:** Ellen Spertus, "Why Are There So Few Female Computer Scientists?," MIT Artificial Intelligence Laboratory Technical Report 1315, 1991, accessed August 18, 2018, http://www.spertus.com/ellen/Gender/pap/pap.html.

204 **equally bleak tales:** *Barriers to Equality in Academia: Women in Computer Science at M.I.T.*, Massachusetts Institute of Technology, Laboratory for Computer Science, Massachusetts Institute of Technology, Artificial Intelligence Laboratory, M.I.T. (1983), accessed August 18, 2018, https://simson.net/ref/1983/barriers.pdf.

205 **rest of the private-sector workforce:** These are my calculations based on figures provided by the US Bureau of Labor Statistics here: "Employed Persons by Detailed Occupation, Sex, Race, and Hispanic or Latino Ethnicity," Bureau of Labor Statistics, accessed August 18, 2018, https://www.bls.gov/cps/cpsaat11.pdf.

205 **4.7 percent in 2016:** "Computer Programmers," Data USA, accessed August 18, 2018, https://datausa.io/profile/soc/151131.

205 **and 4 percent, respectively:** Molla, "It's Not Just Google."

207 **and 8.3 percent identified as LGBTQ:** "Diversity at Slack," *Slack Blog*, updated April 17, 2018, accessed August 18, 2018, https://slackhq.com/diversity-at-slack-2.

208 **no negative material:** Kieran Snyder, "The Abrasiveness Trap: High-achieving Men and Women Are Described Differently in Reviews," *Fortune*, August 26, 2014, accessed August 18, 2018, http://fortune.com/2014/08/26/performance-review-gender-bias.

208 **only 5 percent of them did:** Erin Carson, "When Tech Firms Judge on Skills Alone, Women Land More Job Interviews," *CNET*, August 27, 2016, accessed August 18, 2018, https://www.cnet.com/news/when-tech-firms-judge-on-skills-alone-women-land-more-job-interviews.

209 **"to decide to invest":** Tracey Ross, "The Unsettling Truth about the Tech Sector's Meritocracy Myth," *Washington Post*, April 13, 2016, accessed August 18, 2018, https://www.washingtonpost.com/news/in-theory/wp/2016/04/13/the-unsettling-truth-about-the-tech-sectors-meritocracy-myth.

209 **in his slide deck:** Sarah Mei, "Why Rails Is Still a Ghetto," *Sarah Mei* (blog), April 25, 2009, accessed August 18, 2018, http://www.sarahmei.com/blog/2009/04/25/why-rails-is-still-a-ghetto.

210 **"to girls what we actually do":** Jun Auza, "Why Mark Shuttleworth Owes FOSS-Women an Apology," *TechSource*, September 30, 2009, accessed August 18, 2018, http://www.junauza.com/2009/09/why-mark-shuttleworth-owes-foss-women.html; Chris Ball, "On Keynotes and Apologies," Blog.printf.net, September 25, 2009, accessed August 19, 2018, https://blog.printf.net/articles/2009/09/25/on-keynotes-and-apologies.

210 **ranking women by "hotness":** Andy Lester, "Distracting Examples Ruin Your Presentation," *Andy Lester* (blog), July 26, 2011, accessed August 19, 2018, https://petdance.wordpress.com/2011/07/26/distracting-examples-ruin-your-presentation.

210 **"efficiency and effectiveness":** Arianna Simpson, "Here's What It's Like to Be a Woman at a Bitcoin Meetup," *Business Insider*, February 3, 2014, accessed August 19, 2018, https://www.businessinsider.com/arianna-simpson-on-women-and-bitcoin-2014-2.

210 **to focus while coding:** Rhett Jones, "Lawsuit: VR Company Had a 'Kink Room,' Pressured Female Employees to 'Microdose,'" *Gizmodo*, May 15, 2017, accessed August 19, 2018, https://gizmodo.com/lawsuit-vr-company-had-a-kink-room-pressured-female-e-1795243868. The lawsuit was later settled out of court: Marisa Kendall, "Silicon Valley Virtual Reality Startup Settles 'Kink Room' Lawsuit," *The Mercury News*, September 7, 2017, accessed October 7, 2018, https://www.mercurynews.com/2017/09/07/san-francisco-virtual-reality-startup-settles-kink-room-lawsuit/.

210 **a good example of that:** Adam Fisher, "Sex, Beer, and Coding: Inside Facebook's Wild Early Days," *Wired*, July 10, 2018, accessed August 19, 2018, https://www.wired.com/story/sex-beer-and-coding-inside-facebooks-wild-early-days.

212 **96 percent of the investors are men:** Dan Primack, "Venture Capital's Stunning Lack of Female Decision-makers," *Fortune*, February 6, 2014, accessed August 19, 2018, http://fortune.com/2014/02/06/venture-capitals-stunning-lack-of-female-decision-makers/.

212 **for sex by investors:** Reed Albergotti, "Silicon Valley Women Tell of VC's Unwanted Advances," *The Information*, June 22, 2017, https://www.theinformation.com/articles/silicon-valley-women-tell-of-vcs-unwanted-advances; Sara O'Brien, "Sexual Harassment in Tech: Women Tell Their Stories," *CNN Tech*, https://money.cnn.com/technology/sexual-harassment-tech/; Katie Benner, "Women in Tech Speak Frankly on Culture of Harassment," *New York Times*, June 30, 2017, https://www.nytimes.com/2017/06/30/technology/women-entrepreneurs-speak-out-sexual-harassment.html; all accessed August 19, 2018.

212 **Pao later wrote:** Ellen Pao, *Reset: My Fight for Inclusion and Lasting Change* (New York: Random House, 2017), 78.

213 **gender for Y Combinator:** Cadran Cowansage, "Ask a Female Engineer: Thoughts on the Google Memo," *Y Combinator* (blog), August 15, 2017, accessed August 19, 2018, https://blog.ycombinator.com/ask-a-female-engineer-thoughts-on-the-google-memo/.

213 **as a lawsuit alleged:** Jordan Pearson, "How the Magic Leap Lawsuit Illuminates Tech's Gendered Design Bias," *Motherboard*, February 15, 2017, accessed August 19, 2018, https://motherboard.vice.com/en_us/article/aeygje/how-the-magic-leap-lawsuit-illuminates-techs-gendered-design-bias. The lawsuit was later settled: Adi Robertson, "Magic Leap Settles Sex Discrimination Lawsuit," *The Verge*, May 9, 2017, accessed

October 7, 2018, https://www.theverge.com/2017/5/9/15593578/magic-leap-tannen-campbell-sex-discrimination-settlement.

213 **Heather Gold has written:** Heather Gold, "Video Chat Is Terrible and About to Get Much Worse," October 16, 2018, accessed January 4, 2019, https://medium.com/s/story/video-chat-is-terrible-and-about-to-get-much-worse-174823f3ffb.

214 **Laurie Penny once cracked:** Laurie Penny, "Laurie Penny: A Woman's Opinion Is the Mini-skirt of the Internet," *Independent*, November 4, 2011, accessed August 19, 2018, https://www.independent.co.uk/voices/commentators/laurie-penny-a-womans-opinion-is-the-mini-skirt-of-the-internet-6256946.html.

214 **like Flickr had previously attempted:** Hunter Walk, "Early Employees: Heather Champ & Flickr," LinkedIn post, March 19, 2013, accessed August 19, 2018, https://www.linkedin.com/pulse/20130320042336-7298-early-employees-heather-champ-flickr/.

214 **"company's wholesome family image":** Alex Sherman, Christopher Palmeri, and Sarah Frier, "Disney Said to Have Stopped Twitter Chase Partly over Image," *Boston Globe*, October 19, 2016, accessed August 19, 2018, https://www.bostonglobe.com/business/2016/10/18/disney-said-have-stopped-twitter-chase-partly-over-image/Op40d0HcrsBOXIjWEgLanL/story.html.

214 **"how to behave on the platform":** Dan Primack, "Ex-Twitter CEO: I'm Sorry," *Axios*, February 1, 2017, accessed August 19, 2018, https://www.axios.com/ex-twitter-ceo-im-sorry-1513300246-b0a495a5-f418-415b-9caa-7a32a7fe8d50.html.

215 **"Google's Ideological Echo Chamber":** Kate Conger, "Exclusive: Here's the Full 10-Page Anti-diversity Screed Circulating Internally at Google [Updated]," *Gizmodo*, August 5, 2017, accessed August 19, 2018, https://gizmodo.com/exclusive-heres-the-full-10-page-anti-diversity-screed-1797564320.

216 **to be interested in people:** Simon Baron-Cohen, *The Essential Difference: Male and Female Brains and the Truth about Autism* (New York: Basic Books, 2003).

217 **"is offensive and not OK":** Kara Swisher, "Google Has Fired the Employee Who Penned a Controversial Memo on Women and Tech," *Recode*, August 7, 2017, accessed August 19, 2018, https://www.recode.net/2017/8/7/16110696/firing-google-ceo-employee-penned-controversial-memo-on-women-has-violated-its-code-of-conduct.

217 **their biological makeup was a liability?:** Yonatan Zunger, "So, about This Googler's Manifesto," *Medium*, August 5, 2017, accessed August 19, 2018, https://medium.com/@yonatanzunger/so-about-this-googlers-manifesto-1e3773ed1788.

217 **on their internal forums:** Ashley Feinberg, "Internal Messages Show Some Googlers Supported Fired Engineer's Manifesto," *Wired*, August 8, 2017, accessed August 19, 2018, https://www.wired.com/story/internal-messages-james-damore-google-memo/.

217 **"an egg donor first":** Holly Brockwell, "Recruiter Sends Jaw-droppingly Sexist Email to Female Engineer—then Claims It Was a Stunt," *shinyshiny* (blog), March 16, 2015, accessed August 19, 2018, https://www.shinyshiny.tv/2015/03/recruiter-sends-sexist-email.html.

218 **"dry brush in fire season":** Cynthia Lee, "James Damore Has Sued Google. His Infamous Memo on Women in Tech Is Still Nonsense," *Vox*, January 8, 2018, accessed August 19, 2018, https://www.vox.com/the-big-idea/2017/8/11/16130452/google-memo-women-tech-biology-sexism.

218 **in the halls of tech workplaces:** Angela Saini, *Inferior: How Science Got Women Wrong and the New Research That's Rewriting the Story* (Boston: Beacon Press, 2017); Rosalind

C. Barnett and Caryl Rivers, "We've Studied Gender and STEM for 25 Years. The Science Doesn't Support the Google Memo," *Recode*, August 11, 2017, accessed August 19, 2018, https://www.recode.net/2017/8/11/16127992/google-engineer-memo-research-science-women-biology-tech-james-damore; Megan Molteni and Adam Rogers, "The Actual Science of James Damore's Google Memo," *Wired*, August 15, 2017, accessed August 19, 2018, https://www.wired.com/story/the-pernicious-science-of-james-damores-google-memo; Suzanne Sadedin, "A Scientist's Take on the Biological Claims from the Infamous Google Anti-diversity Manifesto," *Forbes*, August 10, 2017, accessed August 19, 2018, www.forbes.com/sites/quora/2017/08/10/a-scientists-take-on-the-biological-claims-from-the-infamous-google-anti-diversity-manifesto; Tia Ghose, "Google Manifesto: Does Biology Explain Gender Disparities in Tech?," *LiveScience*, August 9, 2017, accessed August 19, 2018, https://www.livescience.com/60079-biological-differences-men-and-women.html.

219 **that girls would code:** Roli Varma and Deepak Kapur, "Decoding Femininity in Computer Science in India," *Communications of the ACM* 58, no. 5 (May 2015): 56–62.

219 **in Kuala Lumpur:** Vivian Anette Lagesen, "A Cyberfeminist Utopia?: Perceptions of Gender and Computer Science among Malaysian Women Computer Science Students," *Science Technology Human Values* 33, no. 1 (2008): 5–27.

220 **"right mentality to do the job":** Abbate, *Recoding Gender*, 67.

221 **unironic, literal concept:** Michael Young, "Down with Meritocracy," *Guardian*, June 28, 2001, accessed August 19, 2018, https://www.theguardian.com/politics/2001/jun/29/comment.

221 **like math or philosophy—they're not:** Sarah-Jane Leslie, Andrei Cimpian, Meredith Meyer, and Edward Freeland, "Expectations of Brilliance Underlie Gender Distributions across Academic Disciplines," *Science* 347, no. 6219 (January 16, 2015): 262–65.

221 **to be the lone genius:** Emilio J. Castilla and Stephen Benard, "The Paradox of Meritocracy in Organizations," *Administrative Science Quarterly* 55 (2010): 543–76.

223 **match those of the men:** Margolis and Fisher, *Unlocking the Clubhouse*, location 1620 of 2083, Kindle.

223 **Using Computational Approaches:** Laura Sydell, "Colleges Have Increased Women Computer Science Majors: What Can Google Learn?," *NPR All Tech Considered*, August 10, 2017, accessed August 19, 2018, https://www.npr.org/sections/alltechconsidered/2017/08/10/542638758/colleges-have-increased-women-computer-science-majors-what-can-google-learn.

223 **found it hard going:** Anya Kamenetz, "A College President on Her School's Worst Year Ever," *nprED*, August 2, 2017, accessed August 19, 2018, https://www.npr.org/sections/ed/2017/08/02/540603927/a-college-president-on-her-schools-worst-year-ever.

223 **the collapse in the mid-'80s:** Linda J. Sax, "Expanding the Pipeline: Characteristics of Male and Female Prospective Computer Science Majors—Examining Four Decades of Changes," *Computing Research News* 29, no. 2 (February 2017): 4–7.

224 **a "high leverage point":** Rushkoff, *Program or Be Programmed*, 133.

224 **"one very, very strange year":** Susan Fowler, "Reflecting on One Very, Very Strange Year at Uber," SusanJFowler.com, February 19, 2017, accessed August 19, 2018, https://www.susanjfowler.com/blog/2017/2/19/reflecting-on-one-very-strange-year-at-uber.

225 **stalk their ex-girlfriends:** Will Evans, "Uber Said It Protects You from Spying. Security Sources Say Otherwise," *Reveal News*, December 12, 2016, accessed August 19, 2018, https://www.revealnews.org/article/uber-said-it-protects-you-from-spying-security-sources-say-otherwise.

225 **after a female journalist:** Sarah Lacy, "Uber Executive Said the Company Would Spend 'A Million Dollars' to Shut Me Up," *Time*, November 14, 2017, accessed August 19, 2018, http://time.com/5023287/uber-threatened-journalist-sarah-lacy.

225 **he calls it "Boob-er":** Mickey Rapkin, "Uber Cab Confessions," *GQ*, February 27, 2014, accessed August 19, 2018, www.gq.com/story/uber-cab-confessions.

225 **had been forced out:** Mike Isaac, "Uber Founder Travis Kalanick Resigns as C.E.O.," *New York Times*, June 21, 2017, accessed August 19, 2018, https://www.nytimes.com/2017/06/21/technology/uber-ceo-travis-kalanick.html.

225 **Chris Sacca and Justin Caldbeck:** Sage Lazzaro, "6 Women Accuse Prominent Tech VC Justin Caldbeck of Sexual Assault and Harassment," *Observer*, June 23, 2017, accessed August 19, 2018, http://observer.com/2017/06/justin-caldbeck-binary-capital-sexual-assault-harssment; Becky Peterson, "'Shark Tank' Judge Chris Sacca Apologizes for Helping Make Tech Hostile to Women—after Being Accused of Inappropriately Touching a Female Investor," *Business Insider*, June 30, 2017, accessed August 19, 2018, https://www.businessinsider.com/chris-sacca-apologizes-after-accusation-of-inappropriate-touching-2017-6; "Dave McClure Quits 500 Startups over Sexual Harassment Scandal," *Reuters*, July 4, 2017, accessed August 19, 2018, http://fortune.com/2017/07/03/dave-mcclure-500-startups-quits; Maya Kosoff, "Silicon Valley's Sexual-harassment Crisis Keeps Getting Worse," *Vanity Fair*, September 12, 2017, accessed August 19, 2018, https://www.vanityfair.com/news/2017/09/silicon-valleys-sexual-harassment-crisis-keeps-getting-worse. McClure resigned from his position and published a post apologizing for his actions: Kaitlin Menza, "Dave McClure's Apology for Sexual Harassment Isn't Applause-worthy—It's the Bare Minimum," *Self*, July 7, 2017, https://www.self.com/story/dave-mcclure-apology-sexual-harassment; Chris Sacca disputed the allegation against him, while writing that he had "sometimes played a role in the larger phenomenon of women not always feeling welcome in our industry": Chris Sacca, "I Have More Work to Do," *Medium*, June 29, 2017, accessed October 7, 2018, https://medium.com/@sacca/i-have-more-work-to-do-c775c5d56ca1; Justin Caldbeck initially denied the charges but later drafted letters of apology to the women who accused him of unwanted sexual advances: Ellen Huet, "After Harassment Allegations, Justin Caldbeck Attempts a Comeback. Critics Want Him to Stay Gone," *Bloomberg Businessweek*, November 13, 2017, accessed October 7, 2018, https://www.bloomberg.com/news/articles/2017-11-13/after-harassment-allegations-justin-caldbeck-attempts-a-comeback-critics-want-him-to-stay-gone.

CHAPTER 8: HACKERS, CRACKERS, AND FREEDOM FIGHTERS

234 **like Google and Yahoo!:** Barton Gellman, "NSA Infiltrates Links to Yahoo, Google Data Centers Worldwide, Snowden Documents Say," *Washington Post*, October 30, 2013, accessed August 19, 2018, https://www.washingtonpost.com/world/national-security/nsa-infiltrates-links-to-yahoo-google-data-centers-worldwide

-snowden-documents-say/2013/10/30/e51d661e-4166-11e3-8b74-d89d714ca4dd_story.html; Barton Gellman, Aaron Blake, and Greg Miller, "Edward Snowden Comes Forward as Source of NSA Leaks," *Washington Post*, June 9, 2013, accessed August 19, 2018, https://www.washingtonpost.com/politics/intelligence-leaders-push-back-on-leakers-media/2013/06/09/fff80160-d122-11e2-a73e-826d299ff459_story.html.

234 **crowdsourced-journalism project:** Andy Greenberg, "Anonymous' Barrett Brown Is Free—and Ready to Pick New Fights," *Wired*, December 21, 2016, accessed August 19, 2018, https://www.wired.com/2016/12/anonymous-barrett-brown-free-ready-pick-new-fights/.

234 **obstruction of justice charge:** Kim Zetter, "Barrett Brown Sentenced to 5 Years in Prison in Connection to Stratfor Hack," *Wired*, January 22, 2015, accessed August 19, 2018, https://www.wired.com/2015/01/barrett-brown-sentenced-5-years-prison-connection-stratfor-hack.

236 **most prolific hackers, later recalled:** Richard Stallman, "My Lisp Experiences and the Development of GNU Emacs," Gnu.org, transcript of speech from October 28, 2002, page last updated April 12, 2014, accessed August 19, 2018, https://www.gnu.org/gnu/rms-lisp.en.html.

237 **"ridiculous concepts as property rights":** Levy, *Hackers*, 95, Kindle.

237 **programmers could learn from it:** Levy, *Hackers*, 436–53, Kindle.

238 **derivative works based on it:** "GNU General Public License," Free Software Foundation, Version 3, June 29, 2007, hosted at Gnu.org, accessed August 19, 2018, https://www.gnu.org/licenses/gpl.txt; Heather Meeker, "Open Source Licensing: What Every Technologist Should Know," *Opensource*, September 21, 2017, accessed August 19, 2018, https://opensource.com/article/17/9/open-source-licensing; Gabriella Coleman, "Code Is Speech: Legal Tinkering, Expertise, and Protest among Free and Open Source Software Developers," *Cultural Anthropology* 24, no. 3 (2009): 420–54, accessed August 19, 2018, https://steinhardt.nyu.edu/scmsAdmin/uploads/005/984/Coleman-Code-is-Speech.pdf.

239 **reprogramming phone systems:** Elinor Mills, "Q&A: Mark Abene, from 'Phiber Optik' to Security Guru," *CNET*, June 29, 2009, accessed August 19, 2018, https://www.cnet.com/news/q-a-mark-abene-from-phiber-optik-to-security-guru/; Michelle Slatalla and Joshua Quittner, "Gang War in Cyberspace," *Wired*, December 1, 1994, accessed August 19, 2018, https://www.wired.com/1994/12/hacker-4; Abraham Riesman, "Twilight of the Phreaks: The Fates of the 10 Best Early Hackers," *Motherboard*, March 9, 2012, accessed August 19, 2018, https://motherboard.vice.com/en_us/article/wnn7by/twilight-of-the-phreaks-the-fates-of-the-10-best-early-hackers.

239 **judge explained at the sentencing:** Julian Dibbell, "The Prisoner: Phiber Optik Goes Directly to Jail," *Village Voice*, January 12, 1994, accessed copy on August 19, 2018, on http://www.juliandibbell.com/texts/phiber.html; Trip Gabriel, "Reprogramming a Convicted Hacker; to His On-line Friends, Phiber Optik Is a Virtual Hero," January 14, 1995, *New York Times*, accessed August 19, 2018, https://www.nytimes.com/1995/01/14/nyregion/reprogramming-convicted-hacker-his-line-friends-phiber-optik-virtual-hero.html; Wired Staff, "Phiber Optik Goes to Prison," *Wired*, April 1, 1994, accessed August 19, 2018, https://www.wired.com/1994/04/phiber-optik-goes-to-prison.

240 **"never met face-to-face":** Joe Mullin, "Newegg Trial: Crypto Legend Takes the Stand, Goes for Knockout Patent Punch," *Ars Technica*, November 24, 2013, accessed August 19, 2018, https://arstechnica.com/tech-policy/2013/11/newegg-trial-crypto-legend-diffie-takes-the-stand-to-knock-out-patent/.

241 **idea of powerful crypto:** Steven Levy, "Prophet of Privacy," *Wired*, November 1, 1994, accessed August 19, 2018, https://www.wired.com/1994/11/diffie; Steve Fyffe and Tom Abate, "Stanford Cryptography Pioneers Whitfield Diffie and Martin Hellman Win ACM 2015 A. M. Turing Award," *Stanford News Service*, March, 1, 2016, accessed August 19, 2018, https://news.stanford.edu/press-releases/2016/03/01/pr-turing-hellman-diffie-030116.

241 **head of the NSA:** Thomas Rid, "The Cypherpunk Revolution," *Christian Science Monitor*, July 20, 2016, accessed August 19, 2018, projects.csmonitor.cypherpunk.

241 **Stewart Baker, recalled:** Gregory Ferenstein, "How Hackers Beat the NSA in the '90s and How They Can Do It Again," *TechCrunch*, June 29, 2013, accessed August 19, 2018, https://techcrunch.com/2013/06/28/how-hackers-beat-the-nsa-in-the-90s-and-how-they-can-do-it-again.

242 **or a missile:** Dan Froomkin, "Deciphering Encryption," *Washington Post*, May 8, 1998, accessed August 19, 2018, https://www.washingtonpost.com/wp-srv/politics/special/encryption/encryption.htm.

242 **First Amendment rights:** John Markoff, "Judge Rules against U.S. in Encryption Case," *New York Times*, December 19, 1996, https://www.nytimes.com/1996/12/19/business/judge-rules-against-us-in-encryption-case.html; "Bernstein v. US Department of Justice," Electronic Frontier Foundation, accessed August 19, 2018, https://www.eff.org/cases/bernstein-v-us-dept-justice.

242 **without explaining why:** John Markoff, "Data-Secrecy Export Case Dropped by U.S.," *New York Times*, January 12, 1996, accessed October 3, 2018, https://www.nytimes.com/1996/01/12/business/data-secrecy-export-case-dropped-by-us.html.

242 **the NSA couldn't break:** Steven Levy, "Battle of the Clipper Chip," *New York Times Magazine*, June 12, 1994, accessed August 19, 2018, https://www.nytimes.com/1994/06/12/magazine/battle-of-the-clipper-chip.html.

243 **"individual's preemptive protection":** "The Cyphernomicon," Nakomoto Institute, accessed August 19, 2018, https://nakamotoinstitute.org/static/docs/cyphernomicon.txt.

243 **hand down that ruling:** Markoff, "Judge Rules"; Rid, "The Cypherpunk Revolution."

244 **"of Big Brother":** Levy, "Battle of the Clipper Chip."

244 **it looked incompetent:** Sharon Begley, "Foiling the Clipper Chip," *Newsweek*, June 12, 1994, accessed August 19, 2018, https://www.newsweek.com/foiling-clipper-chip-188912.

245 **"cultural Dark Ages":** Gabriella Coleman, *Coding Freedom: The Ethics and Aesthetics of Hacking* (Princeton, NJ: Princeton University Press, 2013), 84.

247 **a $2,250,000 fine:** "US v. ElcomSoft & Sklyarov FAQ," Electronic Frontier Foundation, updated February 19, 2002, accessed August 19, 2018, https://www.eff.org/pages/us-v-elcomsoft-sklyarov-faq. The charges against Sklyarov were later

dropped: John Leyden, "Case against Dmitry Sklyarov Dropped," *The Register*, December 14, 2001, accessed October 7, 2018, https://www.theregister.co.uk/2001/12/14/case_against_dmitry_sklyarov_dropped/.

247 **open source software DeCSS:** J. S. Kelly, "Meet the Kid behind the DVD Hack," *CNN*, January 31, 2000, accessed August 19, 2018, http://www.cnn.com/2000/TECH/computing/01/31/johansen.interview.idg.

247 **of Norwegian law:** Declan McCullagh, "Norway Cracks Down on DVD Hacker," *Wired*, January 10, 2002, accessed August 19, 2018, https://www.wired.com/2002/01/norway-cracks-down-on-dvd-hacker/; "DVD Lawsuit Questions Legality of Linking," *New York Times*, January 7, 2000, accessed August 19, 2018, https://www.nytimes.com/2000/01/07/technology/dvd-lawsuit-questions-legality-of-linking.html; Amy Harmon, "Free Speech Rights for Computer Code?," *New York Times*, July 31, 2000, accessed August 19, 2018, https://archive.nytimes.com/www.nytimes.com/library/tech/00/07/biztech/articles/31rite.html.

248 **to criminalize programming:** John Leyden, "2600 Withdraws Supreme Court Appeal in DeCSS Case," *The Register*, July 4, 2002, accessed August 19, 2018, https://www.theregister.co.uk/2002/07/04/2600_withdraws_supreme_court_appeal; "Teen Cleared in Landmark DVD Case," *CNN*, January 7, 2003, accessed August 19, 2018, http://www.cnn.com/2003/TECH/01/07/dvd.johansen/index.html; Carl S. Kaplan, "The Year in Internet Law," *New York Times*, December 28, 2001, accessed August 19, 2018, https://www.nytimes.com/2001/12/28/technology/the-year-in-internet-law.html.

248 **"hacker cultural DNA":** Gabriella Coleman, "From Internet Farming to Weapons of the Geek," *Current Anthropology* 58, no. S15 (February 2017), accessed August 19, 2018, https://www.journals.uchicago.edu/doi/full/10.1086/688697.

250 **super-open form of copyright:** Tim Carmody, "Memory to Myth: Tracing Aaron Swartz through the 21st Century," *The Verge*, January 22, 2013, accessed August 19, 2018, https://www.theverge.com/2013/1/22/3898584/aaron-swartz-profile-memory-to-myth.

250 **he committed suicide:** Justin Peters, "The Idealist," *Slate*, February 7, 2013, accessed August 19, 2018, http://www.slate.com/articles/technology/technology/2013/02/aaron_swartz_he_wanted_to_save_the_world_why_couldn_t_he_save_himself.single.html.

252 **"or sectarian infighting":** Coleman, "From Internet Farming."

254 **buy drugs and guns:** Nick Bilton, *American Kingpin: The Epic Hunt for the Criminal Mastermind behind the Silk Road* (New York: Penguin, 2017), 34.

254 **use Tor to avoid scrutiny:** Robert Graham, "How Terrorists Use Encryption," *CTC Sentinel* 9, no. 6 (June 2016), accessed August 19, 2018, https://ctc.usma.edu/how-terrorists-use-encryption.

254 **of its overall use:** Andy Greenberg, "No, Department of Justice, 80 Percent of Tor Traffic Is Not Child Porn," *Wired*, January 28, 2015, accessed August 19, 2018, https://www.wired.com/2015/01/department-justice-80-percent-tor-traffic-child-porn.

254 **"possible to break the law":** Andy Greenberg, "Meet Moxie Marlinspike, the Anarchist Bringing Encryption to All of Us," *Wired*, July 31, 2016, accessed August 19, 2018, https://www.wired.com/2016/07/meet-moxie-marlinspike-anarchist-bringing-encryption-us/.

254 **"should be permitted":** Moxie Marlinspike, "We Should All Have Something to Hide," *Moxie* (blog), June 12, 2013, accessed August 19, 2018, https://moxie.org/blog/we-should-all-have-something-to-hide.

258 **began cracking open the machines:** Jim Finkle, "Hackers Scour Voting Machines for Election Bugs," *Reuters*, July 28, 2017, accessed August 19, 2018, https://www.reuters.com/article/us-cyber-conference-election-hacking/hackers-scour-voting-machines-for-election-bugs-idUSKBN1AD1BF.

258 **you could take control:** Matt Blaze, Jake Braun, Harri Hursti, Joseph Lorenzo Hall, Margaret MacAlpine, and Jeff Moss, "DEFCON 25 Voting Machine Hacking Village," Defcon.org, September 2017, accessed August 19, 2018, https://www.defcon.org/images/defcon-25/DEF%20CON%2025%20voting%20village%20report.pdf.

259 **admitted in an essay, "we're jerks":** Christian Ternus, "Infosec's Jerk Problem," *Adversarial Thinking* (blog), June 19, 2013, accessed August 19, 2018, http://adversari.es/blog/2013/06/19/cant-we-all-just-get-along.

260 **"couldn't pay in six months":** James Stevenson, "The Who, What, Where, When, and Why of WCry," *Hacking Insider*, May 13, 2017, accessed August 19, 2018, http://www.hackinginsider.com/2017/05/the-who-what-where-when-and-why-of-wcry.

260 **security experts suspected:** "Cyber Attack Hits 200,000 in at Least 150 Countries: Europol," *Reuters*, May 14, 2017, accessed August 19, 2018, https://www.reuters.com/article/us-cyber-attack-europol/cyber-attack-hits-200000-in-at-least-150-countries-europol-idUSKCN18A0FX; Julia Carrie Wong and Olivia Solon, "Massive Ransomware Cyber-attack Hits Nearly 100 Countries around the World," *Guardian*, May 12, 2017, https://www.theguardian.com/technology/2017/may/12/global-cyber-attack-ransomware-nsa-uk-nhs; Thomas P. Bossert, "It's Official: North Korea Is Behind WannaCry," *Wall Street Journal*, December 18, 2017, accessed August 19, 2018, https://www.wsj.com/articles/its-official-north-korea-is-behind-wannacry-1513642537.

261 **in-demand infosec talent is:** Reeves Wiedeman, "Gray Hat," *New York*, February 19, 2018, accessed August 19, 2018, http://nymag.com/selectall/2018/03/marcus-hutchins-hacker.html.

262 **them back to you:** Doug Olenick, "Simple, but Not Cheap, Phishing Kit Found for Sale on Dark Web," *SC Magazine*, April 26, 2018, accessed August 19, 2018, https://www.scmagazine.com/simple-but-not-cheap-phishing-kit-found-for-sale-on-dark-web/article/761520; Kishalaya Kundu, "New Phishing Kit on Dark Web Lets Anyone Launch Cyber Attacks," *Beebom*, April 30, 2018, accessed August 19, 2018, https://beebom.com/new-phishing-kit-dark-web; Ionut Arghire, "New Advanced Phishing Kit Targets eCommerce," *SecurityWeek*, April 25, 2018, accessed August 19, 2018, https://www.securityweek.com/new-advanced-phishing-kit-targets-ecommerce.

262 **of all intrusion groups:** *Internet Threat Security Report: Volume 23* (March 2018), Symantec, accessed August 19, 2018, https://www.symantec.com/security-center/threat-report.

263 **of gray indeed:** Brian Krebs, "Who Is Anna-Senpai, the Mirai Worm Author?," *Krebs on Security*, January 17, 2017, accessed August 19, 2018, https://krebsonsecurity.com/2017/01/who-is-anna-senpai-the-mirai-worm-author; Brian Krebs, "Mirai IoT Botnet Co-authors Plead Guilty," *Krebs on Security*, December 17, 2017, accessed August 19, 2018, https://krebsonsecurity.com/2017/12/mirai-iot-botnet-co-authors

-plead-guilty; Mark Thiessen, "3 Hackers Get Light Sentences after Working with the FBI," Associated Press, September 19, 2018, accessed October 2, 2018, https://apnews.com/b6f03f9a13e04b19afed3375476b4132; Garrett M. Graff, "The Mirai Botnet Architects Are Now Fighting Crime with the FBI," *Wired*, September 18, 2018, accessed October 2, 2018, https://www.wired.com/story/mirai-botnet-creators-fbi-sentencing/.

264 **That's how you learn:** Wiedeman, "Gray Hat."

265 **(if it's a state):** Michael Schwirtz and Joseph Goldstein, "Russian Espionage Piggybacks on a Cybercriminal's Hacking," *New York Times*, March 12, 2017, accessed August 19, 2018, https://www.nytimes.com/2017/03/12/world/europe/russia-hacker-evgeniy-bogachev.html; Garrett M. Graff, "Inside the Hunt for Russia's Most Notorious Hacker," *Wired*, March 21, 2017, accessed August 19, 2018, https://www.wired.com/2017/03/russian-hacker-spy-botnet.

265 **other sites online:** Lisa Vaas, "DNC Chief Podesta Led to Phishing Link 'Thanks to a Typo,'" *Naked Security by Sophos*, December 16, 2016, accessed August 19, 2018, https://nakedsecurity.sophos.com/2016/12/16/dnc-chief-podesta-led-to-phishing-link-thanks-to-a-typo; Eric Lipton, David E. Sanger, and Scott Shane, "The Perfect Weapon: How Russian Cyberpower Invaded the U.S. Image," *New York Times*, December 13, 2016, accessed August 19, 2018, https://www.nytimes.com/2016/12/13/us/politics/russia-hack-election-dnc.html.

265 **phishing attempts each day:** Teri Robinson, "Defense Dept. Blocks 36M Malicious Emails Daily, Fends off 600 Gbps DDoS Attacks," *Cybersecurity Source*, January 19, 2018, accessed August 19, 2018, https://www.scmagazine.com/defense-dept-blocks-36m-malicious-emails-daily-fends-off-600-gbps-ddos-attacks/article/738292.

266 **to name just one example:** Masashi Crete-Nishihata, Jakub Dalek, Etienne Maynier, and John Scott-Railton, "Spying on a Budget: Inside a Phishing Operation with Targets in the Tibetan Community," The Citizen Lab, January 30, 2018, accessed August 19, 2018, https://citizenlab.ca/2018/01/spying-on-a-budget-inside-a-phishing-operation-with-targets-in-the-tibetan-community.

266 **when the NSA itself was hacked:** Lily Hay Newman, "The Leaked NSA Spy Tool That Hacked the World," *Wired*, May 7, 2018, accessed August 19, 2018, https://www.wired.com/story/eternalblue-leaked-nsa-spy-tool-hacked-world.

CHAPTER 9: CUCUMBERS, SKYNET, AND RISE OF THE AI

267 **AlphaGo dominated, 4 to 1:** John Markoff, "Alphabet Program Beats the European Human Go Champion," *New York Times*, January 27, 2016, accessed August 19, 2018, https://bits.blogs.nytimes.com/2016/01/27/alphabet-program-beats-the-european-human-go-champion; Bloomberg News, "How You Beat One of the Best Go Players in the World? Use Google," *Washington Post*, March 14, 2016, accessed August 19, 2018, https://www.washingtonpost.com/national/health-science/how-you-beat-one-of-the-best-go-players-in-the-world-use-google/2016/03/14/1efd1176-e6fc-11e5-b0fd-073d5930a7b7_story.html.

268 **atoms in the universe:** Alan Levinovitz, "The Mystery of Go, the Ancient Game That Computers Still Can't Win," *Wired*, May 12, 2014, accessed August 19, 2018,

https://www.wired.com/2014/05/the-world-of-computer-go; David Silver and Demis Hassabis, "AlphaGo: Mastering the Ancient Game of Go with Machine Learning," *Google AI Blog*, January 27, 2016, accessed August 19, 2018, https://ai.googleblog.com/2016/01/alphago-mastering-ancient-game-of-go.html.

268 **model of the game:** Silver and Hassabis, "AlphaGo."

268 **tales about AlphaGo:** Cade Metz, "The Sadness and Beauty of Watching Google's AI Play Go," *Wired*, March 11, 2016, accessed August 19, 2018, https://www.wired.com/2016/03/sadness-beauty-watching-googles-ai-play-go.

269 **messing around with it:** Tom Simonite, "Google Stakes Its Future on a Piece of Software," *MIT Technology Review*, June 27, 2017, accessed August 19, 2018, https://www.technologyreview.com/s/608094/google-stakes-its-future-on-a-piece-of-software.

269 **"learned to sort cucumbers well":** Amos Zeeberg, "D.I.Y. Artificial Intelligence Comes to a Japanese Family Farm," *New Yorker*, August 10, 2017, accessed August 19, 2018, https://www.newyorker.com/tech/elements/diy-artificial-intelligence-comes-to-a-japanese-family-farm.

270 **to make it better:** Kaz Sato, "How a Japanese Cucumber Farmer Is Using Deep Learning and TensorFlow," *Google Cloud Platform Blog*, August 31, 2016, accessed August 19, 2018, https://cloud.google.com/blog/products/gcp/how-a-japanese-cucumber-farmer-is-using-deep-learning-and-tensorflow.

272 **to get machines to "think":** Stuart Armstrong, Kaj Sotala, and Seán S. Ó hÉigeartaigh, "The Errors, Insights and Lessons of Famous AI Predictions—and What They Mean for the Future," *Journal of Experimental & Theoretical Artificial Intelligence* 26, no. 3 (May 2014): 317–42.

272 **break down and become useless:** Abhishek Anand, "The Problem with Chatbots—How to Make Them More Human?," *Chatbots Magazine*, April 21, 2107, accessed August 19, 2018, https://chatbotsmagazine.com/the-problem-with-chatbots-how-to-make-them-more-human-d7a24c22f51e.

272 **once described it to me:** Clive Thompson, "What Is I.B.M.'s Watson?," *New York Times Magazine*, June 16, 2010, accessed August 19, 2018, https://www.nytimes.com/2010/06/20/magazine/20Computer-t.html.

274 **would be a monster hit:** Clive Thompson, "Hit Song Science," *New York Times Magazine*, December 14, 2003, accessed August 19, 2018, https://www.nytimes.com/2003/12/14/magazine/2003-the-3rd-annual-year-in-ideas-hit-song-science.html.

276 **their own "AI winter":** Alex Kantrowitz, "Meet the Man Who Makes Facebook's Machines Think," *BuzzFeed*, April 17, 2017, accessed August 19, 2018, https://www.buzzfeednews.com/article/alexkantrowitz/meet-the-man-who-makes-facebooks-machines-think; "Yann LeCun: An AI Groundbreaker Takes Stock," *Forbes*, July 17, 2018, accessed August 19, 2018, https://www.forbes.com/sites/insights-intelai/2018/07/17/yann-lecun-an-ai-groundbreaker-takes-stock.

277 **to reboot the crawl:** Steven Levy, *In the Plex: How Google Thinks, Works, and Shapes Our Lives* (New York: Simon & Schuster, 2011), 42.

277 **"Jeff Dean warns compilers":** Will Oremus, "The Optimizer," *Slate*, January 23, 2013, accessed August 19, 2018, http://www.slate.com/articles/technology/doers/2013/01/jeff_dean_facts_how_a_google_programmer_became_the_chuck_norris_of_the_internet.html.

278 **It was an AI moon shoot:** Alex Krizhevsky, Ilya Sutskever, and Geoffrey E. Hinton, "ImageNet Classification with Deep Convolutional Neural Networks," *NIPS '12 Proceedings of the 25th International Conference on Neural Information Processing Systems—Volume 1* (December 2012), 1097–1105, accessed August 19, 2018 via https://www.nvidia.cn/content/tesla/pdf/machine-learning/imagenet-classification-with-deep-convolutional-nn.pdf; Adit Deshpande, "The 9 Deep Learning Papers You Need to Know About (Understanding CNNs Part 3)," adeshpande3.github.io, August 24, 2016, accessed August 19, 2018, https://adeshpande3.github.io/The-9-Deep-Learning-Papers-You-Need-To-Know-About.html.

279 **"automate with AI":** Andrew Ng (@AndrewYNg), "Pretty much anything," Twitter, October 18, 2016, accessed October 2, 2018, https://twitter.com/AndrewYNg/status/788548053745569792.

279 **between their language and English:** Gideon Lewis-Kraus, "The Great A.I. Awakening," *New York Times Magazine*, December 14, 2016, accessed August 19, 2018, https://www.nytimes.com/2016/12/14/magazine/the-great-ai-awakening.html.

280 **"human-level performance," as they noted:** Steven Levy, "Inside Facebook's AI Machine," *Wired*, February 23, 2017, accessed August 19, 2018, https://www.wired.com/2017/02/inside-facebooks-ai-machine/; Yaniv Taigman, Ming Yang, Marc'Aurelio Ranzato, and Lior Wolf, "DeepFace: Closing the Gap to Human-Level Performance in Face Verification," Conference on Computer Vision and Pattern Recognition (CVPR), June 24, 2014, accessed August 19, 2018, https://research.fb.com/publications/deepface-closing-the-gap-to-human-level-performance-in-face-verification.

280 **to navigate roads:** Andrew J. Hawkins, "Inside Waymo's Strategy to Grow the Best Brains for Self-driving Cars," *The Verge*, May 9, 2018, accessed August 19, 2018, https://www.theverge.com/2018/5/9/17307156/google-waymo-driverless-cars-deep-learning-neural-net-interview.

280 **where new rides will emerge:** Nikolay Laptev, Slawek Smyl, and Santhosh Shanmugam, "Engineering Extreme Event Forecasting at Uber with Recurrent Neural Networks," Uber Engineering, June 9, 2017, accessed August 19, 2018, https://eng.uber.com/neural-networks.

280 **detect cancer in CT scans:** Cade Metz, "Using AI to Detect Cancer, Not Just Cats," *Wired*, May 11, 2017, accessed August 19, 2018, https://www.wired.com/2017/05/using-ai-detect-cancer-not-just-cats.

280 **74 minutes a day using it:** Anu Hariharan, "The Hidden Forces Behind Toutiao: China's Content King," *Y Combinator* (blog), October 12, 2017, accessed August 19, 2018, https://blog.ycombinator.com/the-hidden-forces-behind-toutiao-chinas-content-king.

281 **"That's my question":** Thompson, "What Is I.B.M.'s Watson?"

284 **"it's going to work":** Erik Brynjolfsson and Andrew McAfee, "The Business of Artificial Intelligence," *Harvard Business Review* (July 2017), accessed August 19, 2018, https://hbr.org/cover-story/2017/07/the-business-of-artificial-intelligence.

285 **"ocean of math":** Jason Tanz, "Soon We Won't Program Computers. We'll Train Them Like Dogs," *Wired*, May 17, 2016, accessed August 19, 2018, https://www.wired.com/2016/05/the-end-of-code/.

286 **"many, many years":** Tanz, "Soon We Won't Program Computers."

287 **"*that one* came up":** "Why Google 'Thought' This Black Woman Was a Gorilla," WNYC, September 30, 2015, accessed August 19, 2018, https://www.wnycstudios.org/story/deep-problem-deep-learning/; this latter quote appears in the audio of the interview, not in the text of the story on the website.

287 **the performance of their AI:** "Google Apologises for Photos App's Racist Blunder," *BBC News*, July 1, 2015, accessed August 19, 2018, https://www.bbc.com/news/technology-33347866.

287 **only 2 percent black:** Danielle Brown, "Google Diversity Annual Report 2018," Google, accessed August 19, 2018, https://diversity.google/annual-report.

287 **with white people:** Joy Buolamwini, "How I'm Fighting Bias in Algorithms," TEDxBeaconStreet, 8:45, November 2016, accessed August 21, 2018, http://www.ted.com/talks/joy_buolamwini_how_i_m_fighting_bias_in_algorithms/transcript.

289 **associated with *homemaker*:** Tolga Bolukbasi, Kai-Wei Chang, James Zou, Venkatesh Saligrama, and Adam Kalai, "Man Is to Computer Programmer as Woman Is to Homemaker?: Debiasing Word Embeddings," *NIPS '16 Proceedings of the 30th International Conference on Neural Information Processing Systems* (December 2016): 4356–64, accessed August 21, 2018, via https://arxiv.org/pdf/1607.06520.pdf.

290 **jobs of $200,000 and up:** Byron Spice, "Questioning the Fairness of Targeting Ads Online," *Carnegie Mellon University News*, July 7, 2015, accessed August 21, 2018, https://www.cmu.edu/news/stories/archives/2015/july/online-ads-research.html.

290 **removed that autosuggestion:** Carole Cadwalladr, "Google, Democracy and the Truth about Internet Search," *Guardian*, December 4, 2016, accessed August 21, 2018, https://www.theguardian.com/technology/2016/dec/04/google-democracy-truth-internet-search-facebook; Samuel Gibbs, "Google Alters Search Autocomplete to Remove 'Are Jews Evil' Suggestion," *Guardian*, December 5, 2016, accessed August 21, 2018, https://www.theguardian.com/technology/2016/dec/05/google-alters-search-autocomplete-remove-are-jews-evil-suggestion.

290 **how its system makes predictions:** Julia Angwin, Jeff Larson, Surya Mattu, and Lauren Kirchner, "Machine Bias," *ProPublica*, May 23, 2016, accessed August 21, 2018, https://www.propublica.org/article/machine-bias-risk-assessments-in-criminal-sentencing.

291 **machine learning to study:** Cathy O'Neil, *Weapons of Math Destruction: How Big Data Increases Inequality and Threatens Democracy* (New York: Broadway Books, 2016), 23–26.

291 **"invent the future":** O'Neil, *Weapons*, 203.

292 **hungry and depleted:** Kurt Kleiner, "Lunchtime Leniency: Judges' Rulings Are Harsher When They Are Hungrier," *Scientific American*, September 1, 2011, accessed August 21, 2018, https://www.scientificamerican.com/article/lunchtime-leniency.

295 **"learning to be a jerk":** Robyn Speer, "ConceptNet Numberbatch 17.04: Better, Less-stereotyped Word Vectors," *ConceptNet* (blog), April 24, 2017, accessed August 21, 2018, http://blog.conceptnet.io/posts/2017/conceptnet-numberbatch-17-04-better-less-stereotyped-word-vectors.

296 **you get zero results:** Tom Simonite, "When It Comes to Gorillas, Google Photos Remains Blind," *Wired*, January 11, 2018, accessed August 21, 2018, https://www.wired.com/story/when-it-comes-to-gorillas-google-photos-remains-blind.

297 **"and not on alchemy":** Alister, *Ali Rahimi's Talk at NIPS (NIPS 2017 Test-of-Time Award Presentation)*, YouTube, 23:19, December 5, 2017, accessed August 21, 2018, https://www.youtube.com/watch?v=Qi1Yry33TQE.

297 **how exactly their tools are working:** Oscar Schwartz, "Why the EU's 'Right to an Explanation' Is Big News for AI and Ethics," The Ethics Centre, February 19, 2018, accessed August 21, 2018, http://www.ethics.org.au/on-ethics/blog/february-2018/why-eu-right-to-an-explanation-is-big-news-for-ai.

299 **"that man need ever make":** Irving John Good, "Speculations Concerning the First Ultraintelligent Machine," *Advances in Computers* 6 (1966): 33, accessed August 21, 2018, via https://exhibits.stanford.edu/feigenbaum/catalog/gz727rg3869.

299 **all humanity working together:** Nick Bostrom, *Superintelligence: Paths, Dangers, Strategies* (Oxford: Oxford University Press, 2014), location 1603 of 8770, Kindle.

300 **the better to sneak away:** Bostrom, *Superintelligence*, 2243–87.

300 **"the observable universe into paper clips":** Bostrom, *Superintelligence*, 2908.

300 **"a faint ticking sound":** Raffi Khatchadourian, "The Doomsday Invention," *New Yorker*, November 23, 2015, accessed August 21, 2018, https://www.newyorker.com/magazine/2015/11/23/doomsday-invention-artificial-intelligence-nick-bostrom.

301 **a machine that can truly reason:** Kevin Hartnett, "To Build Truly Intelligent Machines, Teach Them Cause and Effect," *Quanta Magazine*, May 15, 2018, accessed August 21, 2018, https://www.quantamagazine.org/to-build-truly-intelligent-machines-teach-them-cause-and-effect-20180515; Gary Marcus, "Deep Learning: A Critical Appraisal," arXiv, January 2, 2018, accessed August 21, 2018, https://arxiv.org/abs/1801.00631.

301 **with surprising speed:** Bostrom, *Superintelligence*, 1723.

301 **rise up to kill us:** "Elon Musk Talks Cars—and Humanity's Fate—with Governors," *CNBC*, July 16, 2017, accessed August 21, 2018, https://www.cnbc.com/2017/07/16/musk-says-a-i-is-a-fundamental-risk-to-the-existence-of-human-civilization.html; Maureen Dowd, "Elon Musk's Billion-dollar Crusade to Stop the A.I. Apocalypse," *Vanity Fair*, April 2017, accessed August 21, 2018, https://www.vanityfair.com/news/2017/03/elon-musk-billion-dollar-crusade-to-stop-ai-space-x.

303 **10 to 25 years from now:** Oren Etzioni, "No, the Experts Don't Think Superintelligent AI Is a Threat to Humanity," *MIT Technology Review*, September 20, 2016, accessed August 21, 2018, https://www.technologyreview.com/s/602410/no-the-experts-dont-think-superintelligent-ai-is-a-threat-to-humanity.

CHAPTER 10: SCALE, TROLLS, AND BIG TECH

308 **exacerbate partisan hatreds:** Eli Rosenberg, "Twitter to Tell 677,000 Users They Were Had by the Russians. Some Signs Show the Problem Continues," *Washington Post*, January 19, 2018, accessed August 21, 2018, https://www.washingtonpost.com/news/the-switch/wp/2018/01/19/twitter-to-tell-677000-users-they-were-had-by-the-russians-some-signs-show-the-problem-continues.

308 **by Trump's followers:** Claire Landsbaum, "Donald Trump's Harassment of a Teenage Girl on Twitter Led to Death and Rape Threats," *The Cut*, December 9, 2016, accessed August 21, 2018, https://www.thecut.com/2016/12/trumps-harassment-of-an-18-year-old-girl-on-twitter-led-to-death-threats.html.

309 **Franklin Foer dubs it:** Franklin Foer, *World without Mind: The Existential Threat of Big Tech* (New York: Penguin, 2017).

313 **over $100 billion:** "Twitter Announces Fourth Quarter and Fiscal Year 2017 Results," *PR Newswire*, February 8, 2018, https://www.prnewswire.com/news-releases/twitter-announces-fourth-quarter-and-fiscal-year-2017-results-300595731.html; "Facebook Reports Fourth Quarter and Full Year 2017 Results," *PR Newswire*, January 31, 2018, https://www.prnewswire.com/news-releases/facebook-reports-fourth-quarter-and-full-year-2017-results-300591468.html; Seth Fiegerman, "Google Posts Its First $100 Billion Year," *CNN Tech*, February 1, 2018, https://money.cnn.com/2018/02/01/technology/google-earnings/index.html; all accessed September 28, 2018.

314 **"some of the responsibility, probably":** Charlie Warzel, "How People Inside Facebook Are Reacting to the Company's Election Crisis," *BuzzFeed*, October 20, 2017, accessed August 21, 2018, https://www.buzzfeednews.com/article/charliewarzel/how-people-inside-facebook-are-reacting-to-the-companys.

315 **he joked on Twitter:** Joseph Milord, "What Data Does Facebook Collect? The Company Can Store Info about Your Phone Calls," *elite daily*, March 28, 2018, accessed August 21, 2018, https://www.elitedaily.com/p/what-data-does-facebook-collect-the-company-can-store-info-about-your-phone-calls-8631437.

316 **wheat from the chaff:** Paresh Dave, "YouTube Sharpens How It Recommends Videos Despite Fears of Isolating Users," *Reuters*, November 29, 2017, accessed August 21, 2018, https://www.reuters.com/article/us-alphabet-YouTube-content/YouTube-sharpens-how-it-recommends-videos-despite-fears-of-isolating-users-idUSKBN1DT0LL; Andrew Hutchinson, "How Twitter's Feed Algorithm Works—as Explained by Twitter," *SocialMediaToday*, May 11, 2017, accessed August 21, 2018, https://www.socialmediatoday.com/social-networks/how-twitters-feed-algorithm-works-explained-twitter; Roger Montti, "Facebook Discusses How Feed Algorithm Works," *Search Engine Journal*, April 10, 2018, accessed August 21, 2018, https://www.searchenginejournal.com/facebook-news-feed-algorithm/248515; Shannon Tien, "How the Facebook Algorithm Works and How to Make It Work for You," *Hootsuite Blog*, April 25, 2018, accessed August 21, 2018, https://blog.hootsuite.com/facebook-algorithm/; Oremus, "Who Controls Your News Feed."

316–317 **"talking about it":** Steve Rayson, "We Analyzed 100 Million Headlines. Here's What We Learned (New Research)," *Buzzsumo*, June 26, 2017, accessed August 21, 2018, https://buzzsumo.com/blog/most-shared-headlines-study.

317 **made-up scandals:** Jackie Mansky, "The Age-old Problem of 'Fake News,'" *Smithsonian*, May 7, 2018, accessed August 21, 2018, https://www.smithsonianmag.com/history/age-old-problem-fake-news-180968945/.

317 **as BuzzFeed described it:** Charlie Warzel and Remy Smidt, "YouTubers Made Hundreds of Thousands Off of Bizarre and Disturbing Child Content," *BuzzFeed*, December 11, 2017, accessed August 21, 2018, https://www.buzzfeednews.com/article/charliewarzel/YouTubers-made-hundreds-of-thousands-off-of-bizarre-and.

317 **conspiracy theories and 9/11 "truthers":** Zeynep Tufekci, "YouTube, the Great Radicalizer," *New York Times*, March 10, 2018, accessed August 21, 2018, https://www.nytimes.com/2018/03/10/opinion/sunday/YouTube-politics-radical.html.

318 **the profit from the clicks:** Jonathan Albright, "Untrue-Tube: Monetizing Misery and Disinformation," *Medium*, February 25, 2018, accessed August 21, 2018, https://medium.com/@d1gi/untrue-tube-monetizing-misery-and-disinformation-388c4786cc3d.

318 **"can just breed and multiply":** Sheera Frenkel, "She Warned of 'Peer-to-Peer Misinformation.' Congress Listened," *New York Times*, November 12, 2017, accessed August 21, 2018, www.nytimes.com/2017/11/12/technology/social-media-disinformation.html.

319 **had a Democratic staffer murdered:** Amanda Robb, "Anatomy of a Fake News Scandal," *Rolling Stone*, November 16, 2017, accessed August 21, 2018, https://www.rollingstone.com/politics/politics-news/anatomy-of-a-fake-news-scandal-125877; Colleen Shalby, "How Seth Rich's Death Became an Internet Conspiracy Theory," *Los Angeles Times*, May 24, 2017, http://www.latimes.com/business/hollywood/la-fi-ct-seth-rich-conspiracy-20170523-htmlstory.html.

319 **they already agreed with:** Clive Thompson, "Social Networks Must Face Up to Their Political Impact," *Wired*, January 5, 2017, accessed August 21, 2018, https://www.wired.com/2017/01/social-networks-must-face-political-impact.

319 **upvoted online memes:** Clive Thompson, "Online Hate Is Rampant. Here's How to Keep It From Spreading," *Wired*, August 2, 2018, accessed August 21, 2018, https://www.wired.com/story/covering-online-hate.

320 **being "false news":** Erich Owens and Udi Weinsberg, "Showing Fewer Hoaxes," Facebook Newsroom, January 20, 2015, accessed September 30, 2018, https://newsroom.fb.com/news/2015/01/news-feed-fyi-showing-fewer-hoaxes/; Caroline O'Donovan, "What Does Facebook's New Tool for Fighting Fake News Mean for Real Publishers?," *NiemanLab*, January 21, 2015, accessed September 30, 2018, http://www.niemanlab.org/2015/01/what-does-facebooks-new-tool-for-fighting-fake-news-mean-for-real-publishers/.

320 **"'want us to do about it?'":** Warzel, "How People Inside Facebook Are Reacting."

320 **and harass journalists:** Siva Vaidhyanathan, *Antisocial Media: How Facebook Disconnects Us and Undermines Democracy* (New York: Oxford University Press, 2018), location 187–92 of 6698, Kindle.

320 **her primary rivals:** "Russia Spent $1.25M per Month on Ads, Acted Like an Ad Agency: Mueller," *Ad Age*, February 16, 2018, accessed August 21, 2018, http://adage.com/article/digital/russia-spent-1-25m-ads-acted-agency-mueller/312424.

321 **"a highly responsive audience":** Dipayan Ghosh and Ben Scott, "Digital Deceit: The Technologies Behind Precision Propaganda on the Internet," New America (policy paper), January 2018, accessed August 21, 2018, https://www.newamerica.org/public-interest-technology/policy-papers/digitaldeceit/.

321 **"But hatred favors Facebook":** Vaidhyanathan, *Antisocial Media*, 195, Kindle.

321 **"says Wernher von Braun":** "Don't Be Evil: Fred Turner on Utopias, Frontiers, and Brogrammers," *Logic* 5 (Winter 2017), accessed August 21, 2018, https://logicmag.io/03-dont-be-evil/.

325 **or "doxing" you:** Taylor Wofford, "Is Gamergate about Media Ethics or Harassing Women? Harassment, the Data Shows," *Newsweek*, October 25, 2014, accessed August 21, 2018, https://www.newsweek.com/gamergate-about-media-ethics-or-harassing-women-harassment-data-show-279736; Brad Glasgow, "A Definition of Twitter Harassment," *Medium*, November 2, 2015, accessed August 21, 2018, https://medium.com/@Brad_Glasgow/a-definition-of-twitter-harassment-f8acfa9ae3a8; Simon Parkin, "Gamergate: A Scandal Erupts in the Video-Game Community," *New Yorker*, October 17, 2014, accessed August 21, 2018, https://www.newyorker.com/tech/elements/gamergate-scandal-erupts-video-game-community.

325 **de facto workplace environment:** Timothy B. Lee, "One Scholar Thinks Online Harassment of Women Is a Civil Rights Issue," *Vox*, September 22, 2014, accessed August 21, 2018, https://www.vox.com/2014/9/22/6367973/online-harassment-of-women-a-civil-rights-issue; Danielle Keats Citron, *Hate Crimes in Cyberspace* (Cambridge, MA: Harvard University Press, 2014), 22.

325 **and Donald Trump:** Whitney Phillips, "The Oxygen of Amplification," Data & Society, May 22, 2018, accessed September 30, 2018, https://datasociety.net/wp-content/uploads/2018/05/FULLREPORT_Oxygen_of_Amplification_DS.pdf.

326 **hijack for electoral chicanery:** Ian Sherr and Erin Carson, "GamerGate to Trump: How Video Game Culture Blew Everything Up," *CNET*, November 27, 2017, accessed August 21, 2018, https://www.cnet.com/news/gamergate-donald-trump-american-nazis-how-video-game-culture-blew-everything-up; Matt Lees, "What Gamergate Should Have Taught Us about the 'Alt-right,'" *Guardian*, December 1, 2016, accessed August 21, 2018, https://www.theguardian.com/technology/2016/dec/01/gamergate-alt-right-hate-trump; Thompson, "Online Hate"; Thompson, "Social Networks."

326 **avalanche of anti-Semitic threats:** Danielle Citron and Benjamin Wittes, "Follow Buddies and Block Buddies: A Simple Proposal to Improve Civility, Control, and Privacy on Twitter," *Lawfare* (blog), January 4, 2017, accessed August 21, 2018, https://www.lawfareblog.com/follow-buddies-and-block-buddies-simple-proposal-improve-civility-control-and-privacy-twitter.

326 **slurs, quit, too:** Anna Silman, "A Timeline of Leslie Jones's Horrific Online Abuse," *The Cut*, August 24, 2016, accessed August 21, 2018, https://www.thecut.com/2016/08/a-timeline-of-leslie-joness-horrific-online-abuse.html.

328 **target political ads:** Matthew Rosenberg, Nicholas Confessore, and Carole Cadwalladr, "How Trump Consultants Exploited the Facebook Data of Millions," *New York Times*, March 17, 2018, accessed August 21, 2018, https://www.nytimes.com/2018/03/17/us/politics/cambridge-analytica-trump-campaign.html; Ian Sherr, "Facebook, Cambridge Analytica and Data Mining: What You Need to Know," *CNET*, April 18, 2018, accessed August 21, 2018, https://www.cnet.com/news/facebook-cambridge-analytica-data-mining-and-trump-what-you-need-to-know/; "Full Text: Mark Zuckerberg's Wednesday Testimony to Congress on Cambridge Analytica," *Politico*, April 9, 2018, accessed August 21, 2018, https://www.politico.com/story/2018/04/09/transcript-mark-zuckerberg-testimony-to-congress-on-cambridge-analytica-509978.

329 **developing-news stories:** Issie Lapowsky, "YouTube Debuts Plan to Promote and Fund 'Authoritative' News," *Wired*, July 9, 2018, accessed August 21, 2018, https://www.wired.com/story/YouTube-debuts-plan-to-promote-fund-authoritative-news/.

329 **sites appeared overall:** Michael J. Coren, "Facebook Will Now Show You More Posts from Friends and Family Than News," *Quartz*, January 12, 2018, accessed August 21, 2018, https://qz.com/1178186/facebook-fb-will-now-show-you-more-posts-from-friends-and-family-than-news-in-an-update-to-its-algorithm; Mike Isaac, "Facebook Overhauls News Feed to Focus on What Friends and Family Share," *New York Times*, January 11, 2018, https://www.nytimes.com/2018/01/11/technology/facebook-news-feed.html.

329 **to be local US newspapers:** Craig Timberg and Elizabeth Dwoskin, "Twitter Is Sweeping Out Fake Accounts Like Never Before, Putting User Growth at Risk," *Washington Post*, July 6, 2018, accessed August 21, 2018, https://www.washingtonpost.com/technology/2018/07/06/twitter-is-sweeping-out-fake-accounts-like-never-before-putting-user-growth-risk.

329 **accounts it doesn't follow:** Del Harvey and David Gasca, "Serving Healthy Conversation," *Twitter Blog*, May 15, 2018, accessed August 21, 2018, https://blog.twitter.com/official/en_us/topics/product/2018/Serving_Healthy_Conversation.html.

330 **death threats *reduced*:** Austin Carr and Harry McCracken, "'Did We Create This Monster?' How Twitter Turned Toxic," *Fast Company*, April 4, 2018, accessed August 21, 2018, www.fastcompany.com/40547818/did-we-create-this-monster-how-twitter-turned-toxic.

331 **10,000 to scour YouTube videos:** April Glaser, "Want a Terrible Job? Facebook and Google May Be Hiring," *Slate*, January 18, 2018, accessed August 21, 2018, https://slate.com/technology/2018/01/facebook-and-google-are-building-an-army-of-content-moderators-for-2018.html.

331 **that's genuinely threatening:** Alexis C. Madrigal, "The Basic Grossness of Humans," *The Atlantic*, December 15, 2017, accessed August 21, 2018, https://www.theatlantic.com/technology/archive/2017/12/the-basic-grossness-of-humans/548330.

332 **whitewash their reputation:** Joseph Cox, "Leaked Documents Show Facebook's Post-Charlottesville Reckoning with American Nazis," *Motherboard*, May 25, 2018, accessed August 21, 2018, https://motherboard.vice.com/en_us/article/mbkbbq/facebook-charlottesville-leaked-documents-american-nazis.

333 **he tells me:** Thompson, "Social Networks."

333 **"inevitably exploit them":** David Greene, "Alex Jones Is Far from the Only Person Tech Companies Are Silencing," *Washington Post*, August 12, 2018, accessed August 21, 2018, https://www.washingtonpost.com/opinions/beware-the-digital-censor/2018/08/12/997e28ea-9cd0-11e8-843b-36e177f3081c_story.html.

333 **"a law on his hands":** Taylor Hatmaker, "Senator Warns Facebook Better Shape Up or Get 'Broken Up,'" *TechCrunch*, April 6, 2018, accessed August 21, 2018, https://techcrunch.com/2018/04/06/facebook-zuckerberg-regulation-wyden.

334 **in the early days, isn't enough:** Max Read, "Does Even Mark Zuckerberg Know What Facebook Is?," *New York*, October 2, 2017, accessed August 21, 2018, http://nymag.com/selectall/2017/10/does-even-mark-zuckerberg-know-what-facebook-is.html.

335 **"infrastructure for the future":** danah boyd, "Google and Facebook Can't Just Make Fake News Disappear," *Wired*, March 27, 2017, accessed August 21, 2018, https://www.wired.com/2017/03/google-and-facebook-cant-just-make-fake-news-disappear/.

339 **ever be used for weaponry:** Scott Shane, Cade Metz, and Daisuke Wakabayashi, "How a Pentagon Contract Became an Identity Crisis for Google," *New York Times*, May 30, 2018, https://www.nytimes.com/2018/05/30/technology/google-project-maven-pentagon.html.

339 **contract worth billions:** Lee Fang, "Leaked Emails Show Google Expected Lucrative Military Drone AI Work to Grow Exponentially," *The Intercept*, May 31, 2018, accessed August 21, 2018, https://theintercept.com/2018/05/31/google-leaked-emails-drone-ai-pentagon-lucrative.

340 **"for the Defense industry":** Shane, Metz, and Wakabayashi, "How a Pentagon Contract Became an Identity Crisis for Google."

340 **they wouldn't renew it:** Kate Conger, "Google Plans Not to Renew Its Contract for Project Maven, a Controversial Pentagon Drone AI Imaging Program," *Gizmodo*, June 2, 2018, accessed August 21, 2018, https://www.gizmodo.com.au/2018/06/google-plans-not-to-renew-its-contract-for-project-maven-a-controversial-pentagon-drone-ai-imaging-program.

341 **"we refuse to be complicit":** Colin Lecher, "The Employee Letter Denouncing Microsoft's ICE Contract Now Has Over 300 Signatures," *The Verge*, June 21, 2018, accessed September 30, 2018, https://www.theverge.com/2018/6/21/17488328/microsoft-ice-employees-signatures-protest.

341 **"people are quick to judge":** Olivia Solon, "Ashamed to Work in Silicon Valley: How Techies Became the New Bankers," *Guardian*, November 8, 2017, accessed August 21, 2018, https://www.theguardian.com/technology/2017/nov/08/ashamed-to-work-in-silicon-valley-how-techies-became-the-new-bankers.

CHAPTER 11: BLUE-COLLAR CODING

343 **a town of 7,000:** "Pikeville, KY," Data USA, accessed August 21, 2018, https://datausa.io/profile/geo/pikeville-ky.

343 **the country away from coal:** Clifford Krauss, "Coal Production Plummets to Lowest Level in 35 Years," *New York Times*, June 10, 2016, accessed August 21, 2018, https://www.nytimes.com/2016/06/11/business/energy-environment/coal-production-decline.html.

343 **dropped to 6,500:** Erica Peterson, "From Coal to Code: A New Path for Laid-off Miners in Kentucky," *All Tech Considered*, May 6, 2016, accessed August 21, 2018, https://www.npr.org/sections/alltechconsidered/2016/05/06/477033781/from-coal-to-code-a-new-path-for-laid-off-miners-in-kentucky.

343 **coal that didn't have any market:** Bill Estep, "Coal Piles Up at Power Plant as Cheap Natural Gas Wrecks Eastern Kentucky's Economy," *Lexington Herald-Leader*, August 3, 2018, accessed August 21, 2018, https://www.kentucky.com/news/state/article215994320.html.

344 **"revolution to Eastern Kentucky":** Lauren Smiley, "Can You Teach a Coal Miner to Code?," *Backchannel* at *Wired*, November 18, 2015, accessed August 21, 2018, https://www.wired.com/2015/11/can-you-teach-a-coal-miner-to-code.

349 **"the Blue-collar Coder":** Anil Dash, "The Blue Collar Coder," *Anil Dash* (blog), October 5, 2012, accessed August 21, 2018, https://anildash.com/2012/10/05/the_blue_collar_coder.

349 **median annual wage for all job types:** "Computer and Information Technology Occupations," Bureau of Labor Statistics, last modified April 13, 2018, accessed August 21, 2018, https://www.bls.gov/ooh/computer-and-information-technology/home.htm.

349 **by 2020 alone:** Vivek Ravisankar, "Unlocking Trapped Engineers," *TechCrunch*, January 12, 2016, accessed September 29, 2018, https://techcrunch.com/2016/01/12/unlocking-trapped-engineers/.

349 **"career track" jobs in 2015:** "Beyond Point and Click: The Expanding Demand for Coding Skills," Burning Glass Technologies, June 2016, accessed August 21, 2018, https://www.burning-glass.com/research-project/coding-skills/.

349 **located in Silicon Valley:** "Six-figure Tech Salaries: Creating the Next Developer Workforce," The App Association (formerly ACT), accessed August 21, 2018, via http://www.arcgis.com/apps/MapJournal/index.html?appid=b1c59eaadfd945a68a59724a59dbf7b1.

350 **a mere four years:** CRA Enrollment Committee Institution Subgroup, "Generation CS: Computer Science Undergraduate Enrollments Surge Since 2006," Computing Research Association, 2017, accessed August 21, 2018, https://cra.org/data/generation-cs/.

350 **most popular major on campus:** Sarah McBride, "Computer Science Now Top Major for Women at Stanford University," *Reuters*, October 9, 2015, accessed August 21, 2018, https://www.reuters.com/article/us-women-technology-stanford-idUSKCN0S32F020151009; Andrew Myers, "Period of Transition: Stanford Computer Science Rethinks Core Curriculum," Stanford Engineering, June 14, 2012, accessed August 21, 2018, https://engineering.stanford.edu/news/period-transition-stanford-computer-science-rethinks-core-curriculum.

350 **a temporary vogue:** Sax, "Expanding the Pipeline."

351 **overcrowded sardine tins:** Monte Whaley, "Colorado Colleges Overflowing with Huge Wave of Computer Science Students," *Denver Post*, November 28, 2017, accessed August 21, 2018, https://www.denverpost.com/2017/11/28/colorado-colleges-overflowing-computer-science-students/.

351 **this was a graduate class:** John Markoff, "Brainlike Computers, Learning from Experience," *New York Times*, December 28, 2013, accessed August 21, 2018, https://www.nytimes.com/2013/12/29/science/brainlike-computers-learning-from-experience.html.

354 **programs, has said:** Kif Leswing, "Google Says Coding School Graduates 'Not Quite Prepared' to Work at Google," *Business Insider*, December 6, 2016, accessed August 21, 2018, https://www.businessinsider.com/google-says-coding-bootcamp-graduates-need-additional-training-2016-12.

354 **noted on Twitter:** Ceej Silverio (@ceejbot), "A very tiny percentage of our industry," Twitter, August 6, 2017, accessed August 21, 2018, https://twitter.com/ceejbot/status/894258853339987968.

355 **looking for jobs:** Silvia Li Sam, "Are Coding Bootcamps the Next 'Big Thing' in Edtech & Marketing?," Thrive Global, August 21, 2017, accessed August 21, 2018,

https://www.thriveglobal.com/stories/11764-are-coding-bootcamps-the-next-big-thing-in-edtech; Liz Eggleston, "2016 Coding Bootcamp Market Size Study," Course Report, June 22, 2016, accessed August 21, 2018, https://www.coursereport.com/reports/2016-coding-bootcamp-market-size-research.

356 **sketchy and under-regulated:** Sarah McBride, "Want a Job in Silicon Valley? Keep Away from Coding Schools," *Bloomberg*, December 6, 2016, accessed August 21, 2018, https://www.bloomberg.com/news/features/2016-12-06/want-a-job-in-silicon-valley-keep-away-from-coding-schools.

356 **find work in the field:** Mitch Pronschinske, "Bootcamps Won't Make You a Coder. Here's What Will," *TechBeacon*, accessed August 21, 2018, https://techbeacon.com/bootcamps-wont-make-you-coder-heres-what-will; Liz Eggleston, "2016 Course Report Alumni Outcomes & Demographics Study," Course Report, September 14, 2016, accessed August 21, 2018, https://www.coursereport.com/reports/2016-coding-bootcamp-job-placement-demographics-report; Liz Eggleston, "2015 Course Report Alumni Outcomes & Demographics Study," Course Report, October 26, 2015, accessed August 21, 2018, https://www.coursereport.com/reports/2015-coding-bootcamp-job-placement-demographics-report.

356 **and the rest freelancing:** "Tech's Strongest Outcomes. Education's Highest Standards," Flatiron School, accessed August 21, 2018, https://flatironschool.com/outcomes.

356 **a since-shuttered school nearby:** Steve Lohr, "As Coding Boot Camps Close, the Field Faces a Reality Check," *New York Times*, August 24, 2017, accessed August 21, 2018, https://www.nytimes.com/2017/08/24/technology/coding-boot-camps-close.html.

358 **operating without licenses:** Rip Empson, "Handcuffs for Hacker Schools? Why a 'Code of Conduct' for Coding Bootcamps Could Actually Be Good for the Ecosystem," *TechCrunch*, February 6, 2014, accessed August 21, 2018, https://techcrunch.com/2014/02/05/bootcamp-regulators-why-a-code-of-conduct-for-coding-academies-in-california-could-be-a-good-thing/; Klint Finley, "California Cracks Down on Hacker Boot Camps," *Wired*, January 31, 2014, accessed August 21, 2018, https://www.wired.com/2014/01/california-hacker-bootcamps.

358 **ordered the school to be shut down:** McBride, "Want a Job."

359 **student who filed a claim:** "A. G. Schneiderman Announces $375,000 Settlement with Flatiron Computer Coding School for Operating without a License and for Its Employment and Salary Claims," NY.gov, October 13, 2017, accessed August 21, 2018, https://ag.ny.gov/press-release/ag-schneiderman-announces-375000-settlement-flatiron-computer-coding-school-operating.

359 **for-profit schools:** Mary Beth Marklein, Jodi Upton, and Sandhya Kambhampati, "College Default Rates Higher Than Grad Rates," *USA Today*, July 2, 2013, accessed September 27, 2018, https://www.usatoday.com/story/news/nation/2013/07/02/college-default-rates-higher-than-grad-rates/2480295/; "2018–2019 Consumer Information Guide," University of Phoenix, November 2018, accessed November 15, 2018, https://www.phoenix.edu/content/dam/altcloud/doc/about_uopx/Consumer-Information-Guide.pdf.

359 **she told *Logic* magazine:** "Teaching Technology: Tressie McMillan Cottom on Coding Schools and the Sociology of Social Media," *Logic* 3 (Winter 2017), accessed August 21, 2018, https://logicmag.io/03-teaching-technology/.

360 **a well-paid technical field:** Alyssa Mazzina, "Do Developers Need College Degrees?"; Paul Krill, "Stack Overflow Survey: Nearly Half of Developers Are Self-taught," *InfoWorld*, April 10, 2015, accessed August 21, 2018, https://www.infoworld.com/article/2908474/application-development/stack-overflow-survey-finds-nearly-half-have-no-degree-in-computer-science.html.

361 **teaching themselves to code:** Quincy Larson, "We Asked 20,000 People Who They Are and How They're Learning to Code," freeCodeCamp, May 4, 2017, accessed August 21, 2018, https://medium.freecodecamp.org/we-asked-20-000-people-who-they-are-and-how-theyre-learning-to-code-fff5d668969.

362 **socio-economic backing:** "Teaching Technology," *Logic.*

364 **estimates of the industry's average:** "Employed Persons by Detailed Occupation," Bureau of Labor Statistics.

365 **an "uneducated imagination":** Northrop Frye, *The Educated Imagination* (Toronto: Anansi, 2002).

365 **on the side or retrained later:** David Kalt, "Why I Was Wrong about Liberal-Arts Majors," *Wall Street Journal*, June 1, 2016, accessed August 21, 2018, https://blogs.wsj.com/experts/2016/06/01/why-i-was-wrong-about-liberal-arts-majors.

367 **less than back-end work:** Miriam Posner, "JavaScript Is for Girls," *Logic* 1 (Spring 2017), accessed August 21, 2018, http://logicmag.1./01-javascript-is-for-girls.

368 **"beyond hope of regeneration":** Edsger W. Dijkstra, "How Do We Tell Truths That Might Hurt?," *ACM SIGPLAN Notices* 17, no. 5 (May 1982): 13–15.

368 **JavaScript, HTML, CSS:** Cecily Carver, "Things I Wish Someone Had Told Me When I Was Learning How to Code," freeCodeCamp, November 22, 2013, accessed August 21, 2018, https://medium.freecodecamp.org/things-i-wish-someone-had-told-me-when-i-was-learning-how-to-code-565fc9dcb329; Nico Koenig, "CSS Isn't Real Programming—Just Like JavaScript," *Medium*, September 22, 2017, accessed September 27, 2018, https://medium.com/@TheNicoKoenig/css-isnt-real-programming-just-like-javascript-70c8c0fe70b0.

369 **"meritocracy that does not exist":** Posner, "JavaScript Is for Girls."

370 **at the elementary-school level:** Valerie Strauss, "All Students Should Learn to Code. Right? Not So Fast," *Washington Post*, January 30, 2016, accessed August 21, 2018, https://www.washingtonpost.com/news/answer-sheet/wp/2014/05/29/all-students-should-learn-to-code-right-not-so-fast; Jean Dimeo, "Re-coding Literacy," *Inside Higher Ed*, September 6, 2017, accessed August 21, 2018, https://www.insidehighered.com/digital-learning/article/2017/09/06/professor-writes-everyone-should-code-movement-re-coding.

370 **which is, essentially, computer programming:** Papert, *Mindstorms*, 6, 45–54.

371 **and robotics competitions:** Anderson Silva, "Is Scratch Today Like the Logo of the '80s for Teaching Kids to Code?," opensource, March 29, 2017, accessed August 21, 2018, https://opensource.com/article/17/3/logo-scratch-teach-programming-kids; Code.org, "The Hour of Code: An International Movement," *Medium*, February 16, 2017, https://medium.com/anybody-can-learn/the-hour-of-code-an-international-movement-66702e388d35; D. Frank Smith, "Rise of the Robots: STEM-fueled Competitions Gaining Traction Nationwide," *EdTech Magazine*, March 4, 2015, accessed August 21, 2018, https://edtechmagazine.com/k12/article/2015/03/rise-robots-stem-fueled-competitions-gaining-traction-nationwide.

371 **their school systems follow suit:** Stefani Cox, "China Is Teaching Kids to Code Much, Much Earlier Than the U.S.," Big Think, November 19, 2015, accessed August 21, 2018, https://bigthink.com/ideafeed/china-is-already-teaching-coding-to-the-next-generation; Kathy Pretz, "Computer Science Classes for Kids Becoming Mandatory," *The Institute*, November 21, 2014, accessed August 21, 2018, http://theinstitute.ieee.org/career-and-education/education/computer-science-classes-for-kids-becoming-mandatory.

371 **"Are you *sure*?":** Jeff Atwood, "Please Don't Learn to Code," *Coding Horror* (blog), May 15, 2012, accessed August 21, 2018, https://blog.codinghorror.com/please-dont-learn-to-code/.

373 **Ian Bogost once noted:** Clive Thompson, "The Minecraft Generation," *New York Times Magazine*, April 14, 2016, accessed August 21, 2018, https://www.nytimes.com/2016/04/17/magazine/the-minecraft-generation.html.

374 **"around me are programmers":** Thompson, "The Minecraft Generation."

374 **"made the game for ourselves":** Thompson, "The Minecraft Generation."

375 **"in front of a bull's face":** Smiley, "Can You Teach."

Index

MAR 2019